Investment, Profit, and Tenancy

Investment, Profit, and Tenancy

The Jurists and the Roman Agrarian Economy

Dennis P. Kehoe

Ann Arbor

THE UNIVERSITY OF MICHIGAN PRESS

Copyright © by the University of Michigan 1997
All rights reserved
Published in the United States of America by
The University of Michigan Press
Manufactured in the United States of America
⊗ Printed on acid-free paper

2000   1999   1998   1997   4   3   2   1

No part of this publication may be reproduced,
stored in a retrieval system, or transmitted
in any form or by any means, electronic,
mechanical, or otherwise, without the written
permission of the publisher.

A CIP catalog record for this book is available from the British Library.

Library of Congress Cataloging-in-Publication Data

Kehoe, Dennis P.
    Investment, profit, and tenancy : the jurists and the Roman
  agrarian economy / Dennis P. Kehoe.
      p.   cm.
  Includes bibliographical references and index.
  ISBN 0-472-10802-6 (cloth)
    1. Agriculture—Economic aspects—Rome.   2. Agriculture—Rome—
Finance.   3. Farm tenancy—Rome.   4. Farm management—Economic
aspects—Rome.   5. Agricultural laws and legislation (Roman law).
6. Digesta.   I. Title.
HD139.A36K439   1997
338.1'0937—dc21
                                                  97-24345
                                                      CIP

For Connie

Preface and Acknowledgments

My purpose in this study is to clarify the basic relationships characterizing the agrarian economy of the early Roman Empire by analyzing the economic mentality that guided upper-class Romans in managing their wealth. My conviction is that the type of agrarian economic planning based on upper-class conceptions of investment and profit played a fundamental role in shaping the Roman economy as a whole. However, the lack of evidence available to the historian of the Roman economy makes it difficult to determine how upper-class Romans' understanding of economic issues affected the choices that they made in managing their agricultural property. In this study, I propose to address this problem by analyzing the assumptions that the Roman jurists in the Digest of Justinian and in other classical sources made about investment and profit in agriculture as they addressed legal issues involving private property. I base my analysis on the legal sources because, as I shall demonstrate, they offer a more comprehensive picture of the economic interests of upper-class Romans than we can gain from other types of evidence, including literary evidence, documentary papyri, and inscriptions.

The present study builds upon previous studies that I have undertaken on Roman economic planning. These studies have involved examining diverse evidence, including the inscriptions from the imperial estates in Africa during the second century, the Letters of Pliny the Younger, and documentary papyri attesting estate management in early Roman Egypt. All of these sources allow us to trace important aspects of Roman economic planning. The African inscriptions attest how the imperial government managed extensive properties in a key grain-producing province, and they demonstrate the factors that limited the imperial government's ability to develop a long-term economic policy. Pliny reveals assumptions about economic conditions that he shares with his correspondents and readers. Consequently, Pliny's experiences as a landowner provide the basis for general conclusions about the Roman estate economy, even if we cannot be certain that other landowners adopted the same solutions to economic problems that Pliny devised for his own estates. Finally, the documentary papyri from Egypt allow us to trace in great detail how individual landowners in that province managed estates. Although the documentary papyri rarely permit us to see the personal perspective of the landowner, this evidence does allow us to infer the general principles influencing landowners in the methods that they developed to manage their estates.

Examination of the Roman agrarian economy from these three different perspectives shows how the imperial government and private landowners in Italy and the provinces all struggled with the same basic constraints. Even so, the analyses that I have undertaken of the Roman economy have been based on restricted types of evidence, each with its difficult problems of interpretation. There are of course competing models for describing the fundamental relationships in the Roman economy. It is therefore desirable to investigate the Roman agrarian economy from a different perspective, in particular on the basis of evidence that allows us to define how upper-class Romans interpreted their own economic situation. The Roman jurists provide an appropriate source for such an investigation. As members of the aristocracy, they can reasonably be expected to have developed legal rules for private property with the interests of the upper classes in mind. In my view, the economic assumptions of the Roman jurists allow us to trace in a manner not possible with other types of evidence how the Roman aristocracy understood the economic conditions of the Roman Empire and the means that they had at their disposal to achieve economic goals. My purpose in examining the legal material is not to provide a comprehensive account of the economic interests and activities of upper-class Romans but rather to determine the most important relationships affecting their decision making. At the very least, the writings of the Roman jurists provide for the historian of the Roman economy a large amount of material that deserves to be investigated in a systematic fashion. The present study represents an attempt to bring this material more fully into the continuing debate about the structure of the Roman economy.

I have a number of institutions and individuals to thank for supporting my work on this project. I began much of the research during a year at the Institut für Papyrologie of the University of Heidelberg in

1989–90, under the auspices of a fellowship from the Alexander von Humboldt Foundation. Summer grants from the Humboldt Foundation enabled me to return to Heidelberg in 1991 and 1992. I also benefited from a summer fellowship from the National Endowment for the Humanities in 1994 and spent part of that summer in Heidelberg. I am very grateful to Professor Dieter Hagedorn, Professor Bärbel Kramer, and Wolfgang Habermann for their friendship and hospitality at the Institut, and to Professors Géza Alföldy and Fritz Gschnitzer for welcoming me to the Seminar für Alte Geschichte in Heidelberg. In addition, the Committee on Research at Tulane University provided me with a summer grant for 1993. Several scholars active during these times at the Institut für Papyrologie, including James Cowey, Ruth Duttenhöffer, Bernhard Palme, Brian McGing, Fritz Mitthof, and Amphilochios Papathomas, have made the Institut a truly ideal place to work. My research on Roman law and the Roman economy greatly benefited from my participation in two seminars in Italy, one sponsored by the Danish Academy in Rome in January 1993, and one sponsored by the University of Naples in October 1995. I thank the organizers of these conferences, Professors Jesper Carlsen, P. Ørsted, and J.E. Skydsgaard at the Danish Academy, and Professor Elio Lo Cascio at the University of Naples, for providing me with the opportunity to participate. In addition, I have benefited from comments made at oral presentations of papers based on this book at the Classical Association of the Middle West and South, the Roman Law Society, the University of Marburg (at the kind invitation of Professor H.-G. Leser), and New York University.

My interest in Roman law draws its inspiration from my teacher Bruce Frier, and I owe him and another of my teachers, Ludwig Koenen, a debt of gratitude for supporting my scholarly endeavors throughout my career. I also thank several scholars for providing me with copies of their work, including Jesper Carlsen, Luigi Capogrossi Colognesi, Andrea Jördens, James Keenan, Jerzy Kolendo, Elio Lo Cascio, David Mattingly, Pasquale Rosafio, Walter Scheidel, Karl Strobel, and Domenico Vera. I have benefited from discussing my work with Jens-Uwe Krause, Susan Martin, Thomas McGinn, and Shael Herman. I especially thank my friend Michael Peachin for reading the entire manuscript with painstaking care and for offering many helpful suggestions and references; his criticisms have helped me to make all my work better. The referees of the University of Michigan Press provided very helpful comments, and I thank Ellen Bauerle, James Laforest, Christina Milton, and the copyeditor of the press for answering many questions and expediting the publication of this book.

My wife, Constance Mui, edited large portions of the manuscript and has helped me to focus my arguments. Above all, her love and support are beyond measure, and to her I dedicate this book.

Contents

Abbreviations	xiii
Introduction: The Use of Legal Sources as Evidence for Economic History	1
Chapter 1. Investment Goals and the Law of Tutorship	22
Chapter 2. Profit, Security, and the Law of Legacy	77
Chapter 3. The Juristic Farm Tenant and the Management of Estates	137
Chapter 4. The Allocation of Resources in Farm Tenancy	181
Conclusion: The Jurists and the Roman Agrarian Economy	237
Bibliography	241
List of Ancient Sources	253
General Index	261

Abbreviations

CIL	*Corpus Inscriptionum Latinarum.*
CJ	*Corpus Iuris Civilis.* Vol. II, *Codex Justinianus.* Ed. and rev. P. Krueger. Berlin, 1954. Reprint, Hildesheim, 1989.
Coll.	*Mosaicarum et Romanarum Legum Collatio.* In *FIRA* II 541–89.
CTh	*Codex Theodosianus.*
D.	*Corpus Iuris Civilis.* Vol. I, *Digesta.* Ed. T. Mommsen. Rev. P. Krueger. Berlin, 1963. Reprint, Hildesheim, 1988.
FIRA II	*Fontes Iuris Romani Antejustiniani.* Part 2, "Auctores." Ed. J. Baviera and J. Furlani. Florence, 1968.
FIRA III	*Fontes Iuris Romani Antejustiniani.* Part 3, "Negotia." 2d ed. Ed. V. Arangio-Ruiz. Florence, 1969.
Frag. Vat.	*Fragmenta quae dicuntur Vaticana.* In *FIRA* II 461–540
ILS	H. Dessau, ed. *Inscriptiones Latinae Selectae.* 3 vols. Berlin, 1892–1916. Reprint, 1962.
Kaser, *RP* I	M. Kaser. *Das römische Privatrecht.* 2d ed. Vol. I. Munich, 1971.
[Paul.] *Sent.*	*Sententiarum receptarum libri quinque qui vulgo Iulio Paulo adhuc tribuuntur.* In *FIRA* II 317–417.
PIR²	E. Groag, A. Stein, and L. Petersen. *Prosopographia Imperii Romani.* 2d ed. Berlin, 1933. Reprint, 1965.
Sherwin-White, *Commentary*	A. N. Sherwin-White, *The Letters of Pliny: A Social and Historical Commentary.* Oxford, 1966. Rev. ed., 1985.

References to papyri follow the conventions of J.F. Oates, W.H. Willis, R.S. Bagnall, and K.A. Worp, *Checklist of Editions of Greek and Latin Papyri, Ostraca, and Tablets*, 4th ed., *BASP* Suppl. 7 (Atlanta, 1992). Abbreviations of journals follow those of *l'Année philologique*. Other common abbreviations are also used. Translations of quoted passages are my own unless otherwise noted.

Introduction: The Use of Legal Sources as Evidence for Economic History

The approaches that upper-class Roman landowners took toward formulating and achieving long-term economic goals had important consequences in shaping the basic relationships that defined the agrarian economy of the early Roman Empire.[1] Upper-class Romans depended on their ability to achieve sufficient income from agriculture to maintain their social privileges. Because they controlled a substantial portion of the empire's wealth, the ways in which they understood their estates as investments and accordingly sought to profit from them affected the situation of all classes involved in the Roman agrarian economy, from the large landowners themselves to the farmers of humbler means who comprised a majority of the empire's population.[2] The

---

1. For a balanced assessment of the Roman imperial economy, see E. Lo Cascio, "Forme dell'economia imperiale," in A. Schiavone, ed., *Storia di Roma*, vol. II, pt. 2 (Turin, 1991), 313–65. Recent discussions of the basic characteristics of the Roman agricultural economy include P. Garnsey and R. Saller, *The Roman Empire* (Berkeley and Los Angeles, 1987), 43–103, and W. Jongman, *The Economy and Society of Pompeii* (Amsterdam, 1988), especially 97–154, 187–203. See also K. Hopkins' introduction to P. Garnsey, K. Hopkins, and C.R. Whittaker, eds., *Trade in the Ancient Economy* (London, 1983) ix–xxv. R. Duncan-Jones, *Structure and Scale in the Roman Economy* (Cambridge, 1990), 121–42, analyzes the quantitative evidence for landholding patterns in the Roman empire. For a survey of the archaeological evidence for Roman estates, see K. Greene, *The Archaeology of the Roman Economy* (Berkeley and Los Angeles, 1986), 67–141. S. Alcock, *Graecia Capta* (Cambridge, 1993), especially chaps. 2–3, analyzes the archaeological evidence for the agrarian economy in Roman Greece. For a survey discussion of Roman agricultural history, see D. Flach, *Römische Agrargeschichte* (Munich, 1990).

2. For a comparative assessment of the general features of the Roman economy, see H.W. Pleket, "Wirtschaft," in *Handbuch der Europäischen Wirtschafts- und Sozialgeschichte*, vol. I (Stuttgart, 1990), 25–160, as well as R.W. Goldsmith, *Premodern Financial Systems* (Cambridge, 1987), 34–59, and idem, "An Estimate of the Size and Structure of the National Product of the Early Roman Empire," *Review of Income and Wealth* 30, no. 3 (1984): 263–88. W.V. Harris, "Between

objective of the present study is to determine how Roman landowners' general conceptions of investment and profit affected their choices in managing their land, given the constraints imposed upon them by a primarily agrarian economy with limited prospects for growth. By constructing a model for understanding upper-class economic planning, we can better identify the most important characteristics of the ancient economy, which will help us to distinguish it from the economies of other periods of history.

Investment and *profit,* however, are elusive terms in the Roman economy, given the restricted range of evidence available to the modern historian. One potentially rich source for examining key questions about the nature of investment and profit in the Roman economy is provided by the literature of classical Roman private law, particularly the Digest of Justinian and the Justinian Code. The Digest preserves the writings of the classical Roman jurists addressing a broad range of legal questions concerning agricultural property, while the Code contains rescripts by emperors to actual petitions concerning the same legal questions. In particular, we can investigate upper-class Roman approaches toward managing agricultural property by examining areas of the law in which the jurists' conception of agriculture as an investment was crucial for their formulation of legal rules. Two areas of the law seem especially appropriate for this type of investigation: (1) the duties of tutors and curators in administering the property belonging to wards and minors (see especially *D.* 26.7, 27.9) and (2) the bequeathing of agricultural property (see especially *D.* 33.7). In areas of private law, such as tutorship and bequest, which directly impinged on the financial interests of the upper classes, one could expect the jurists to have demonstrated a particular concern to adapt legal rules to what

---

Archaic and Modern: Problems in Roman Economic History," in W.V. Harris, ed., *The Inscribed Economy* (Ann Arbor, 1993), 11–29, addresses the important issues in defining the essential characteristics of the Roman economy. The basic demographic features of the Roman Empire are discussed in R.S. Bagnall and B.W. Frier, *The Demography of Roman Egypt* (Cambridge, 1994), as well as in T. Parkin, *Demography and Roman Society* (Baltimore, 1992), and E. Lo Cascio, "The Size of the Roman Population: Beloch and the Meaning of the Augustan Census Figures," *JRS* 84 (1994): 23–40. For the importance of demography for agrarian relationships, see E. Lo Cascio, "Fra equilibrio e crisi," in *Storia di Roma,* vol. II, pt. 2, 701–31, at 707–16.

they perceived as concrete social and economic conditions.[3] In so doing, the Roman jurists frequently made assumptions about the economic interests of the groups affected by their rulings, and these assumptions, based on the jurists' perception of the economic world around them, offer a broad and, as I shall argue, reliable framework in which to analyze the individual pieces of evidence for investing in and managing agricultural property.

At this point it would be helpful to clarify what is meant in this study by the terms *investment* and *profit*. The limited growth in the Roman economy certainly restricted the options available to people in managing their wealth. The question that I seek to explore in this study is how the Roman jurists understood the economic goals of upper-class Romans and the means that they had at their disposal to achieve them. In other words, in the view of the jurists, given the social demands imposed on them, what options did upper-class Romans have in managing their wealth so as to maintain their social position? Could upper-class Romans realistically choose among alternative strategies in managing wealth? The answers to these questions offer plausible explanations of the economic mentality of upper-class Romans, which are crucial for understanding the overall structure of the Roman economy and the place of agriculture, commerce, and other industries within it.

The relationship between landowner and tenant was especially important in defining the economy and the social structure of the Roman Empire. Certainly, there were numerous ways for upper-class Roman landowners to manage estates, both in Italy and in the provinces, and it is beyond doubt that, in Italy at least, many estates continued to be exploited by their landowners throughout the early empire without recourse to farm tenancy. These estates were tended by gangs of slave laborers under the management of *vilici*. Even so, it is also clear that other estates in Italy often consisted of diverse and scattered holdings and that many upper-class landowners derived their incomes in large part from *fundi* leased out to tenants. The holdings belonging to a large landowner in a given locality might not form a contiguous estate but might instead be comprised of a number of separate farms, often scattered and fragmented. Such estates will have grown

---

3. On this issue, see later in this introduction.

over time as landowners gradually added individual parcels to their holdings. This pattern of landownership is indicated by the Veleian alimentary table and certainly characterized large estates in many regions of Italy. Indeed, a recent investigation has shown that, in all likelihood, Pliny's estates in the area around Tifernum Tiberinum were not contiguous but consisted of a number of individual farms of varying sizes scattered among similar holdings belonging to other landowners.[4] Such estates shared significant characteristics with estates in Egypt, which have recently been the focal point of investigation; and it seems highly likely that estates belonging to individual landowners in other areas of the empire might also consist of scattered and diverse holdings. Under this circumstance, it was probably very difficult for the owners of large holdings to manage these as unified enterprises, unless they were able to create a complex system of farm management, such as the one attested on the estate of Aurelius Appianus, a third-century Egyptian magnate.[5]

A characteristic way to solve such management problems was to rely on farm tenancy. It may be hypothesized that, in managing their estates, landowners attempted to minimize their own risk and involvement by imposing on their tenants the costs and risks involved in keep-

---

4. For a thorough discussion of the structure and organization of estates in Italy, see D. Vera, "Dalla 'villa perfecta' alla villa di Palladio: sulle trasformazioni del sistema agrario in Italia fra principato e dominato," pts. 1 and 2, *Athenaeum* 83, no. 1 (1995): 189–211, no. 2 (1995): 331–56; cf. J.R. Patterson, "Crisis: What Crisis? Rural Change and Urban Development in Imperial Appennine Italy," *PBSR* 55 (1987): 115–46. The process of estate building is also analyzed in L. Capogrossi Colognesi, *Ai margini della proprietà fondiaria* (Rome, 1995), 245–74. Chapter 4 of Capogrossi Colognesi's book is published in a slightly different version as "Il regime degli affitti agrari," *Scienze dell'antichità, Storia, Archeologia, Antropologia* 6–7 (1992–93): 163–253 (for estate building see 221–37). See also F.G. de Pachtère, *La table hypothécaire de Veleia* (Paris, 1920), 78–97, W.E. Heitland, *Agricola* (Cambridge, 1921; reprint, Westport, Conn., 1970), 203–12; and P.W. de Neeve, *Colonus* (Amsterdam, 1984), 224–30. For the physical situation of Pliny's estates, see idem, "A Roman Landowner and His Estates: Pliny the Younger," *Athenaeum* 78 (1990): 363–402, at 373, referring to *Ep.* 3.19.1; and Garnsey and Saller, *The Roman Empire*, 68–69.

5. For a full analysis of the estate of the third-century Alexandrian magnate Aurelius Appianus, see D. Rathbone, *Economic Rationalism and Rural Society in Third-Century A.D. Egypt* (Cambridge, 1991). I analyze Appianus' estates in *Management and Investment on Estates in Roman Egypt during the Early Empire* (Bonn, 1992), 92–118, 120–24.

ing an estate productive for the long term. Although it is not possible to quantify this, it seems clear that farm tenancy was an important form of land tenure throughout Roman antiquity. This conclusion can be drawn for the case of Italy, and it also seems apparent that tenancy was the principal method of exploiting estates in most provinces of the empire.[6] Although no positive proof can be offered from legal, literary, or archaeological evidence, recent investigations suggest that Roman landowners in Italy may have increasingly relied on farm tenancy as their holdings grew and it was no longer convenient to exploit estates through gangs of slave laborers.[7] In the Roman law of legacy, we can

---

6. Recent studies of Roman farm tenancy, which concentrate primarily on early imperial Italy, include M.I. Finley, "Private Farm Tenancy in Italy before Diocletian," in M.I. Finley, ed., *Studies in Roman Property* (Cambridge, 1976), 103–21; B.W. Frier, "Law, Technology, and Social Change: The Equipping of Italian Farm Tenancies," *ZRG* 96 (1979): 204–28; K.-P. Johne, V. Weber, and J. Köhn, *Die Kolonen in Italien und den westlichen Provinzen des Römischen Reiches* (Berlin, 1983); De Neeve, *Colonus*; L. Capogrossi Colognesi, "Grandi proprietari, contadini e coloni nell'Italia Romana (I–III d.C.)," in A. Giardina, ed., *Società romana e impero tardo-antico*, vol. I (Rome and Bari, 1986), 325–65, 703–23; idem, *Ai margini della proprietà fondiaria*, 143–301 and "Il regime," 163–253; L. Foxhall, "The Dependent Tenant: Land Leasing and Labour in Italy and Greece," *JRS* 80 (1990): 97–114; E. Lo Cascio, "Considerazioni sulla struttura e sulla dinamica dell'affitto agrario in età imperiale," in H. Sancisi-Weerdenburg et al., eds., *De Agricultura* (Amsterdam, 1993), 296–316; idem, "L'affitto agrario in Italia nella prima età imperiale: A proposito di alcuni lavori recenti," *Scienze dell'antichità, Storia, Archeologia, Antropologia* 6–7 (1992–93): 257–68, who emphasizes the importance of demographic changes for the development of Roman farm tenancy; W. Scheidel, *Grundpacht und Lohnarbeit in der Landwirtschaft des römischen Italien* (Frankfurt, 1994); and F. De Martino, "Coloni in Italia," *Labeo* 41 (1995): 35–65. I discuss tenancy on the estates of Pliny the Younger in "Allocation of Risk and Investment on the Estates of Pliny the Younger," *Chiron* 18 (1988): 15–42, and "Approaches to Economic Problems in the 'Letters' of Pliny the Younger: The Question of Risk in Agriculture," *ANRW* II 33, no.1 (1989): 555–90. For a contrasting analysis of the relationship between Pliny and his tenants, see De Neeve, *Athenaeum* 78 (1990): 363–402. J. Rowlandson, *Landowners and Tenants in Roman Egypt* (Oxford, 1996), especially 202–79, provides a thorough analysis of farm tenancy in Roman Egypt. I also discuss economic aspects of tenancy in Roman Egypt in *Management and Investment*, chap. 4.

7. On the decline of the villa exploited by gangs of slave labor, see Vera, *Athenaeum* 83, no.1 (1995): 189–211, 83, no. 2 (1995): 331–56; and idem, "Schiavitù rurale e colonato nell'Italia imperiale," *Scienze dell'antichità, Storia, Archeologia, Antropologia* 6–7 (1992–93): 291–339. Vera argues that slave labor continued

trace an assumption made by the Roman jurists of the overall importance of farm tenancy to the economic interests of upper-class Romans. In regulating bequests of land, the jurists often assumed that an estate would be leased out to tenants and that the income from the estate consisted solely of the rent paid by the tenants (as discussed in chap. 2).

Accordingly, a careful analysis of the legal and economic relationships between landowner and tenant is crucial for understanding the fundamental relationships defining both the agrarian economy of the early Roman Empire and its social structure. The need for such analysis can certainly be seen in the later Roman Empire, when tenants tended to fall increasingly under the economic and even the legal control of powerful landowners.[8] (This tendency was one of the major historical themes characterizing that period.) Although changes in their situation during the principate may have been less drastic, the conditions under which tenants cultivated their land were no less important as a historical factor then than in the later period. In my judgment, any attempt to investigate the situation of the empire's farming population is possible only if we understand the economic motives of the Roman upper classes and the constraints that limited their scope for action as they pursued their economic goals.

In light of this, I shall argue that an estate represented to the jurists the resource from which an upper-class Roman obtained an income and that the desirability of an agricultural estate as an investment did not depend on external and changing economic conditions, since there were really very few alternatives for investing wealth. Certainly, a landowner might have to undertake substantial expenditures to make an estate resourceful enough to produce an income. But an estate represented first and foremost an asset for enhancing financial and social security, rather than an enterprise that could be evaluated among alternative potential investments in terms of the likelihood of gain balanced against associated risks. The emphasis on this function of an estate implies that in studies of the Roman agrarian economy the term *profit* should also be understood in a restricted sense: in general, profit repre-

---

to play an important role in Italian agriculture alongside tenancy throughout the imperial period.

8. For the continuity between tenancy in the early empire and the colonate of late antiquity, see the papers in E. Lo Cascio, ed., *Dall'affitto agrario al colonato tardo-antico: Continuità o frattura?* (Naples, 1997), the proceedings of a conference hosted in October 1995 by the University of Naples.

sented simply the income that a landowner might gain from the land at his or her disposal. Thus profit in agriculture represented not a return on a given level of investment but an income from a resource whose realization depended on the landowner's implementation of managerial techniques suitable to keeping production costs under control. My thesis is that farm tenancy made a crucial contribution to the efforts of Roman landowners to derive an income from their holdings in agriculture. Farm tenancy provided landowners with a reliable means of obtaining a stable income from diverse and scattered holdings, while at the same time maintaining the productivity of their land for the long term.

In view of these considerations, chapters 1 and 2 of my study address the ways in which upper-class Romans understood investment and profit in agriculture, by analyzing the jurists' treatment of the responsibilities of tutors in administering the property of pupils and the bequest of agricultural property. In examining the relevant legal texts, I seek to demonstrate that the jurists' understanding of the Roman economy informed their formulation of rules in the areas of private law under scrutiny.[9] To be sure, my approach places the jurists' treatment of these two subjects in a broader social context by examining relevant literary, epigraphical, and papyrological evidence. Such evidence gives us examples of individuals acting on the basis of economic concerns that I have inferred from the legal sources.

Since, as I shall argue, tenancy played a crucial role in enabling upper-class Romans to pursue the financial stability that I have hypothesized as their major economic concern, I devote a large part of this study to an examination of how Roman landowners contended with farm tenancy as an institution as they pursued this security. Here again, my analysis is based on the assumptions that the Roman jurists made about the economic interests of landowners and tenants. My working hypothesis is that the conservative approach toward managing wealth that was characteristic of upper-class economic planning had important implications for the relationships between landowners and tenants. To test my hypothesis, I devote the second half of my study to an examination of the ways in which the Roman legal author-

---

9. I have published a preliminary version of these two chapters in "Approaches to Profit and Management in Roman Agriculture: The Evidence of the Digest," in J. Carlsen et al., eds., *Landuse in the Roman Empire* (Rome, 1994), 45–58.

ities understood these relationships as they formulated regulations for farm tenancy. In chapter 3, I introduce my analysis of farm tenancy by examining how the Roman legal authorities dealt with managerial questions connected with this institution. The basis for my analysis is the idealized normative tenant that stands behind much of the juristic discussion of farm tenancy. The tenant of the Digest is often the product of an abstract conception by the jurists of a specific type of contract for the lease of a particular type of *fundus*. The jurists' discussion about the rules surrounding this contract offer us a way of tracing their general conception of the economic advantages that farm tenancy provided upper-class Roman landowners. As we will see, however, the jurists subsumed a variety of land tenure arrangements under this standard form of farm tenancy. The jurists' effort to accommodate Roman private law to the realities of the Roman economy is a focal point of this study, particularly in chapter 4, in which I examine the allocation of resources and risk between landowner and tenant. In my view, one of the fundamental characteristics of farm tenancy in the Roman world was that it allowed the landowner to share with the tenant the costs of investment and the risks involved in agriculture. At the same time, each party sought to use whatever economic and legal leverage available to impose on the other as many as possible of these costs and risks. The allocation of resources and risk between landowner and tenant was a subject whose importance the Roman legal authorities recognized. Accordingly, chapter 4 considers how the jurists and the Roman imperial government developed rules by which complex economic relationships between landowner and tenant were regulated. I supplement my investigation of the Digest with comparative evidence for farm tenancy from other types of sources, in particular the Letters of Pliny the Younger and documentary papyri from Roman Egypt. One cannot directly apply to the rest of the empire conclusions drawn on the basis of the papyri, but these texts allow us to trace how certain economic relationships played themselves out in the real world of Roman Egypt.[10]

---

10. For the importance of papyrological evidence to general questions about the Roman economy, see Rathbone, *Economic Rationalism,* and idem, "The Ancient Economy and Graeco-Roman Egypt," in L. Criscuolo and G. Geraci, eds., *Egitto e storia antica dall'ellenismo all'età araba* (Bologna, 1989), 159–76.

Using Legal Sources as Evidence

An initial difficulty in using legal sources as evidence for economic conditions is one common to any study of Roman law, namely, the ambiguity concerning the extent to which the Digest and the Justinian Code genuinely preserve the writings of the classical Roman legal authorities. Both were compiled in the sixth century under the emperor Justinian, whose purpose was not to preserve classical Roman law but to select from existing imperial rescripts and from the vast writings of the classical Roman jurists texts that would serve as a basis for the private legal relationships in the Byzantine Empire. The commission charged with the task of excerpting from the classical legal literature, then, was not primarily concerned with offering a representative picture of the writings of the Roman jurists. Consequently, it is very likely that much of value of these writings for the historian of Roman law in general and for the purposes of this study in particular has been lost to posterity.

The task of interpreting what is preserved in the Digest and the Code is itself fraught with difficulties. In their concern to serve contemporary needs, Justinian's commissioners suppressed much from the classical sources that was completely obsolete in their own day, and they were also charged with the task of eliminating contradictions among the various sources excerpted. The commissioners fulfilled this task only to some degree, and the major debate today is centered on the extent to which Justinian's compilers are believed to have substantially altered the classical texts that they excerpted. In addition, the legal historian must reckon with the possibility of postclassical interpretation of the texts used by Justinian's compilers. While it is difficult to make any general assertion on this issue, it seems clear that contemporary scholarship is justified in moving away from the suspicion that the Digest is so thoroughly interpolated that the opinions of the classical jurists can only be recovered after exhaustive reconstruction of the texts. Instead, many Romanists agree that, although interpolation is widespread in these documents, it is not so rampant as to distort completely what the classical legal authorities wrote.[11] But interpolation can take many

---

11. For full discussion of the important issues, see M. Kaser, *Zur Methodologie der römischen Rechtsquellenforschung* (Vienna, 1972), as well as F. Wieacker, "Textkritik und Sachforschung: Positionen in der gegenwärtigen Romanistik," *ZRG* 91 (1974): 1–40, and *Römische Rechtsgeschichte* (Munich, 1988), 154–82.

forms, so that we must always contend with the possibility that the surviving sources distort the reasoning of a classical jurist beyond recovery. Even so, it still seems reasonable to assert that the problem posed by the possibility of interpolation does not make it impossible to evaluate the economic assumptions of the Roman jurists, even if in the case of certain interpolated passages we cannot be absolutely certain about the soundness of the classical text. In this study, I will deal with problems of interpolation as they arise in connection with individual texts, but my study is premised on the belief that the existing collections of legal texts do allow us to trace the thinking, in particular the economic thinking, of the classical Roman legal authorities.[12]

Even if we can be sure about the validity of this premise, the legal sources present additional difficulties of interpretation to the economic historian, since it is not at all clear to what extent the writings of the jurists inform us about actual historical conditions. Roman legal historians have vigorously debated whether the jurists, in formulating legal rules, consciously addressed the social concerns of their contemporaries.[13] The alternative understanding is that the jurists operated essentially within a closed legal system, subject to its own rules, but largely immune from influences of economic and social considerations.[14] According to this hypothesis, the classical jurists were not at all concerned to adapt the law in response to changing social needs but instead persisted in developing rules for a legal system in which social and economic concerns found little expression. As such, one might plausibly claim that descriptions of economic conditions present in the jurists' writings need not bear any resemblance to actual historical conditions but merely represent traditional hypothetical situations that remained fixed in the juristic tradition. Hence, any text of the jurists that seems to describe economic conditions in the Roman

---

12. For discussion of this issue with reference to the legal sources on farm tenancy, see De Neeve, *Colonus*, 27–30, who bases his optimistic position on Kaser, *Zur Methodologie der römischen Rechtsquellenforschung*, and idem, *RP* I, vi.

13. This approach to examining the jurists is applied to the law of urban leases in B.W. Frier, *Landlords and Tenants in Imperial Rome* (Princeton, 1980); for private farm tenancy, see idem, *ZRG* 96 (1979): 204–28, and "Law, Economics, and Disasters down on the Farm: 'Remissio Mercedis' Revisited," *BIDR*, 3d ser., 31–32 (1989–90): 237–70. For the Roman building industry, see S.D. Martin, *The Roman Jurists and the Organization of Private Building in the Late Republic and Early Empire* (Brussels, 1989).

14. See A. Watson, *The Spirit of Roman Law* (Athens, Ga., 1995).

Empire could be dismissed as simply the product of a very conservative legal tradition.

Indeed, it is clear that the jurists typically based their decisions on abstract conceptions of the Roman economy that at times simplify and idealize reality, without seeming to take into account actual social concerns. The responses in the Digest, when they are based on cases from the real world, are generally formulated in such a way as to reduce to a minimum the description of the actual circumstances surrounding a case. And even these minimal descriptions are often generalized to accord a given response the widest legal authority.

Examples of this type of abstraction can be found right in the Roman jurists' treatment of tenancy. In addressing legal questions involving farm tenancy, the jurists generally assumed the existence of a normative tenant who leased a *fundus* for a five-year period in exchange for a cash rent. As a result, they often overlooked, or at least did not seem to take into account, the wide variety of contractual arrangements that undoubtedly characterized Italian farm tenancy in early imperial Italy, let alone in the rest of the empire.[15] For example, the jurists paid relatively little attention to sharecropping. There is a ready explanation for this: classical Roman lease law required that the rent, or *merces,* be paid in cash. Gaius made a casual reference to sharecropping when discussing the rights of the tenant to a remission of rent in the event of *vis maior,* an unforeseeable disaster making his farm impossible to cultivate (*D.* 19.2.25.6, 10 *ad ed. prov.*). In this passage, the jurist only mentioned sharecropping to emphasize that the normative tenant, that is, the one paying a cash rent, had a right to a remission under the conditions that he was describing.[16] It is noteworthy that Gaius referred to sharecropping in terms strongly implying that it was a widespread

---

15. On the "ideal" farm tenant in the Digest, see Frier, ZRG 96 (1979): 204–28; for the methodology of investigating the "legal culture" in the regulation of the Roman building industry, see Martin, *The Roman Jurists and the Organization of Private Building,* especially 11–14, 138–40. For the jurists' "hypothetical casuistry," see B.W. Frier, *The Rise of the Roman Jurists* (Princeton, 1985), 163–71. De Neeve, *Colonus,* 25–26, views the writings of the Roman jurists as directly reflecting society, albeit a conservative view of society. But cf. the comments of Frier, ZRG 100 (1983): 667–76, at 673–76.

16. Gaius *D.* 19.2.25.6: "apparet autem de eo nos colono dicere, qui ad pecuniam numeratam conduxit: alioquin partiarius colonus quasi societatis iure et damnum et lucrum cum domino fundi partitur."

form of land tenure in the Roman world. Sharecropping was indeed the lease arrangement that Pliny introduced on his Tuscan estates, and Pliny's account of his actions indicates that sharecropping was a form of land tenure well known to his reader and his audience (*Ep.* 9.37). It also represented the form of land tenure on the second-century imperial estates in the Medjerda Valley in North Africa, and it was a frequent lease arrangement in Roman Egypt, particularly for vineyards. The jurists' persistence in describing tenancy in terms of a cash rent aided them in developing a manageable set of rules for farm tenancy, but it should warn us about equating the decisions in the Digest concerning farm tenancy with descriptions of actual historical conditions in Italy or in any other part of the empire.

It is therefore not immediately clear to what extent legal evidence can be used to describe historical conditions of the early Roman Empire. But, as I will argue, we can overcome this inherent difficulty in the juristic sources to some degree by analyzing the assumptions that the jurists made about the Roman economy as they addressed legal issues involving tutorship, legacy, and farm tenancy. Tutorship and legacy offer particularly fertile ground for formulating hypotheses about upper-class Roman approaches to economic problems, since they represent two areas of the law where we could most expect the jurists to have displayed a concern for contemporary social and economic practices.[17] Both tutorship and the bequest of estates were institutions basic to Roman upper-class life. In Roman law, the affairs of all fatherless children under the age of puberty had to be managed by a tutor, and it was also common for minors (males under the age of twenty-five) to have their affairs supervised by a curator. According to the demographic model recently developed by Saller, a very considerable proportion of the empire's upper-class population had to have their

---

17. For discussion of the limits on the ability of the jurists to take economic considerations into account when formulating legal rules for the bequeathing of estates, see A. Steinwenter, *Fundus cum instrumento* (Vienna and Leipzig, 1942), 9–10, discussed further in chap. 2 in connection with the bequest of estates. For additional discussion of the economic thinking of the Roman jurists, see G. Tozzi, *Economisti greci e romani* (Milan, 1961), 383–498, G. Melillo, *Economia e giurisprudenza a Roma* (Naples, 1978), and C. Nicolet, "La pensée économique des Romains, République et Haut-Empire," in *Rendre à César* (Mesnil-sur-l'Estrée, 1988), 117–219, = "Il pensiero economico dei Romani," in L. Firpo, ed., *Storia delle idee politiche, economiche e sociali,* vol. I (Turin, 1982), 877–960.

property administered in this way. As many as one-third of all children lost their fathers before puberty, while another third of the population lost their fathers before reaching majority at the age of twenty-five. Consequently, the rules developed by the jurists for the institutions of tutorship and curatorship had a very real effect on how private property was managed in the Roman Empire.[18] Likewise, the bequeathing of property through legacy was also a practice of fundamental importance to the finances of upper-class Romans. Upper-class Romans regularly used the institution of legacy as a means of providing for loved ones and other dependents in their wills.[19] As a result, we should expect that, if they ever did so in any area of the law, the Roman legal authorities developed legal rules for tutorship and bequest on the basis of their understanding of economic conditions in the world around them.[20]

The rules that the jurists developed in these two areas of the law reveal their assumptions about the restricted range of investment options open to a landowner. The question, then, is whether the assumptions made by the jurists correspond in any significant degree to the actual practices of upper-class Romans. I shall argue that other nonlegal evidence for upper-class Roman economic planning, especially literary and papyrological evidence when available, corroborates the view that the assumptions made by the jurists were drawn from more broadly held conceptions of the Roman economy. Thus the economic assumptions made by the jurists are not just to be considered within the framework of developing rules within a legal system subject to its own logic and hence divorced from the real world. Rather, these assumptions present us with an idealized view of the overall conditions

---

18. For discussion of the importance of tutorship as a Roman social institution, see R.P. Saller, *Patriarchy, Property and Death in the Roman Family* (Cambridge, 1994), 181–203, especially 189. Saller (202, 229) emphasizes the importance of the rules for *tutela* for understanding the economic mentality of upper-class Romans. On the situation of pupils, see also J.-U. Krause, *Witwen und Waisen im römischen Reich*, vol. III, *Rechtliche und soziale Stellung von Waisen* (Stuttgart, 1995), especially 85–112.

19. On this subject, see E. Champlin, *Final Judgments* (Berkeley and Los Angeles, 1991), and D. Johnston, *The Roman Law of Trusts* (Oxford, 1988), as well as Saller, *Patriarchy, Property and Death in the Roman Family*, 161–80.

20. For the observation that the jurists, in developing legal rules for legacy, were concerning themselves with an institution affecting the financial interests of their own class, see Kaser, *RP* I, 742.

to which all economic activities on the part of upper-class Romans were subject.

Certainly the economic assumptions made by the jurists cannot tell us directly how individual upper-class Romans managed their wealth. As we will see, the jurists were concerned, in the case of tutorship, to prevent fraud, so they imposed many more restrictions on the tutor in managing the pupil's property than upper-class Romans would have been likely to observe in managing their own holdings. Similarly, in the case of testamentary law, the jurists were concerned not with making any prescription about how agriculture was to be practiced but rather with settling areas of dispute among beneficiaries of wills. Even so, in imposing restrictions on tutors or in settling disputes among heirs and legatees, the jurists made a consistent set of assumptions both about the uses to which upper-class Romans put their wealth and about their ability to maintain financial security and social status not only for themselves but for future generations. These assumptions paint a picture that helps us to understand the economic world in which all upper-class Romans operated, even though we must recognize that it is an idealized picture that does not reveal how particular individuals managed their wealth.

An analysis of the economic assumptions in the juristic sources, then, should allow us to construct a model describing the general principles that guided upper-class Romans in managing their wealth. The point is not to predict or account for the economic planning on the part of particular upper-class Romans but rather to define the most important relationships characterizing the Roman economy. I shall focus primarily on the classical Roman jurists, who wrote from roughly the late republic until the early third century A.D., as well as on the works of later jurists, most notably Hermogenianus, who served as petitions secretary under Diocletian and wrote an epitome of Roman law at this time.[21] In addition, imperial constitutions preserved in the Justinian Code down to the time of Diocletian indicate how the imperial administration grappled with the problems faced by the jurists in developing

---

21. For Hermoginianus' service as *magister libellorum* under Diocletian, see T. Honoré, *Emperors and Lawyers*, 2d ed. (Oxford, 1994), 163–81, no. 20; D. Liebs, *Die Jurisprudenz im spätantiken Italien (240–640 n.Chr.)* (Berlin, 1987); 36–52; W. Kunkel, *Herkunft und soziale Stellung der römischen Juristen*, 2d ed. (Graz, Vienna, and Cologne, 1967), 263; and A. Berger, *Encyclopedic Dictionary of Roman Law* (Philadelphia, 1953; reprint, 1980), 487.

the legal regulations that I will be examining. These constitutions, in particular the rescripts to petitions, addressed real-life problems experienced by individuals and complement the evidence provided by the jurists. Even though rescripts are often formulated in a general way to enhance their legal authority, they do draw us away from the hypothetical world of the jurists into the real world of concrete legal problems.[22]

Models of the Roman Economy

To generalize at this point from the evidence that I will examine, agriculture provided the only reliable form of investment on which upper-class Romans could generally depend for a stable long-term income. But the "profit" that could be generated from estates was severely limited, and for many landowners long-term stability was a far more important concern than maximizing immediate revenues. To understand the structure of the economy of the early empire, it is crucial that we look at the particular strategies that this concern for stability led Roman landowners to adopt. In particular, the constraints that upper-class landowners perceived in the general conditions of the Roman economy made tenancy an attractive and logical system by which to manage their agricultural property. But the efforts that landowners undertook to make this system work to their advantage complicated their relationships with their tenants and were hampered by a chronic shortage of suitable tenants and by the tenants' resulting bargaining power.[23]

---

22. See W. Turpin, "Imperial Subscriptions and the Administration of Justice," *JRS* 81 (1991): 101–18, as well as D. Nörr, "Zur Reskriptenpraxis in der hohen Prinzipatszeit," *ZRG* 98 (1981): 1–46. Some responses in the *CJ* may have been issued to judges: see M. Peachin, "Consultation with a Magistrate in Justinian's *Code*," *CQ* 42, no. 2 (1992): 448–58. In general on the administration of justice in the Roman Empire, see idem, *Iudex vice Caesaris* (Stuttgart, 1996), 10–91. On the emperor's handling of petitions and judging in civil cases, see F. Millar, *The Emperor in the Roman World (31 B.C.–A.D. 337)* (Ithaca, 1977), 240–52, 528–49.

23. In his statistical analysis of wages, wheat prices, and rents for grain land in Roman Egypt, "Real Land Rentals in Early Roman Egypt," *Explorations in Economic History* 31 (1994): 210–24, R.F. Muth argues that real wages and real land rents declined in the first three centuries of Roman rule. It is not clear whether one can accept Muth's explanation of an apparent simultaneous

In making this argument, I seek to develop hypotheses about the Roman economy that I have formulated on the basis of other sources of evidence. Since other historians have adhered to very different models, it is appropriate that I state explicitly my basic working assumptions about the nature of the Roman agrarian economy. In general, I argue that Roman landowners had a very specific understanding of the economic conditions affecting their ability to pursue their livelihoods. Certainly the Roman economy imposed severe constraints on them, limiting their options for investing large amounts of wealth. The scope for decision making in economic matters was constrained by the inescapable fact that the Roman economy, largely agrarian, was characterized by very limited growth.[24] To be sure, the Roman economy was largely monetized, even in rural areas, and money played a significant role not only in urban commercial transactions but also in the rural economy.[25] The agrarian economy of the Roman Empire thus had an important commercial sector, and the growth of agrarian fortunes

---

decline of both wages and rents as resulting from increasing imperial taxation, but a clear decline in rents would suggest that tenants gained increasing bargaining power during the early empire. It is difficult to assess rent levels in any period: see Rowlandson, *Landowners and Tenants in Roman Egypt*, 247–52.

24. For the limited growth in the Roman economy, see Goldsmith, *Premodern Financial Systems*, 36, and *Review of Income and Wealth* 30, no. 3 (1984): 275–76. For the importance to economic growth of a financial infrastructure capable of converting savings into investment in medieval Europe, see C.M. Cipolla, *Before the Industrial Revolution*, 3d ed. (New York and London, 1994) 160–208.

25. See C. Howgego, "The Supply and Use of Money in the Roman World, 200 B.C. to A.D. 300," *JRS* 82 (1992): 1–31, who argues that the use of money was widespread even in the countryside in the Roman Empire and that the use of crops to make certain types of rental or tax payments does not imply that coinage was not available. The Heroninos archive suggests strongly that the rural economy of third-century Egypt was highly monetized: see Rathbone, *Economic Rationalism*, 318–30, 393, 397–98. For the importance of the Roman imperial government's exaction of taxes in cash as a stimulus to commerce, see K. Hopkins, "Taxes and Trade in the Roman Empire (200 B.C.–A.D. 400)," *JRS* 70 (1980): 101–25. For more general discussion of the importance of coinage to the Roman economy, see K.W. Harl, *Coinage in the Roman Economy, 300 B.C. to A.D. 700* (Baltimore and London, 1996), 207–89. For a contrasting view about the degree of monetization in the Roman Empire and about the relationship between imperial fiscal policy and commerce, see Duncan-Jones, *Structure and Scale*, 30–47, and idem, *Money and Government in the Roman Empire* (Cambridge, 1994), especially 3, 20–32.

depended on the possibility of selling agricultural surpluses to urban markets.²⁶ Indeed, it seems clear that the commercial sector of the Roman economy grew substantially under the imperial peace. But even though certain individuals might make great fortunes from commerce and other pursuits—indeed many more people made modest fortunes—the backbone of the Roman economy was agriculture, so most upper-class Romans had to invest the bulk of their wealth in land.²⁷ Given the limited possibilities for growth in the Roman economy, the primary concern of most landowners was to maintain stability, rather than to seek to increase their wealth. But I argue that, even within these constraints, upper-class Romans could carefully weigh the options available to them and make calculated economic decisions that would, in their mind, safeguard this desired economic stability.

Consequently, it seems questionable whether the pessimistic and static view of Roman agriculture recently presented by Jongman in his analysis of the economy and society of Pompeii can provide an adequate basis for analyzing the Roman agrarian economy in general. According to Jongman, the Roman economy offered little scope for growth, and the elite landowners of Pompeii, in their pursuit of eco-

---

26. For the role of olive oil as an export crop increasing the wealth of the elite, see the arguments of D.J. Mattingly, "Oil for Export? A Comparison of Libyan, Spanish and Tunisian Olive Oil Production in the Roman Empire," *JRA* 1 (1988): 33–56, and "The Olive Boom: Oil Surpluses, Wealth and Power in Roman Tripolitania," *Libyan Studies* 19 (1988): 21–41.

27. See Lo Cascio, *Storia di Roma*, vol. II, pt. 2, 327–30, 344–58. For upper-class profits from commerce, see J.H. D'Arms, *Commerce and Social Standing in Ancient Rome* (Cambridge, Mass., 1981), as well as Harris, "Between Archaic and Modern," 14–25. For the general thesis that commerce occupied a structurally similar position in the Roman economy to that it would hold in early modern Europe, see Pleket, "Wirtschaft," 25–160, especially 31–55, with discussion of comparative evidence, as well as T. Schleich, "Überlegungen zum Problem senatorischer Handelsaktivitäten," pts. 1 and 2, *MBAH* 2, no.2 (1983): 65–90, 3, no.1 (1984): 37–76. M. Rostovtzeff, *The Social and Economic History of the Roman Empire*, 2d ed., rev. P.M. Fraser (Oxford, 1957), 130–91, had argued that large fortunes in the Roman Empire were primarily based in commerce. See also H. Pavis-d'Escurac, "Aristocratie sénatoriale et profits commerciaux," *Ktema* 2 (1977): 339–55. C.R. Whittaker, "Trade and the Aristocracy in the Roman Empire," *Opus* 4 (1985): 49–75, argues that trade in the Roman Empire was not driven by the free market as much as by state needs and the effort of elites to satisfy their own needs.

nomic stability, relied on small tenants cultivating their estates with a bare minimum of resources and eking out at best a precarious living.[28] Jongman's view of the Roman economy has much in common with that of Finley, who has argued that the very limited growth and opportunities for enrichment outside of agriculture in the Roman economy limited the scope for decision making for upper-class Romans. In Finley's view, Roman landowners did not have sufficient knowledge about the Roman economy to engage in any sort of rational, long-term economic planning.[29] The decisions that Roman landowners made in economic matters were more likely to be the result of customary practice than of any careful and sustained evaluation of the costs and benefits involved. On the other side of the debate, some economic historians believe that Roman landowners engaged in economic planning in many respects comparable to that of a modern capitalistic entrepreneur. Indeed, De Neeve, in his study of tenancy on the estates of Pliny the Younger, distinguishes between "capitalistic" and "peasant" sectors in the Roman economy, and he sees upper-class landowners as operating largely within the capitalistic sector.[30] Capogrossi Colognesi has now questioned the usefulness of this dichotomy for understanding the Roman economy and has offered a more nuanced analysis. He suggests that a continuum linked the practices of the wealthiest tenants—the group most likely to have displayed elements of capitalistic economic planning—and those of the humble tenants of much more modest resources. Capogrossi Colognesi further emphasizes that all tenants, even humble farmers with few resources of their own, were involved with the market, if only to sell a small surplus to be able to pay a cash rent. Whereas we should not relegate the farmers representing the most modest levels of wealth to a purely subsistence economy, neither should we overestimate the "capitalistic" aspect of the economic planning of wealthier tenants and landowners.[31]

---

28. Jongman, *The Economy and Society of Pompeii*, especially 97–154, 187–203.

29. M.I. Finley, *The Ancient Economy*, 2d ed. (Berkeley and Los Angeles, 1985), especially 95–122, with the review of M.W. Frederiksen, "Theory, Evidence and the Ancient Economy," *JRS* 65 (1975): 164–71, especially 168–70.

30. See De Neeve, *Athenaeum* 78 (1990): 364–65; see also idem, "The Price of Agricultural Land in Roman Italy and the Problem of Economic Rationalism," *Opus* 4 (1985): 77–109, and *Colonus*.

31. See Capogrossi Colognesi, *Ai margini della proprietà fondiaria*, chap. 4, especially 235–37, and "Il regime," 163–253, especially 215–16. On whether the Roman economy can be analyzed in terms of the modern notion of capitalism, see J.R. Love, *Antiquity and Capitalism* (London and New York, 1991).

A different perspective on upper-class economic planning is offered by Rathbone in his thorough study of the large estate in third-century Egypt mentioned earlier, that of Aurelius Appianus. The management of this estate is attested in a degree of detail unparalleled in the Roman world by the so-called Heroninos archive, a collection of hundreds of documents, primarily accounts and correspondence, kept by a manager of an individual division of the estate.[32] As Rathbone demonstrates, the methods of accounting and cost control practiced on the estate of Appianus were quite sophisticated, and the managers had sufficient information at their disposal to assess the relative profitability of the various crops produced on the estate. Under this circumstance, the managers of the estate were able to engage in "rational" long-term planning and to make the most remunerative use of the resources of labor and capital that were available to them. Still, in my view, the scope for economic planning by large landowners was somewhat more constrained than is argued by Rathbone's painstaking analysis of this archive.[33]

Specifically, what would any method of accounting, no matter how sophisticated, permit a landowner in the Roman Empire to accomplish? On the one hand, the central administration of the estate of Appianus could keep accurate records about costs of production and could decide on this basis which crops to cultivate. The central administration of the estate, moreover, would be in a position to reallocate resources in response to changing market conditions. On the other hand, Appianus and estate owners like him had few alternatives, if any, for investing large amounts of wealth, and they usually had no choice but to invest the bulk of their wealth in producing crops for the market, no matter how the market for farm products might change. Given these constraints, their objective was to stabilize income by keeping the costs of production under control. The sophisticated methods of accounting attested in the Heroninos archive allowed Appianus to achieve this objective. Insofar as he could keep costs under a tight control, he could invest his resources in an optimal manner that balanced the risks of investing in more remunerative crops, such as wine, against the desirability of keeping his income stable.

---

32. See Rathbone, *Economic Rationalism*, especially 244–330, on the sharing of resources within the estate.

33. I discuss Rathbone's analysis in "Economic Rationalism in Roman Agriculture," *JRA* 6 (1993): 476–84.

Nevertheless, no system of management, no matter how carefully conceived, could alter the basic constraints under which landowners like Appianus had to operate. All landowners depended for their income on marketing a relatively limited range of crops, and the scope for introducing new cash crops was certainly restricted. Put simply, they had no choice but to seek their fortunes in an economy with little short-term growth and few ways to invest their wealth safely. This planning as exhibited in the Heroninos archive, then, can be regarded as "rational" only in a narrow sense. What was rational about the economic planning of landowners was the way in which they sought to adapt existing social and legal institutions to safeguard a stable yearly income.[34]

In undertaking this study, I must acknowledge at the outset that a model of the Roman economy based primarily on legal sources is no more sufficient for explaining the economic planning by all Roman landowners than is a model based on other sources, such as documentary papyri or the correspondence of Pliny the Younger. But the legal sources do provide one important advantage to the historian of the ancient economy; namely, they represent the most extensive body of literature revealing the way upper-class Romans understood and undertook to solve economic problems affecting their class. The problems, of course, are formulated in legal terms, but the solutions that the jurists and the Roman government prescribed allow us to see, however indirectly, the economic planning that upper-class Romans considered feasible or desirable in the Roman economy. Hence, an analysis of the legal evidence can help us to trace what upper-class Romans saw as their basic and common financial goals, as well as their assessment of the major constraints that limited their capacity to achieve economic security. Such a model cannot be expected to describe or predict the behavior of individual landowners, but it portrays the general relation-

---

34. The achievement of this goal can usefully be understood in terms of what H.A. Simon characterizes as "bounded rationality"—that is, the limits on available knowledge and the ability to implement a decision—and "satisficing" solutions, which do not necessarily make the most optimal possible use of resources but rather achieve a desired goal. See H.A. Simon, *Administrative Behavior*, 3d ed. (New York, 1976), 35–41, 80–81, 240–44, and *The Sciences of the Artificial*, 2d ed. (Cambridge, Mass., 1981), 31–61, as well as *Reason in Human Affairs* (Stanford, 1983), 84–85, and his earlier *Models of Man, Social and Rational* (New York, 1957), 196–206.

ships that characterized the Roman economy. Analyzing the legal sources in the manner proposed thus gives us a better grasp of the circumstances affecting all landowners, even if individual landowners adopted many different solutions to the common economic problems facing them. The model that I seek to develop here will have a heuristic function, in that it offers a useful reference point for analyzing the economic planning by individual landowners, to the extent that it is possible to do so.[35]

---

35. For this conception of the usefulness of a generalizing model to understanding fundamental relationships in the Roman economy, see Jongman, *The Economy and Society of Pompeii*, 16–62. See also M.I. Finley, *Ancient History: Evidence and Models* (London, 1985), passim, especially 60, cited by S. Hodkinson, "Animal Husbandry in the Greek Polis," in C.R. Whittaker, ed., *Pastoral Economies in Classical Antiquity* (Cambridge, 1988), 35–74, at 69 n. 2.

1
Investment Goals and the Law of Tutorship

In this chapter, I begin my investigation of economic planning by upper-class Romans by examining the assumptions that the Roman legal authorities made about investment and income when they developed rules for tutorship and curatorship. As we will see, in the jurists' thinking, the careful administration of a pupil's property served to make sure that the pupil would have sufficient income to maintain a standard of living appropriate to his or her class. This aspect of the law of tutorship is very significant, since the perception on the part of the jurists about the financial needs of pupils had important consequences for the degree to which they would allow tutors to take risks in managing pupils' wealth. My argument is that, in developing rules concerning the management of pupils' property, Roman legislators and the jurists applied a consistent conception of the preeminent value of productive land. The Roman legal authorities were faced with the task of formulating general rules to protect the financial interests of pupils and minors, and for this purpose agricultural property provided the only resource that could generally be relied on to provide an upper-class Roman with a secure long-term income. By implication, the uninterrupted and undisturbed ownership of productive land was the single most important measure of an upper-class Roman's financial security. As I will show, this conception of agricultural property as a form of investment was not chiefly a product of the internal logic of the law but was rooted in a more general upper-class approach to economic problems.

Tutorship, or *tutela*, was one of the venerable institutions of Roman civil law, dating back to the Twelve Tables. Like the institution of inheritance, tutorship was a product of the Roman concern to protect a family's interest in maintaining control over its property.[1] In general terms,

---

1. For the general importance of tutorship in the Roman world, see R.P. Saller, *Patriarchy, Property and Death in the Roman Family* (Cambridge, 1994),

a fatherless child (whether through the death of the father or through emancipation) required the supervision of a tutor until the age of puberty. The tutor was responsible for the pupil's education and upbringing, and he was also required to administer the pupil's property. The pupil could not make any significant transaction without the tutor's *auctoritas*.

An institution in many respects similar to tutorship was *cura minorum*, which involved the supervision of the business affairs of minors (males who had not yet reached the age of twenty-five). The two institutions, originally distinct from one another, were completely assimilated by Justinian's time, and this fact is reflected in the legal sources, which to a large extent treat the duties of the tutor and the curator as identical. The assimilation of tutorship and curatorship in the legal sources was long considered to be the result of Justinianic interpolation, but more recently legal historians have recognized that the classical legal authorities had already begun defining the duties of the curator to make them resemble those of the tutor.[2] In classical times, then, minors under certain circumstances were treated much like pupils, in that they required the assistance of a curator to manage their property and transact major business.

In analyzing the Roman law of tutorship, we should recognize at the outset that the jurists concerned themselves to a large degree with the interests of the upper classes. The reason for this focus of concern is straightforward; the administration of property became increasingly complex as the parties involved became wealthier, and controversies about the financial affairs of the wealthy were fertile ground for lawsuits.[3] Although pupils and tutors represented all ranks of society, and

---

181–203, as well as J.A. Crook, "Women in Roman Succession," in B. Rawson, ed., *The Family in Ancient Rome* (London and Ithaca, 1986), 58–82, at 62–63, and Kaser, *RP* I, 85–86, 352–54. For the duties of tutors and curators, see Kaser, *RP* I, 83–91, 352–72; for tutors, see also E. Sachers, "Tutela," *RE* 7 (1948): 1497–1599, and J.-U. Krause, *Witwen und Waisen im römischen Reich*, vol. III, *Rechtliche und soziale Stellung von Waisen* (Stuttgart, 1995), 85–112, especially 97–103 on the duties of tutors in managing wealth of pupils.

2. For discussion of the scholarly debate, see G. Cervenca, "Studi sulla *Cura Minorum*," pt. 1, "*Cura Minorum* e *Restitutio in Integrum*," *BIDR*, 3d ser., 14 (1972): 235–317, at 235–44.

3. I owe the observation about the frequency of law suits involving upper-class pupils to Professor Shael Herman. See also Krause, *Witwen und Waisen*, 87. For a similar judgment about the upper-class concerns of the jurists in deciding

although the same law applied irrespective of social class, the jurists in many cases were clearly concerned with treating issues exclusively affecting the interests of wealthier pupils. As a consequence, the jurists' assumptions about the economic interests of pupils offer a means of analyzing how upper-class Romans in general evaluated their own economic interests.

This preoccupation on the part of the jurists is indeed made explicit in a number of their responses in the Digest that address issues that are solely the concern of upper-class Romans. These responses deal with the legal responsibility of tutors when the pupil's property to be administered was distributed over a wide geographical area. Thus in several responses, the jurists posit that a pupil or a minor will be represented by a number of tutors or curators, each representing a *regio* in which the individual held property (e.g., Ulp. D. 26.7.4, 9 *ad Sab.*, Ven. D. 26.7.51, 6 *stipul.*). The pupil might also own property both in Italy and in the provinces (Scaev. D. 26.7.47.2, 2 *resp.*, Papin. D. 26.7.39.3, 26.7.39.8, 5 *resp.*). In the understanding of the mid-second century-jurist Venuleius, for example, it was a common practice to separate the administration of the pupil's property in Rome from that of his or her property in the provinces (loc. cit.), while Hermoginianus, writing at the time of Diocletian, discussed legal complications arising from a case involving a pupil with property in Rome as well as in the provinces (D. 26.5.27 pr., 2 *iuris epit.*). The imperial government also answered petitions involving wealthy pupils. By way of illustration for a different legal point, Ulpian referred to a rescript, issued by Septimius Severus and Caracalla, concerning a pupil with property in Italy and Africa (D. 26.7.3.4, 35 *ad ed.*). These same emperors addressed in another rescript the possibility that curators responsible for overseeing the management of property in Italy might be relieved of this responsibility for property in the provinces (CJ 5.62.2, A.D. 204). In another rescript, Caracalla provided a mechanism to select tutors for pupils who only owned property outside of the province in which they lived (CJ 5.32.1, A.D. 215). Later, Alexander Severus addressed the problem of assigning curators to assist a tutor administering widely dispersed properties (" . . . late diffusum patrimonium," CJ 5.36.3 = 5.62.11.1, A.D. 231). This sampling

---

cases in testamentary law, see E. Champlin, *Final Judgments* (Berkeley and Los Angeles, 1991), 41–63. For discussion of many of the passages examined in this section, see S. Solazzi, "Tutele e Curatele," in *Scritti di diritto romano*, vol. II (Naples, 1957), 1–66, originally published in two parts in RISG 53 (1913): 263–97; 54 (1914): 17–70, 273–94.

of cases suggests the high financial stakes involved in a tutor's administration of a pupil's property.[4]

The Uses of Income for Upper-Class Romans

Before we evaluate how the jurists regulated the duties of tutors in administering the wealth of pupils, it will be helpful to consider the purposes that wealth served for upper-class Romans in the thinking of the jurists. We can appreciate the thinking of the Roman legal authorities on this matter by considering first how the Roman senator Pliny the Younger conceived of his own financial situation. Pliny, whose published correspondence allows us to trace the economic concerns of an upper-class Roman landowner in some detail, took an extremely conservative approach to his finances, seeking a steady income with as little risk as possible. Pliny's senatorial career and lofty social position imposed a number of expenses on him, including those associated with public office, as well as the obligation to engage in acts of public and private generosity, and Pliny seems to have been concerned with stabilizing his income as much as possible to make sure of his ability to meet these various obligations.[5]

Pliny reveals a conservative attitude toward his income in a letter to a certain Calvina (*Ep.* 2.4).[6] In this letter, Pliny excuses the debts that her

---

4. On the jurist Veneleius Saturninus, see W. Kunkel, *Herkunft und soziale Stellung der römischen Juristen*, 2d ed. (Graz, Vienna, and Cologne, 1967), 181–84. The range of social classes with which the jurists dealt when making rules about tutorship is made more specific in two passages. Gaius referred to a senatus consultum authorizing the appointment of a curator to oversee, in order to pay off debts, the alienation of property belonging to a "clara persona, veluti senatoris vel uxoris eius" (*D.* 27.10.5, 9 *ad ed. prov.*). A much lower social status is involved in the case of a centurion who assigned a curator to his "impubes filius" (Papin. *D.* 26.7.40 pr., 6 *resp.*). The issue of administering property dispersed among provinces was also handled in the later empire. Thus Constantine and Licinius discussed the division along provincial boundaries of the risk in administering a tutorship (*CJ* 5.40.2, A.D. 319 = *CTh.* 2.4.1.1).

5. For Pliny's finances, see R. Duncan-Jones, *The Economy of the Roman Empire*, 2d ed. (Cambridge, 1982), 17–32, as well as P.W. de Neeve, "A Roman Landowner and His Estates: Pliny the Younger," *Athenaeum* 78 (1990): 363–402, and A. Chastagnol, *Le sénat romain à l'époque imperiale* (Paris, 1992), 145–53.

6. My discussion of this letter is based on my treatment of it in "Investment in Estates by Upper-Class Landowners in Early Imperial Italy: The Case of Pliny the Younger," in H. Sancisi-Weerdenburg et al., eds., *De Agricultura* (Amsterdam, 1993), 214–37, at 215–16.

now late father had owed him, so that she can accept her inheritance. This letter reveals how Pliny's social relationships involved him in a complex array of financial obligations. For reasons now unknown, Pliny had earlier maintained long-standing ties with Calvina's father, repeatedly helping him financially. When Calvina married, Pliny saved her apparently impoverished father from a great embarrassment by providing the needed dowry in the amount of one hundred thousand sesterces (*Ep.* 2.4.2). The dowry was just one of a series of gifts or loans that Pliny made to Calvina's father, and the excusing of debts in the present letter was the final benefit that Pliny was able to confer on a family now dependent on his generosity.

It is significant for our purposes that Pliny valued his own income chiefly because it enabled him to perform the type of generosity exemplified in this letter, and he does not omit reminding Calvina of this principle. Thus Pliny describes to Calvina how his ability to bear the expenses imposed by his social position and public career depended directly on the security of the income that he received from his estates.

> Sunt quidem omnino nobis modicae facultates, dignitas sumptuosa, reditus propter condicionem agellorum nescio minor an incertior; sed quod cessat ex reditu, frugalitate suppletur, ex qua velut fonte liberalitas nostra decurrit. (*Ep.* 2.4.3)

> [I have altogether modest resources, an expensive social rank, and an income that, on account of the condition of my little fields, is either lower or more uncertain. But what is lacking from income is made up for by frugality, from which our generosity flows, as if from a spring.]

Pliny, like most members of his class, derived the bulk of his income from his estates. He may have had particular reasons to be concerned about his estates at the time that he wrote this letter, since throughout the period of his correspondence he experienced problems with indebtedness on the part of his tenants. These difficulties led him to take some drastic measures, including granting remissions of rent. When that did not succeed in alleviating these problems, he replaced with sharecropping the traditional system of leasing for cash rents (*Ep.* 3.19, 7.30.3, 9.37, 10.8.5).[7] It does not matter whether Pliny's finances were really as

---

7. Pliny's problems with his tenants are discussed in chap. 4.

"modest" as he characterizes them.⁸ Rather, it is more to the point that Pliny links his requirement for an income directly to the financial demands imposed by his social position. Pliny emphasizes this principle when he describes his finances to Calvina, but it was no doubt well understood by Pliny's contemporaries. Forgiving the debts of Calvina's father, like his earlier gift of one hundred thousand sesterces to her as a dowry, represented just one of those many discretionary expenses.⁹ According to his own description, when his resources did not prove sufficient to provide the income that his social obligations required, Pliny neither changed his obligations nor sought new sources of income. Rather, he saw as his only choice reducing his discretionary

---

8. Pliny's level of wealth was dwarfed by that of his rival, the praetorian senator M. Aquilius Regulus, who according to Pliny had achieved a fortune of sixty million sesterces (*Ep.* 2.20.13). Duncan-Jones, *Economy,* 17–20, estimates Pliny's holdings in land at ca. seventeen million sesterces, perhaps twice what was needed to provide an income appropriate for a senator; cf., however, De Neeve, *Athenaeum* 78 (1990): 369–70, who is skeptical about estimates of Pliny's fortune. For a lower assessment of the minimum senatorial fortunes, see R. Talbert, *The Senate of Imperial Rome* (Princeton, 1984), 495–97. For the varying financial resources of senators, see Talbert, 47–53; see also Duncan-Jones, 343–44, for a list of the known sizes of private fortunes during the principate, with sources. Cf., however, W. Scheidel, "Finances, Figures and Fiction," *CQ* 46, no. 1 (1996): 222–38, at 230–32, who argues that the sizes of private fortunes reported in literary sources are generally stylized and so cannot be pressed

9. For Pliny's gifts, see Duncan-Jones, *Economy,* 27–31, who notes that Pliny made total public gifts of about five million sesterces, both while living and through his will. This figure does not include the numerous gifts that Pliny made to private individuals. Such gifts will have represented only part of Pliny's socially imposed financial obligations, which included expenses associated with officeholding, social functions, and so on. For the financial obligations imposed by a senatorial career, see Talbert, *The Senate of Imperial Rome,* 54–66; cf. also K. Hopkins and G. Burton, "Political Succession in the Late Republic (249–50 B.C.)," in K. Hopkins, *Death and Renewal* (Cambridge, 1983), 31–119. The difficulties encountered by senatorial landowners in the later empire in achieving incomes sufficient to meet their immense social and political expenses is emphasized by D. Vera, "Strutture agrarie e strutture patrimoniali nella tarda antichità: L'aristocrazia romana fra agricoltura e commercio," *Opus* 2 (1983): 489–533; cf. also idem, "Simmaco e le sue proprietà: Struttura e funzionamento di un patrimonio aristocratico del quarto secolo d.C," in F. Paschoud et al., eds., *Colloque Génevois sur Symmaque* (Paris, 1986), 231–76, at 231, 265–66. See also A. Bürge, "Vertrag und personale Abhängigkeiten im Rom der späten Republik und der frühen Kaiserzeit," *ZRG* 97 (1980): 105–56, at 139–43.

expenses. Here we see Pliny revealing a basic attitude toward his wealth. His overriding concern was to maintain the level of income adequate for his social obligations.[10] No doubt, like many of his contemporaries, he wanted to avoid winding up like Calvina's father, who had an estate so encumbered by debt that his daughter would not have been able, without Pliny's timely generosity, to take up her inheritance.[11]

A comparable concern for maintaining a standard of living appropriate to one's class underlies the jurists' treatment of the discretionary expenses that a tutor was expected to authorize for his pupil. The tutor not only oversaw the management of the pupil's property but provided for the pupil's sustenance and education. In fact, the tutor might have to provide for the full range of expenses that characterized aristocratic life, and the level of expenditures that the tutor might authorize was expected to be based on the social rank of the pupil concerned, as well as on the resources available to that individual. This fact made the responsibilities of the tutor broad, imposing on him, in the Severan jurist Paul's formulation, the obligation to provide teachers in accordance with the wealth and social rank of the pupil and to authorize on the same basis other discretionary expenses, such as providing for the sustenance of the pupil's freedmen and slaves or sending gifts to the pupil's relatives (Paul. D. 26.7.12.3, 38 *ad ed.*).[12]

---

10. Cf. Pliny's description of the finances of his friend Atilius Crescens: "homo est alieni abstinentissimus sui diligens; nullis quaestibus sustinetur, nullus illi nisi ex frugalitate reditus" (*Ep.* 6.8.5–6). Pliny sought to manage his income so as to maintain a standard of living, not to increase wealth for its own sake. This approach to property and the concomitant concepts of profit from agriculture are analyzed in an important article by J. Love, "The Character of the Roman Agricultural Estate in the Light of Max Weber's Economic Sociology," *Chiron* 16 (1986): 99–146, especially 138–41, with comparative discussion of the Renaissance humanist Leon Batista Alberti and of M. Weber's analysis of the economic activities of that figure. See also idem, *Antiquity and Capitalism* (London, 1991), 101–102.

11. For the legal situation in this letter, see J.W. Tellegen, *The Roman Law of Succession in the Letters of Pliny the Younger*, vol. I (Zutphen, 1982), 18–29, as well as Sherwin-White, *Commentary,* 149. For the social significance of the type of generosity exhibited by Pliny, see R.P. Saller, *Personal Patronage under the Early Empire* (Cambridge, 1982), 119–43.

12. Paul. D. 26.7.12.3: "Cum tutor non rebus dumtaxat, sed etiam moribus pupilli praeponatur, imprimis mercedes praeceptoribus, non quas minimas poterit, sed pro facultate patrimonii, pro dignitate natalium constituet, ali-

The tutor entrusted with such an important duty was to apply the same standard of care in administering discretionary expenditures for the pupil as he was in general to apply in administering the pupil's property (see the next section of this chapter). The standard of care to which the tutor was held, then, was that which a pater familias would exhibit toward his own affairs (Ulp. D. 26.7.10, 49 [59 Lenel] *ad ed.*).[13] Consequently, the tutor was not to authorize any expenditure from the pupil's property that he would not have authorized from his own property: "nimium est licere tutori respectu existimationis pupilli erogare ex bonis eius, quod ex suis non honestissime fuisset erogaturus" (Paul. D. 26.7.12.2, 38 *ad ed.*). The tutor was apparently given some latitude in deciding on expenditures for the pupil, however. Paul referred to rescripts of Trajan and Hadrian upholding actions done in good faith by the tutor (D. 26.7.12.1, 38 *ad ed.*). Later, Caracalla allowed tutors to make expenditures in good faith even without the authorization of the praetor, as long as they were necessary and arising "from honorable and just causes" ("sumptus . . . necessario et ex honestis iustisque causis . . . ," CJ 5.37.3, A.D. 212).[14] Apparently, there was some general agreement about what a level of daily expenses appropriate for a pupil might be, since Scaevola asserted that this was to be assessed "boni viri arbitrio" (D. 26.7.47.1, 2 *resp.*).

It will be useful to consider specific examples of the types of expenditures that tutors were expected to authorize. One of the principal responsibilities of the tutor was to provide for feeding, clothing, and education of the pupil; the term used for this type of expense in the legal sources is *alimenta*. Failure to provide *alimenta* would constitute a grievous violation on the part of the tutor; a tutor so failing might be *suspectus*, that is, liable to be removed from his post and made subject to penalties (Ulp. D. 26.10.3.14., 35 *ad ed.*, 26.10.7.1, 1 *de omn. trib.*). The pupil, moreover, could be placed in possession of the property belonging to a tutor found to be *suspectus* (Ulp. D. 26.10.7.2, referring to a letter of Septimius Severus and Caracalla).

---

menta servis libertisque, nonnumquam etiam exteris, si hoc pupillo expediet, praestabit, sollemnia munera parentibus cognatisque mittet."

13. Ulp. D. 26.7.10: "Generaliter quotienscumque non fit nomine pupilli quod quivis pater familias idoneus facit, non videtur defendi: sive igitur solutionem sive iudicium sive stipulationem detrectat, defendi non videtur."

14. Cf. ibid.: " . . . nam quod a tutoribus sive curatoribus bona fide erogatur, potius iustitia quam aliena auctoritate firmatur."

Most of the discussion by the jurists about this responsibility concerns the right of the tutor to compensation for expenses that he had to bear in providing for his pupil's *alimenta*. The tutor was entitled to be paid back for these expenses from the pupil's funds or property (cf. Ulp. *D.* 27.2.2 pr., 36 *ad ed.,* Ulp. *D.* 27.4.1 pr., 36 *ad ed.*), but at the same time he was obliged to provide for the pupil's upbringing and could be compelled by the magistrate to do so (Caracalla, *CJ* 5.50.1 pr., A.D. 215).[15] The issue might arise at the *iudicium tutelae*, the judgment at which the tutor was to render an accounting for his administration of the pupil's property. The difficulty was to determine what constituted an appropriate level of spending for which the tutor would have a right to compensation. The praetor or provincial governor had the duty and authority to establish a level for the *alimenta* (Ulp. *D.* 27.2.2.1, 36 *ad ed.,* 27.2.3 pr.-1, 1 *de omn. trib.*; cf. Alexander Severus, *CJ* 5.50.2 pr., A.D. 223). The praetor was to establish an appropriate level of expenditure based on the size of the pupil's property.[16] This official was also to take into consideration the social needs of the pupil, including the number of slaves serving him (or her), the salaries of his teachers, his clothing, his housing, and his age (Ulp. *D.* 27.2.3.2; cf. *CJ* 5.50.2.1).[17] The level of *alimenta* might be lowered if the property of the pupil declined, and likewise the *alimenta* might be raised as the size of the property of the pupil increased (Ulp. *D.* 27.2.3.6, 1 *de omn. trib.*). Controversy might arise if the tutor provided for the needs of the pupil without first obtaining a decree from the praetor setting the level of the *alimenta*. The tutor was still entitled to compensation, but only if his expenditures were made in accordance with the principles I have outlined. The determination

---

15. Ulp. *D.* 27.2.2 pr.: "Officio iudicis, qui tutelae cognoscit, congruit reputationes tutoris non improbas admittere, ut puta [se *ins. Ruckerus*] impendisse in alimenta pupilli vel disciplinas."

16. The *modus patrimonii* (Ulp. *D.* 27.2.3.1); cf. the *pro modo facultatum* (Alexander Severus *CJ* 5.50.2 pr.).

17. Ulp. *D.* 27.2.3.2: "Ante oculos habere in decernendo et mancipia, quae pupillis deserviunt, et mercedes pupillorum [praeceptorum *Leoninus*] et vestem et tectum pupilli: aetatem etiam contemplari, in qua constitutus est cui alimenta decernuntur." Cf. Ulp. *D.* 27.2.2.5: "Idem ad instructionem quoque . . . pupillorum . . . pupillarum [<ve> *ins. Mo.*] . . . solet decernere respectu facultatium et aetatis eorum qui instruuntur." Ulpian (*D.* 27.2.3.3) exhorted the praetor to observe frugality in setting the level of *alimenta:* see the critical apparatus in the Mommsen and Krueger edition.

was to be made by the judge trying the case, who was to take into account the same factors that the praetor would in issuing a decree concerning the pupil's *alimenta* (Ulp. D. 27.2.2 pr.-2, 36 *ad ed.*).[18] A conception that *alimenta* had to be appropriate to a social class finds expression in a ruling of the second-century jurist Valens, according to which *alimenta* whose level was not specified in the testator's will were to be established in accordance with the social rank of the recipient in conjunction with the generosity and also the wealth of the testator (D. 34.1.22 pr., 1 *fideicommis.*).

The tutor was also to provide for other types of expenditures associated with upper-class life. An upper-class Roman had to have slaves, and accordingly Gaius required the tutor to establish the number of slaves in attendance on the pupil in accordance with the pupil's social rank and wealth: "secundum dignitatem facultatesque pupilli" (D. 26.7.13 pr., 12 *ad ed. prov.*). The question of providing gifts on solemn occasions was also one that long occupied the jurists. Gaius set a limit on wedding gifts and other discretionary expenses to be paid for with

---

18. For discussion of this type of expenditure, see Saller, *Patriarchy, Property and Death*, 198–200. The papyri provide examples of documentation of the expenditures that guardians made toward the support of their wards. For example, *P.Fam.Tebt.* 53 (A.D. 208 [209?]-219/20, *Pap.Lugd.Batav.* VI), concerning Antinoite citizens, is a series of receipts issued by a ward's mother to the guardian, for payments toward the ward's maintenance, including τρόφια, ὀψώνια, and ἱματισμός. *P.Oxy.* XLVIII 3921–22 (A.D. 219) is a guardian's accounting for investments and expenditures for two wards. In *P.Oxy.* VI 898 (A.D. 123), a minor complained about the refusal of his guardian (in this case his mother) to provide for his needs. *PSI* XII 1258 (III c.) represents a claim made by a widow to the guardian of her daughter, for payment of an allowance established in the late husband's will. By contrast, *P.Oxy.* XLIII 3113 (A.D. 264/65) is a fragmentary petition to the prefect by a guardian who claimed to have insufficient means to provide for his wards: see J.R. Rea, introd., *P.Oxy.* XLIII 3113, pp. 160–61. Finally, *M.Chr.* 86 (*BGU* I 136, A.D. 135, Memphis) records a trial before the *archidikastes* involving an underage daughter and the two heirs of her brother, with whom she had split the estate. For discussion of the papyrological evidence for tutorship, with more examples, see Krause, *Witwen und Waisen*, 104–7. For the legal institution of tutorship in Roman Egypt, see R. Taubenschlag, *The Law of Greco-Roman Egypt in Light of the Papyri, 332 B.C.–640 A.D.*, 2d ed. (Warsaw, 1955), 168; and idem., "Die Alimentationspflicht im Licht der Papyri," in *Opera Minora*, vol. II (Warsaw, 1959), 539–55, and the earlier discussion by L. Mitteis in L. Mitteis and U. Wilcken, *Grundzüge und Chrestomathie der Papyruskunde*, vol. II (Leipzig, 1912; reprint, Hildesheim, 1963), 1, 248–56.

the money belonging to the pupil (D. 26.7.13.2, 12 *ad ed. prov.*).[19] Another category of expenses was dowries, which aristocratic women required for marriage. A female pupil, of course, might need a dowry for herself, and in determining how large the dowry was to be, the jurists applied the same principle applied to other expenditures. Thus according to Labeo, the size of a dowry to be provided on the judgment of a daughter's tutors could readily be set on the basis of the testator's social standing and wealth: ". . . ex dignitate, ex facultatibus, ex numero liberorum testamentum facientis" (apud Celsus, D. 32.43, 15 *dig.*).[20] The provision of dowries by tutors might engender a wide range of legal problems. For example, in a largely interpolated passage, Paul (D. 26.7.43.1, 7 *quaest.*) discussed the obligations of a tutor representing his niece; the tutor promised his niece a substantial dowry, but then he sought to withdraw from his promise when increasing debts meant that the promised dowry exceeded her resources. In the present question, the degree of the tutor's liability and the pupil's subsequent obligations toward him depended on whether he promised the dowry out of his own property or out of the property belonging to his pupil.

The jurists were also careful to define the precise circumstances under which a pupil could be expected to shoulder the expenses of a dowry for a female connection.[21] Hence Paul (D. 26.7.12.3, 38 *ad ed.*), concerned to prevent the pupil's property from being frittered away, acknowledged the obligation of the pupil to provide a dowry for a sister, but only for one born from the same father; any other type of expense, such as a dowry for a half sister who otherwise could not marry, was considered *liberalitas*, not a justifiable expenditure. Neratius had earlier addressed the obligations on the part of a curator of a minor

---

19. Gaius D. 26.7.13.2: "In solvendis legatis et fideicommissis attendere debet tutor, ne cui non debitum solvat, nec nuptiale munus matri pupilli vel sorori mittere."

20. Cf. Pap. D. 23.3.69.4 (4 *resp.*): the size of a dowry that a father-in-law had promised could be assessed "pro modo facultatium patris et dignitate mariti." On the assessment of dowries in accordance with social station, see Saller, *Patriarchy, Property and Death*, 215–16.

21. On the obligation on the part of upper-class Romans to provide dowries for their relations, see Bürge, *ZRG* 97 (1980): 141 and n. 167, referring to Kaser, *RP* I, 335 n. 22. For the size of the dowry, see S. Treggiari, *Roman Marriage* (Oxford, 1991), 340–48. It is not clear under what circumstances this obligation might be enforced as a legal requirement.

to provide for the expenses of both a dowry and a wedding (D. 26.7.52, 1 *resp.*). The requirement that a curator provide for a woman's dowry was reasserted by Alexander Severus. In answering a petition from a woman named Melitia seeking a dowry, the emperor ruled that the provincial governor could compel curators to provide a dowry appropriate to the social station of the woman: "... quod moderatum est honestae personae" (*CJ* 5.37.9, A.D. 230).

Despite these strictures, the tutor was expected to provide for certain types of expenses not directly affecting the pupil. For example, Julian approved compensation for a tutor who provided for the upkeep and education of a sister of his pupil; the sister had been left only a legacy to be paid out as a dowry by the heir (the pupil of the tutor in question), and the tutor had been importuned by the relatives of the woman to cover these expenses (Julian. D. 27.2.4, 21 *dig.*). The principle was that the tutor was to pay for the basic upkeep of near relatives of his pupil if no other provision had been made for that type of expense. Ulpian generalized on this principle, noting that, although the tutor could not lawfully reduce his pupil's property through gifts, there were certain types of expenditures that the tutor had to authorize, such as the support of an indigent mother or sister. In fact, the tutor might be liable if he failed to carry out this duty (Ulp. D. 27.3.1.2, 36 *ad ed.*). But to avoid liability for authorizing such expenditures, the tutor had to be sure that stringent criteria were met, namely, that the mother was in fact in need and that the pupil had sufficient resources (Ulp. D. 27.3.1.4).[22] Even Gaius, who, as we have seen, took a very conservative view about the discretionary expenditures that he would allow the tutor to make, permitted the tutor to provide for the sustenance of an otherwise destitute mother or sister (D. 26.7.13.2, 12 *ad ed. prov.*).[23] Providing for these social needs was one of the basic duties of a tutor; he was to make sure that the pupil continued to enjoy the privileges and to fulfill the responsibilities of the social class into which he or she was born.[24]

---

22. Again, Ulpian refined a more generous opinion of Labeo (apud loc. cit.), who simply approved such expenditures.

23. Gaius D. 26.7.13.2: "... aliud est, si matri forte aut sorori pupilli tutor ea quae ad victum necessaria sunt praestiterit, cum semet ipsa sustinere non possit: nam ratum id habendum est...."

24. The social rank of the tutor was not always equal to that of the pupil; see Saller, *Patriarchy, Property and Death*, 196–98. Freedmen were often expected to perform this role. This conception of income as maintaining a standard of

The Management of the Pupil's Property

The question now to be addressed concerns what options, in the view of the jurists, Romans might have to achieve the income on which the social position of upper-class pupils depended. Concerned that the pupil achieve a safe income with a minimum degree of risk, the Roman legal authorities required tutors to exercise a great deal of caution in managing the wealth of their pupils. Accordingly, they held the tutor to the same standard of care that a pater familias would be expected to display in managing his own property: "a tutoribus et curatoribus pupillorum eadem diligentia exigenda est circa administrationem rerum pupillarium, quam pater familias [rebus suis *Iust. Eisele*] ex bona fide debet" (Callistr. D. 26.7.33 pr., 4 *de cognitionibus*).[25] The tutor was responsible to see that the pupil fulfilled all legal obligations (Ulp. D. 26.7.10), but the tutor was also restricted in how he might manage the pupil's property.[26] The standard of care to which the jurists held tutors did not mean that they were responsible for all losses. Indeed, if he administered the pupil's property diligently, a tutor could not be held liable for losses (Ulp. D. 27.4.3.7, 36 *ad ed.*).[27]

---

living for a pupil finds expression in a deposition made by the Nabataean Babatha on behalf of her orphaned son Jesus (Joshua): see *P.Yadin* 15, lines 26–28, with H. Cotton, "The Guardianship of Jesus Son of Babatha: Roman and Local Law in the Province of Arabia," *JRS* 83 (1993): 94–108, at 103–4 and n. 113.

25. Cf. Paul. D. 26.7.12.3, quoted in n. 12 in this chapter. For discussion of the social and economic implications of the tutor's duties in managing the pupil's property, see Saller, *Patriarchy, Property and Death*, 202–3.

26. The tutor was still responsible for the pupil's property in a *iudicium tutelae* even if he had authorized no transaction for the pupil (Julian. D. 26.7.18 pr., 21 *dig.*). Ulpian (D. 26.7.9.3, 36 *ad ed.*) stated that the tutor was to be slightly more rigorous in enforcing his own obligations toward the pupil than the obligations of other parties. Pomponius (D. 27.5.4, 16 *ad Quint. Muc.*) required a substitute tutor, or *pro tutore*, to exhibit the same diligence as a tutor.

27. Ulp. D. 27.4.3.7: "In contrario iudicio sufficit tutori bene et diligenter negotia gessisse, etsi eventum adversum habuit quod gestum est." D. 27.4 concerns the *contrarium tutelae iudicium*, in which the tutor was entitled to claim compensation from the pupil for expenses; see Kaser, *RP* I, 367. See also Ulp. D. 27.3.1 pr. (36 *ad ed.*). The tutor was apparently subject to varying degrees of liability, with the degree of liability depending both on the nature of the duty to be performed and on the circumstances surrounding the tutor's inability or failure to perform it. On this question, see G. MacCormack, "The Liability of the Tutor in Classical Roman Law," *Irish Jurist* 5 (1970): 369–90; on this particular passage, see ibid., 375–76. The Roman government wanted to avoid situa-

Outright fraud was one matter, but the interests of the pupil could also be seriously compromised if the tutor was negligent in administering the pupil's property. Thus the jurists were careful to develop guidelines by which the tutor was to invest any funds that the pupil might have.[28] We shall examine later in this chapter how the jurists treated the management of landed property already belonging to the pupil. Two concerns seem to have guided the development of rules concerning the administration of funds belonging to the pupil. The first is that no funds belonging to the pupil should be allowed to sit idle, because any available money could be expected to provide some income. Accordingly, as we shall see, the jurists imposed substantial penalties on a tutor who failed to make timely investment of any extra funds available to the pupil. But the second principle is equally important. In establishing guidelines for the management of property, the jurists operated on the general assumption that it was quite difficult to make an investment of the pupil's funds safe enough to provide a steady income for the long term. Accordingly, the jurists carefully defined under what circumstances the tutor became liable when investments went bad.

We can gain some insight into the difficulty of investing a given amount of capital to provide a secure income from a case, treated by the second-century jurist Scaevola, concerning the efforts of a senior mili-

---

tions like the one that arose in a legal dispute from third-century Oxyrhynchus, in which two orphans, aged four and two, respectively, and both citizens of Alexandria, were allegedly defrauded of their inheritance (*PSI* X 1102). The petitioners were bringing legal action against the heirs of the alleged despoilers to recover their property. On a more humble level, a woman named Aurelia Didyme petitioned the prefect Aristius Optatus for recovery of an inheritance (*P.Oxy.* XXXIV 2713, ca. A.D. 297) Cf. also *P.Mich.* IX 525 (Karanis, A.D. 119–24), in which a woman petitioned the prefect in seeking relief against the brother of her deceased husband, who had seized the property inherited by her two orphaned children. For the victimization of underage heirs, cf. J. Rowlandson, *Landowners and Tenants in Roman Egypt* (Oxford, 1996), 151–52.

28. In this connection, we might consider a case commented on by Scaevola at *D.* 34.3.28 pr. (16 *dig.*). A certain Aurelius Symphorus had served as surety for the tutors managing the affairs of two boys, Arellius Latinus and Arellius Felix. Apparently, the tutors had managed the affairs of the boys badly, and Symphorus was forced to spend a considerable amount of his own money on the boys' upbringing. On this case, see B.W. Frier, "Subsistence Annuities and Per Capita Income in the Early Roman Empire," *CP* 88 (1993): 222–30, at 225–26.

tary officer, described as a *praefectus legionis*, to provide for the investment of money that he left his son (*D.* 26.7.47.4, 2 *resp.*). The testator, concerned that the money that he was leaving his son not be dissipated, offered the son's tutors an option not usually found in wills. Normally the tutors would be required to invest funds belonging to the pupil, either by purchasing property or by lending it out at interest (as will be discussed later in this chapter), but at least the latter possibility clearly was not attractive to the testator, who felt that a risky series of loans might cause his son not only to lose any income from the funds in question but also the capital itself. Accordingly, the testator allowed the tutors to use the money themselves, on the condition that they pay interest of 1 percent each year: "'volo, ut sit in arbitrio tutorum filii mei, si voluerint, huius summae uncias inferre usurarum nomine ita, ne nummi dispargantur'."[29] The testator clearly offered this option because he had no confidence that conventional loans would be safe enough to maintain his son's capital, let alone to provide his son with any regular income. As a result, to guarantee the preservation of the sum of money, he turned it over to the tutors, charging them a nominal interest of 1 percent, well below the conventional market rate, which might range between 6 and 12 percent or could be even higher.[30]

The father in effect renounced a substantial amount of income for his son, and the drastic measure that he took to preserve his son's capital is only understandable in an economy in which the prospects for a safe income from conventional investments were at best uncertain. Placing the money in the hands of the tutors gave the pupil greater legal protection than possible under conventional forms of investment. The tutors would now be liable to return the sum in full when the son reached majority, and at the *iudicium tutelae* the state could proceed against the property of the tutors to gain repayment of the principal. Confiscating the property of the tutors gave the pupil more immediate

---

29. The term *unciae* denotes an interest rate of one-twelfth of 1 percent per month, or 1 percent annual interest; see the *Oxford Latin Dictionary*, s.v., citing this passage. On this case, see the discussion of A. Bürge, "Fiktion und Wirklichkeit: Soziale und rechtliche Strukturen des römischen Bankwesens," *ZRG* 104 (1987): 465–558, at 541.

30. Marcellus (apud Paul. *D.* 35.2.3.2, *lib. sing. ad leg. Falc.*) assumed an annual interest rate of 4 percent when calculating the capital value of an annuity paid to a town; on this passage, see B.W. Frier, "Roman Life Expectancy: Ulpian's Evidence," *HSCP* 86 (1982): 213–51, at 222.

access to his money than would have been possible if the money had been invested in conventional loans, since in the case of conventional investment tutors would have needed time to exact the interest and to proceed against the borrowers. Assigning the money to the tutors eliminated these steps, and the greater security gained as a result compensated for the income lost.

Although this case represents only a single example of how a father might provide for the financial security of a son, it suggests the general background of uncertain prospects for investment against which the jurists developed their rules about the responsibilities of tutors. A similar assumption about the difficulty of providing for a safe income explains another case decided by Scaevola (*D.* 34.1.15 pr., 17 *dig.*). In this case a testator sought to provide sustenance for a foster child (*alumnus*) until the latter reached his twentieth birthday. The testator's method of solving this problem was to set up a trust imposing the obligation of providing that safe income upon a trustee; the trust required the trustee to take charge of a fund (comprising four hundred thousand sesterces), which she could use as she saw fit, under the condition that she provide a 5 percent annual income to the foster child and return the principal to him after he reached the age of twenty. In the meantime, the trustee could keep whatever profits she made over and above the 5 percent annual stipend. The trustee was also enjoined to take charge of the foster child's upbringing. The legal difficulties that Scaevola addressed are envisioned as arising in the event that the trustee might be unwilling to take charge of the fund set aside by the testator.

A testator making such provisions as these clearly was seeking a solution to the continuing problem of providing for a stable long-term income in an economy with few opportunities for safe investment. Clearly, if the trustee charged to invest the money and pay the interest to the foster child refused to take responsibility for the designated fund, she would be displaying just as little confidence in her prospects for achieving a stable income as that displayed by the testator. The trustee presumably might refuse to take charge of the fund because she could not be certain of achieving every year the needed 5 percent return. Imposing the duty to manage funds on a third party may have been a common way for providing for the payment of such fixed annuities, for Scaevola discusses a second case in which a testator seeking to provide *alimenta* for a freedwoman left a fund with two other freedmen, themselves charged to pay 5 percent yearly interest and to restore the fund on the beneficiary's twenty-fifth birthday (*D*. 34.1.16.2, 18 *dig*.). A similar procedure is envisioned in a case ruled on by Papinian (*D*. 34.1.9 pr., 8 *resp*.). In this case, a testator placed ten aurei (ten thousand sesterces) at the disposal of an individual legatee, with the request that the latter provide from that amount *alimenta* for the testator's foster children.[31] Very often, of course, a testator might set up a trust to provide agricultural property as a means for the third party to pay for the obligations imposed on him or her, but I reserve discussion of this topic for chapter 2.

The economic assumptions in these cases emphasize how important the rules imposed on tutors in investing their pupils' property were for maintaining the financial security of a vulnerable social group. Tutors were, as I already stated, required to make appropriate investments of their pupils' funds, and they were financially liable if they failed to carry out this task in a timely manner or if they made unnecessarily risky investments.[32] In general, this meant that the tutor was to invest funds available to the pupil in agricultural property, *praedia* or *agri* (Ulp. *D*. 26.7.3.2, 35 *ad ed*.; cf. Ulp. *D*. 26.7.5 pr.).[33] The failure on the part

---

31. For the ratio of aurei to sesterces in the legal sources, see A. Berger, *Encyclopedic Dictionary of Roman Law* (Philadelphia, 1953; reprint, 1980), s.v. "aureus." See also Frier, *CP* 88 (1993): 223–34.

32. On this topic, with discussion of the passages analyzed here, see J. Crook, "Classical Roman Law and the Sale of Land," in M.I. Finley, ed., *Studies in Roman Property* (Cambridge, 1976), 71–83, 180 n. 3, and Krause, *Witwen und Waisen*, 99–103. Cf. also M.I. Finley, *The Ancient Economy*, 2d ed. (Berkeley and Los Angeles, 1985), 121; M.W. Frederiksen, "Theory, Evidence and the Ancient Economy," *JRS* 65 (1975): 164–71, at 168; and Sachers, *RE* 7 (1948): 1544–51. In general terms, the obligation of the tutor to administer the property of the tutor assiduously was reinforced by his obligation to provide a *cautio*, that is, a binding legal guarantee that the pupil's property would remain intact. This guarantee is the so-called *cautio rem pupilli salvam fore* (Ulp. *D*. 27.8.1.15, 36 *ad ed*., Nerat. *D*. 46.6.11, 4 *membran*.), on which see Kaser, *RP* I, 361 n. 9, 362, 364–65, 539 n. 9. This *cautio* was required of *tutores legitimi*, that is, individuals required by law to take up a tutorship. It was not required of a *tutor testamentarius*, that is, a tutor named in the will, or of a *tutor datus*, that is, a tutor appointed by a magistrate. Apparently, it could also be provided by a *curator minoris* (Kaser, 371). See also Solazzi, "Tutele e curatele," 1–22.

33. Ulp. *D*. 26.7.3.2 discusses the obligation of *tutores honorarii*, that is, tutors with a fiduciary responsibility to oversee the performance of the tutors actually administering the pupil's property, to enforce the requirement that the administering tutor invest the pupil's funds: ". . . et si pecunia sit, quae deponi possit,

of the tutor to invest available funds was a serious offense. Thus the praetor or provincial governor could issue a decree compelling the tutor to purchase property, and a tutor ignoring such a degree was subject to severe penalties (Ulp. *D.* 26.7.7.3, 26.7.7.7). The governor, in fact, could use his power of coercion to enforce the tutor's fulfillment of this obligation.[34] Caracalla ruled that a tutor of freedman status should be turned over to the urban prefect for appropriate punishment if he failed to invest in loans or in the purchase of property money that he owed a *pupilla* (*CJ* 5.37.4, A.D. 213). This particular response was issued to the mother of the *pupilla* in question. But the tutor's obligation to invest in property did not depend on direct intervention on the part of the praetor, so a tutor was always liable to pay interest, *usurae pupillares,* for any funds that he did not invest (Ulp. *D.* 26.7.7.7). The tutor was also subject to pay interest penalties if he delayed in investing his pupil's money (Ulp. *D.* 26.7.7.3, 26.7.7.10). The importance that the Roman legal authorities attached to investing in land is underscored by a rescript of Caracalla, dealing with a minor's curator who had used the minor's funds to purchase property for himself (*CJ* 5.51.3, A.D. 215). The minor was given the option of either taking over the property that the curator had purchased or exacting interest for the time that the curator had held his money.[35] The requirement to purchase property obtained, however, only if funds available were sufficient to allow the purchase of land in economically rational units (Ulp. *D.* 26.7.5 pr.).[36] As Ulpian

---

curare, ut deponatur ad praediorum comparationem." On this passage and the degree to which various types of tutors were liable for the administration of the pupil's property, see Solazzi, "Tutele e Curatele," 1–62, quoting on p. 2 Lenel's reconstruction of this passage.

34. This doctrine is also enunciated at Paul. *D.* 26.7.49 (2 *sent.*): "Ob faenus pupillaris pecuniae per contumaciam non exercitum aut fundorum omissam comparationem tutor, si non ad damnum resarciendum idoneus est, extra ordinem coercebitur"; cf. also Ulp. *D.* 26.10.3.16 (35 *ad ed.*).

35. On this passage, see Solazzi, "Tutele e Curatele," 19–20. For the similarity of the rules governing the administration of a minor's property to those governing that of the pupil's property, see later in this chapter under "The *Oratio Severi* and Restrictions on Alienating the Pupil's Property." Finally, Ulpian (*D.* 42.1.15.12, 3 *de offic. consulis*) referred to the confiscation *in causam iudicati* of a pupil's funds deposited for the purchase of property. On the legal enforcement of the tutor's obligation, see Solazzi, 17–21, with discussion of these passages.

36. According to Ulpian (*D.* 27.4.3.6, 36 *ad ed.*), one circumstance in which the tutor was authorized *not* to invest money deposited for the purchase of property occurred when the tutor was owed money by the pupil for expenses

emphasized, it was impossible to establish a general rule for determining what constituted an amount of money sufficient to allow the purchase of land, so the praetor was authorized to determine this on a case-by-case basis. The emperor Alexander Severus gave the tutor the benefit of the doubt when he released him from his liability to pay interest if it was simply not possible to invest the pupil's money or to find suitable borrowers (*CJ* 5.56.3, A.D. 228). At a much later date, the emperors Honorius and Arcadius acted on a similar assumption in a rescript concerning the duties of tutors. They required the tutor either to invest the pupil's funds in property or, if no property was available, to lend that money out at interest at his own risk (*CJ* 5.37.24.1, A.D. 396).[37] In this constitution lending money out at interest is clearly envisioned as a less preferable alternative to investing in property, but the emperors recognized the difficulty that the tutor would be likely to encounter in finding property to purchase at any particular time.

In defining this responsibility, the jurists took a very broad approach toward investment in land, for all intents and purposes assuming that, as long as no chicanery was involved, any investment in land justified itself. As we shall see, the jurists were quite careful to define the liability of the tutor when he invested the pupil's money in loans, but the tutor's responsibility in an investment in land was much more generally outlined. It is noteworthy that the jurists nowhere define what constituted an appropriate investment in land. Ulpian held tutors liable if they invested dishonestly, "per sordes aut gratiam," in "unsuitable properties" [praedia . . . non idonea] (*D.* 26.7.7.2).[38] When the tutors purchased property in poor condition, Ulpian continued, they were liable for "broad negligence" [lata neglegentia].[39] But to my know-

---

in connection with the tutorship. The tutor was entitled to request the praetor to assign these funds over to himself: the tutor's right to compensation for legitimate expenses took preference over the obligation to invest the pupil's property.

37. *CJ* 5.37.24.1: " . . . ita tamen, ut ex mobilibus aut praedia idonea comparentur aut, si forte (ut adsolet) idonea non potuerint inveniri, iuxta antiqui iuris formam usurarum crescat accessio, quarum exactio ad periculum tutoris pertinet."

38. For the problem of bribery of legal authorities, see M. Peachin, *Iudex vice Caesaris* (Stuttgart, 1996), 76–79.

39. Ulp. *D.* 26.7.7.2: "Competet adversus tutores tutelae actio, si male contraxerint, hoc est si praedia comparaverint non idonea per sordes aut gratiam. quid ergo si neque sordide neque gratiose, sed non bonam condicionem elegerint? recte quis dixerit solam latam neglegentiam eos praestare in hac

ledge, no source provides a positive definition of what constituted an appropriate investment in land.[40] Apparently, the tutor was primarily responsible for making sure that the pupil received the income that in good faith could be realized from an estate. Modestinus used this principle in a response concerning the responsibility of a tutor to render an accounting to his *pupilla* of the revenues from a certain estate (*D.* 26.7.32.2, 6 *resp.*).[41]

The assumption underlying this response seems to be that the income that could in good faith be realized from an estate was readily determined, so it must have fallen within a relatively restricted range. As I intend to show, this assumption figures in other legal rules about landed property. Indeed, in the present case, Modestinus emphasized that the tutor remained liable for a good-faith income from an estate even when he turned its management over to a slave (*D.* 26.7.32.3). The tutor, then, was liable to pay what he could salvage from the *peculium* of the slave.[42] In this case, we should imagine that the jurists envisioned no great difficulty in determining what a reasonable income from an estate might be, so the rent the tutor could be expected to have exacted from the slave tenant was readily ascertained.

In many cases, this reasonable level of income was precisely the rent for which the estate was leased out, again with the jurists envisioning little difficulty in determining what a fair rent was. This principle seems to underlie a response by the jurist Paul concerning the obliga-

---

parte debere." For analysis of this passage, with discussion of the suspected interpolation of the expression *lata neglegentia*, see MacCormack, *IJ* 5 (1970): 369–90, at 385 and n. 61.

40. The Roman legal authorities were more concerned that the pupil be given clear title to the land than with the quality of the land purchased; the point was to assure the pupil of undisturbed possession. Accordingly, tutors were liable to restore to the pupil the purchase price if they knowingly bought property from an individual whose property was subject to confiscation by the state (Scaev. *D.* 26.7.57.1, 10 *dig.*). In this particular case, the seller had appealed his sentence at the time of the sale, only to see his appeal overturned.

41. Mod. *D.* 26.7.32.2: "Modestinus respondit tutorem eorum redituum nomine rationem pupillae reddere debere, qui ex fundo bona fide percipi potuerunt."

42. Mod. *D.* 26.7.32.3: "Item respondit, si minus a servo tutor percepit [recepit *Hal.*], quam bona fide ex fundo percipi potuit, ex eo, de quo pupillae sit obstrictus, quantum ex peculio servi servari possit, eidem tutori proficere debere, scilicet si non perdituro servo administrationem crediderit." The slave was functioning as a *servus quasi colonus*; this institution is discussed in chap. 3.

tions of a curator who had leased out an estate belonging to the pupil. The tenant had fallen in arrears, and these arrears remained outstanding when the pupil, after reaching majority, made the former tenant his business manager (procurator). The curator remained financially liable for the arrears that had accumulated as a result of the tenant's stay on the estate (*D.* 26.7.46 pr., 9 *resp.*). The doctrine, mentioned earlier, whereby the tutor was not held accountable for acts undertaken in good faith (Ulp. *D.* 27.4.3.7) certainly meant that the tutor could not be considered obligated to guarantee any level of income from an investment in an estate. The tutor could dispense with his obligations in managing a pupil's property by leasing it out to a tenant, while the jurists were confident that the judicial authority responsible for enforcing the tutor's responsibility could determine if the lease terms were consistent with the pupil's financial interests.

To carry this line of reasoning further, because the jurists envisioned only a restricted range of possible incomes from a given investment in agricultural land, it seems clear that they also envisioned the tutor as investing the pupil's income in estates with a restricted range of characteristics. This estate would have been a fully productive one and in full operation, and the purchase price would have been a direct function of its income. This very broad approach to investing in land finds a parallel, which I will later examine in more detail, in Pliny's contemplated purchase of an estate at Tifernum Tiberinum (*Ep.* 3.19). A landowner like Pliny, if we can draw general conclusions from this single example, might commonly invest in a fully productive estate, where the purchase price was a function of the estate's current income. This strategy is to be contrasted with the more risky but potentially more profitable strategy of investing in improving a property.[43]

We must not overlook the likelihood that climatic conditions might cause considerable variation in the actual production of an estate. But at what level would these variations be especially felt, and when would the jurists or the judge at trial take them into account? Presumably, the calculated income that could be realized from an estate was based on an assumption of good weather and no unusual conditions that might compromise the yearly income. Certainly a poor harvest resulting from

---

43. Cato had assumed that the landowner would purchase a fully productive estate (*Agr.* 1); on this characteristic type of purchase, see A. Steinwenter, *Fundus cum instrumento* (Vienna and Leipzig, 1942), 29–30.

adverse weather conditions would not present an obstacle to the ability of the legal authorities to determine whether the tutor had fulfilled his obligations in managing the pupil's wealth. If the pupil's estate was leased out for a cash rent, as must often have been the case, then the risk of variations in the harvest and the market price for the crops was passed on to the tenants. As we will see, landowners often had to grant concessions to their tenants in the event of poor harvests, but the strict rules imposed on tutors surely gave them much less flexibility in dealing with the pupil's tenants than many landowners might have displayed. That the situation that real-life landowners faced was more complicated than the idealized picture of estate management offered by the jurists is no surprise, and it does not undermine the conclusion that, in general, upper-class Roman landowners could expect a relatively narrow range of income from a given level of investment in agriculture.

Several papyri from Roman Egypt attesting purchases of land for minors suggest that, in that province at least, it was a common practice for landowners to try to provide the same type of security for their underage children as prescribed by the jurists in regulating the institution of tutorship.[44] These landowners tried to promote the financial security of their children by purchasing for them fully productive land. In one case, a father named Aurelius Hermias provided one thousand drachmas to purchase for his underage daughter, Tetseiris, a one-half share of her half brother's interest in a vineyard (*P.Oxy.* LI 3638, A.D. 220). This vineyard came with a reed plantation (καλαμεία), which supplied materials for supporting the vines, as well as with date palms and other fruit trees, irrigation equipment, and facilities for pressing the grapes (lines 7 f.).[45] The relatively high status of the parties involved in this transaction is indicated by the identity of the seller, M. Aurelius Chairemon, a town councillor at Oxyrhynchus who had served as *agoranomos*. He had previously shared ownership of the vineyard with other siblings, having acquired it through inheritance

---

44. For general conditions surrounding the market for agricultural land in Roman Egypt, see Rowlandson, *Landowners and Tenants in Roman Egypt*, 176–201.

45. For the terminology used to describe the equipment of agricultural properties in Egypt and the legal status of the equipment, see Steinwenter, *Fundus cum instrumento*, 40–61. For the background to this document, see Rowlandson, *Landowners and Tenants in Roman Egypt*, 193–94.

from his mother. The father, in acquiring for his young daughter a share in a fully productive property, was apparently seeking to provide his daughter with a stable source of income, possibly for her to use when she married.

In another example, we see how a member of the provincial aristocracy undertook to provide for the financial security of his young son (*P.Oxy.* XXXIV 2723, III c.). In this case, an individual named C. Calpurnius Firmus purchased a vineyard at Oxyrhynchus for his son, L. Calpurnius Firmus. The father was a member of the council of Alexandria, where he had served as *eutheniarch* and *kosmetes*, and he was currently *gymnasiarch* and *prytanis* at Oxyrhynchus. He thus was one of many wealthy Alexandrians who also owned properties outside the Alexandrian *chora* in the Egyptian nomes.[46] The seller, one Aurelia Apollonia, alias Harpokratiaine, was likewise a citizen of Alexandria. The vineyard was relatively large in size (for Egypt), consisting of eleven and three-fourths *arourae*, and it came with all of the equipment and appurtenances needed to keep a vineyard productive, including a reed plantation, equipment for irrigating the vineyard, facilities for pressing the vintage, and date palms and other fruit trees (lines 7 f.). Some peculiar circumstances surround this purchase, however. The vineyard is described as being in a state of neglect (ἐν ἀμελείᾳ, line 9), and it was bounded on three sides by property belonging to C. Calpurnius Firmus. This purchase certainly enabled the father to consolidate some of his holdings at Oxyrhynchus, and presumably he could gain some economies by cultivating the vineyard in conjunction with his other property in the vicinity. The purchase may have been made in the son's name to give him a source of income and also to allow the father to be more flexible with his will. In setting up the purchase in this way, he was already, in effect, passing this property on to his son. Whatever the motivation for purchasing this particular prop-

---

46. For this family, see Rowlandson, *Landowners and Tenants in Roman Egypt*, 110–11, 196–97. Cf. *P.Oxy.* XXXVIII 2848 (A.D. 225), in which a Calpurnius Firmus, apparently this same individual, is recorded as a benefactor of Oxyrhynchus together with the wealthy Aurelius Horion; this record is discussed further in chap. 2. The family apparently included C. Valerius Firmus, prefect of Egypt ca. 245–47, and Claudius Firmus, prefect ca. 264/65: see G.M. Browne, *P.Oxy.* XXXVIII 2848, introd., pp. 57–58, and, for further discussion of the family's activities, D.W. Rathbone, "Italian Wines in Roman Egypt," *Opus* 2 (1983): 81–98, at 90.

erty, it is clear that the father intended to restore the vineyard to full production and to pass the property on to his son, providing his son with a resource that would be capable of yielding a stable income over the long run. It may well be that the present document attests only one of many such purchases for the son at Oxyrhynchus and elsewhere in Egypt.[47] Although we can only infer the motivation behind purchases of land attested in contracts, it seems clear that these landowners in Egypt, in seeking to provide their children with productive land, were responding to the same general need envisioned by the jurists when regulating tutorship. The general conditions of the Roman economy led them to take the same sorts of steps in making investments for children that the jurists prescribed for pupils.

The Difficulties of Investing in Loans

Apparently only when it was impossible to purchase property were tutors expected to have recourse to the single major alternative form of investing the pupil's money, that is, lending the money out at interest. We should imagine that such loans, denoted by the terms *faenerare*, *credere*, and *nomina*, encompassed a wide variety of transactions. These loans might include investment in businesses, but they might also include purely private consumer loans. Lending money was apparently a basic part of any upper-class Roman's portfolio. Seneca describes money lending as a charcteristic activity of upper-class Romans (*Ep.* 41.7), and Tacitus, in his discussion of judicial proceedings in A.D. 33 against lenders charging usurious interest rates, emphasizes that, from the senate's point of view at least, lending money at interest was commonplace: "neque enim quisquam tali culpa vacuus" (*Ann.*

---

47. See also *P.Oxy.* IX 1208 (A.D. 291), in which a father purchased on behalf of his underage daughter a one-fifth share of four *arourae* of grain land with a corresponding share of irrigation equipment (for a now inflated purchase price of one talent and three thousand drachmas), as well as *P.Turner* 24, in which a mother made a bid to purchase sixteen *arourae* of confiscated catoecic, or private, land on behalf of her daughter; the price offered is three thousand two hundred drachmas. The mother and daughter already owned all the land surrounding the property for which they were bidding. On the legal capacity of the child in the first document, presumably under the *patria potestas* of her father, to own the land purchased, see E. Seidl, *Rechtsgeschichte Ägyptens als römischer Provinz* (St. Augustin, 1973), 137–38. For additional examples of purchases of this type, see Krause, *Witwen und Waisen*, 260–61.

6.16.3).[48] We can gain some appreciation of the role that lending money played in the management of an upper-class Roman's finances by examining the situation of Pliny the Younger.

If Pliny is a representative example of how a Roman senator invested his wealth, it seems doubtful whether lending money out at interest could have generated much wealth in aggregate for upper-class Romans. Rather, it seems to have been a sideline, offering a means for investing spare amounts of capital that could not be invested in land. Pliny describes his own money lending as a form of investment secondary to landowning: "sum quidem prope totus in praediis, aliquid tamen fenero . . ." (*Ep.* 3.19.8). The evidence provided by Pliny's letters suggests that the loans that Pliny might make or take out, even if contracted toward some business purpose, were personal in nature. Formally, such loans would have been classified as *mutuum*, that is, a loan that in and of itself involves neither the payment of interest nor the pledging of property as security.[49] The best example is the loan that Pliny anticipated taking from his mother-in-law, Pompeia Celerina, to finance the purchase of an estate. The purchase price for this estate was three million sesterces, a substantial sum representing perhaps one-fifth of Pliny's overall wealth, but Pliny felt that he would have little difficulty in finding funds to purchase the estate. Pliny could raise some of the money from his own loans, and at the same time he could borrow from his mother-in-law: "nec molestum erit mutuari; accipiam a socru, cuius arca non secus ac mea utor" (*Ep.* 3.19.8).

The loan that Pliny envisioned taking out was to finance a very significant business venture, since the purchase of the estate in *Epistulae* 3.19 would have a considerable and permanent impact on his finances. Pliny no doubt secured more advantageous terms from his mother-in-law than he might have obtained from another lender; a personal loan from a close connection will have given him more flexibility in arrang-

---

48. On lending at interest by upper-class Romans, see Duncan-Jones, *Economy*, 21 and n. 4. See P. Veyne, "Mythe et réalité de l'autarcie à Rome," *REA* 81 (1979): 261–80, at 276–77, on the loans made by Trimalchio (Petr. 76.9).

49. On *mutuum*, see Kaser, *RP* I, 530–32. Interest payments and pledging of security could be contracted for by stipulation. Bürge, *ZRG* 97 (1980): 126–29, 133–38, argues that loans among members of the Roman upper classes were generally contracted without interest. The ability to lend money enhanced the creditor's social prestige; it was a sign of the individual's generosity, and the debtor was socially obliged to the creditor.

ing repayment. It seems reasonable to draw some general conclusions from Pliny's efforts to finance his purchase of the estate in *Epistulae* 3.19. Pliny's correspondent, Calvisius Rufus, was an equestrian friend from Comum who Pliny consulted several times on business and legal matters.[50] In the present letter, nothing in the way in which Pliny describes his arrangements to Calvisius Rufus indicates that either this correspondent or the general reader could have been expected to find anything unusual in them. Since Pliny himself considered lending money at interest to be a basic part of his finances, it was probably common for one upper-class landowner to finance the purchase of an estate by borrowing from another. There was no institutionalized credit market to which Pliny had recourse. Rather, to judge by this example, it was apparently common for upper-class Romans to finance significant business ventures through purely personal loans.[51]

The loans that Pliny himself made were, in all likelihood, equally personal in nature, even though our evidence for them is far from complete. In a case discussed earlier in this chapter, Pliny had lent money

---

50. On Calvisius Rufus, *PIR*² C 349, see Sherwin-White, *Commentary*, 202.

51. H. Pavis d'Escurac, "Aristocratie sénatoriale et profits commerciaux," *Ktema* 2 (1977): 339–55, in arguing that commercial interests had an important place in the finances of Roman senators, suggests that many of the loans that upper-class Romans made were for commercial purposes. Such loans would have been especially characteristic of wealthy patrons financing business undertakings by their freedmen, much as Trimalchio is apparently depicted as doing (Petr. 76.9). The loan that Pliny was to receive from Pompeia Celerina hardly represented a loan for agricultural investment. It was rather a special example of the "consumer" loans that characterized money lending among aristocratic Romans, to some extent coincidentally applied to the purchase of agricultural land. Pompeia Celerina presumably did not figure to profit from this loan, and she hardly had a lien against any of Pliny's property offered as collateral. For general discussion of the nature of the credit market in the Roman Empire as compared with other preindustrial economies, see R.W. Goldsmith, *Premodern Financial Systems* (Cambridge, 1987), 43–47, as well as H.W. Pleket, "Wirtschaft," in *Handbuch der Europäischen Wirtschafts- und Sozialgeschichte*, vol. I (Stuttgart, 1990), 37–38, and W.V. Harris, "Between Archaic and Modern: Problems in Roman Economic History," in W.V. Harris, ed., *The Inscribed Economy* (Ann Arbor, 1993), 11–29, at 19–21. See also the analysis of Bürge, *ZRG* 104 (1987): 465–558, and C. Howgego, "The Supply and Use of Money in the Roman World, 200 B.C. to A.D. 300," *JRS* 82 (1992): 1–31, especially 27–29. Howgego, who discusses the relatively limited development of banking in the Roman Empire, argues that members of the Roman upper classes did not use banks for credit.

out to the father of Calvina and then had excused those debts so that Calvina could accept an inheritance from her father (*Ep.* 2.4). Thus although we cannot be sure of the amount of money involved or the purpose, Pliny's loans to Calvina's father were clearly personal in nature, presumably to help him meet some social obligation, so the loans would have served a purpose similar to the gift of one hundred thousand sesterces that Pliny made toward Calvina's dowry (2.4.2). Pliny, moreover, was at first only one of several creditors to whom Calvina's father owed money, and all of these individuals had probably extended personal loans to him.[52] The loans that Pliny made to Calvina's father were probably little different than the loans that the philosopher Artemidorus had contracted. Artemidorus found himself in embarrassing position when Domitian expelled the philosophers from Rome in A.D. 93. He had substantial debts, contracted apparently to meet some social obligation. Thus, in Pliny's description, Artemidorus had taken out loans for strictly personal uses, that is, for "most seemly reasons": "ut aes alienum exsolveret contractum ex pulcherrimis causis" (3.11.2). Pliny came to the rescue by borrowing the money himself and then giving it to Artemidorus.

From the evidence of Pliny's letters emerges a picture of upper-class Romans lending one another money as part of their general network of mutual social obligations. Admittedly, Pliny offers in his letters a somewhat idealized picture of his financial life, and underneath a veneer of civility there may have lurked some hard-nosed financial transactions. But whether Pliny or others like him remained civil or were ruthless in their financial dealings, there is no hint in the correspondence that anyone from Pliny's circle was involved in a formal financial market that made capital available on a large scale for investment in business ventures. Although they might on occasion be put to what we could term economic uses, such as Pliny's purchase of the estate in *Epistulae* 3.19, the contracting of loans in Pliny's circle remained a personal business. The nature of the circumstances under which a loan might be contracted and collected is suggested by the intervention on the part of Pliny on behalf of a senatorial friend Atilius Crescens (*Ep.* 6.8). Crescens was owed money by a certain Maximus, the heir of the original debtor Valerius Varus. Crescens had had difficulty in collecting his debt,

---

52. *Ep.* 2.4.2: "Cum vero ego ductus adfinitatis officio, dimissis omnibus qui non dico molestiores sed diligentiores erant, creditor solus exstiterim. . . ."

which was overdue for a number of years, and Pliny wrote to a certain Priscus, a friend of Maximus, to ask that Priscus put gentle pressure on Maximus to pay off the debt (*Ep.* 6.8.5).[53] We learn nothing about the original purposes of the loan, but this letter suggests that in the Roman Empire the collection of an outstanding debt might have been as much a question of personal considerations as was the original contracting of a loan. Although the details about the social standing of these individuals escape us, the episode makes clear how the enforcement of a debt must often have been a delicate matter of diplomacy rather than simply a question of legal rights.[54]

The restricted set of examples used in this discussion point out some of the circumstances under which investment in loans might be conducted in the Roman economy, and such circumstances provided a background against which the jurists developed their rules for tutorship. Lending often depended on personal relationships, with the result that the credit market was somewhat haphazard. Collecting debts originally contracted with personal considerations in mind might have involved significant difficulties as well.

Lending the Pupil's Money

The jurists carefully regulated the responsibilities of tutors when lending out money belonging to their pupils or when administering already existing loans, because such loans were assumed to involve a much higher degree of risk than investment in land. In addition to the risk, the jurists also had to reckon with several other factors complicating

---

53. Plin. *Ep.* 6.8.5: "Rogo ergo, exigo etiam pro iure amicitiae, cures ut Atilio meo salva sit non sors modo verum etiam usura plurium annorum." On the circumstances surrounding this letter and possible identities of the individuals involved, see Sherwin-White, *Commentary,* 363–65.

54. In the view of Bürge, *ZRG* 97 (1980): 105–156, the social relationship of creditor and debtor was crucial for determining the ability of the creditor to enforce the debtor's obligations. Bürge argues that creditors often resorted to self-help to exact repayment of a loan. The creditor's ability to move against a debtor in arrears and to collect what was due clearly increased as the social gulf separating creditor and debtor grew wider. As an illustration of this, cf. the case in which Aulus Gellius acted as a private *iudex*: that the distinguished plaintiff had no proof about the debt that he was trying to collect would not necessarily let the defendant of questionable character off (14.2). On this case, see Peachin, *Iudex vice Caesaris,* 66–67.

the enforcement of the tutor's fulfillment of his duties. The jurists had two major enforcement responsibilities: making sure that the tutor was diligent in actually investing the pupil's money and then in exacting the repayment of loans; and making sure that the tutor did not abuse his position of supervising loans of the pupil's funds by diverting those funds and using them for his own purposes.

The tutor was accordingly required to lend out money available to the pupil in a timely manner, being required to pay interest if he failed to do so, and he was also expected to be diligent in exacting full payment, including interest (Scaev. D. 26.7.58.1, 11 *dig.*, Paul. D. 26.7.49, 2 *sent.*, Ulp. D. 26.7.7.10, 35 *ad ed.*).[55] The principle that the jurists applied here is the same as Ulpian enumerated when defining the obligation of the pupil to pay interest on uncompensated expenses that the tutor had to bear in connection with administrating the tutorship. According to Ulpian, the tutor was entitled to interest on any unpaid principal until he received his full compensation, since his money was not to lie idle: "nec enim debet ei sterilis esse pecunia" (D. 27.4.3.4, 36 *ad ed.*).[56] According to Paul, the tutor was liable to pay interest if he failed to press claims against the pupil's debtors and thereby allowed them to become insolvent, or "minus idonei." Likewise, he was given six months to lend out the pupil's money, after which time he was liable to pay interest (D. 26.7.15, 2 *sent.*). A tutor might also be held liable for failing to be diligent in exacting loans contracted with the pupil's father and therefore existing at the time of his taking up the tutorship (Caracalla, CJ 5.51.2, A.D. 213). For existing loans, the tutor was in fact granted a period of two months to exact the interest for loans of the pupil's money (Ulp. D. 26.7.7.11, 35 *ad ed.*).[57] It was of course illegal for the tutor to divert the pupil's money and use it for his own purposes, and the danger of this happening was sufficient to provoke repeated constitutions by the emperors (cf. Paul. D. 26.7.46.2, 9 *resp.*). A tutor guilty of

---

55. See Scaev. D. 26.7.58.3 (11 *dig.*), concerning the tutor's obligation to pay interest for a legacy of two thousand aurei that was not yet invested, and see Ulp. D. 27.4.3.2–3 (36 *ad ed.*, discussed in the subsequent text), concerning a tutor's use of his own funds to support the pupil. P.Oxy. LXVIII 3921–2 (A.D. 219, discussed already) is an annual account in which a guardian reports how he invested money left to the two boys in his charge.

56. For the *contrarium tutelae iudicium*, see n. 27 in this chapter.

57. See Scaev. D. 26.7.57 pr. (10 *dig.*), concerning the liability of tutors who alleged the destruction by fire of records as a reason for not enforcing loans of the pupil's funds.

such an infraction was subject to severe penalties for the interest, and he was of course liable from his own property for the principal (Ulp. D. 26.7.7.4, 35 *ad ed.;* cf. Caracalla, *CJ* 5.56.1, A.D. 213). Indeed, a tutor who for any reason failed to invest his pupil's money fell under the suspicion of taking it for his own use (Ulp. D. 26.7.7.4). The tutor was no more allowed to use the interest from the pupil's money for his own purposes than he was the principal (Ulp. D. 26.7.7.12, 35 *ad ed.*).[58]

Under certain circumstances, the line dividing appropriate and inappropriate use of the pupil's funds was fine. For example, a tutor who loaned out the pupil's money in his own name did not violate the imperial constitutions outlawing the diversion of the pupil's funds, because that tutor would still be liable to pay whatever interest he had been able to achieve for that money (Paul. D. 26.7.46.2–3, 9 *resp.;* cf. Ulp. D. 26.7.7.4, 6). Tryphoninus allowed the tutor to borrow the pupil's funds, as long as such action was taken with the consent of his cotutors and as long as the tutor borrowing the money was to pay the same interest as other debtors (D. 26.7.54, 2 *disput.*). The tutor was also allowed a certain latitude in using the pupil's funds to compensate himself for expenditures made in connection with the tutorship (see Ulp. D. 27.4.3.2–3, 36 *ad ed.*), and he was not liable to pay interest to the pupil for money used in this way.[59]

But there were limits on the freedom of action accorded to the tutor, and in general tutors were held to a very specific and exacting standard of conduct. The strict approach that the jurists took in interpreting the duties of the tutor resulted in a certain inflexibility; the tutor had no capacity, for example, to renegotiate a loan that otherwise could not be repaid. Thus Paul emphasized that the tutor had the right only to exact money from the pupil's debtors, not to make them any concessions; a debtor accorded any concession still remained liable to make full pay-

---

58. In a constitution, the emperor Constantine gave the pupil or minor even greater recourse against a tutor or curator lending his money out at interest. The property of the tutor or curator was subject to the pupil or minor as if it were pledged as a security in a loan; in effect, the tutor or curator pledged his own property as security when he made loans with the money belonging to the pupil or minor (*CJ* 5.37.20, A.D. 314 = *CTh.* 3.30.1).

59. The problem of determining when a tutor diverted the pupil's money is also evidenced in Ulpian's ruling that a tutor who had also been a debtor to the father of his pupil did not commit this infraction if he neglected to pay back his debt; the tutor was still liable as debtor to pay the interest originally contracted for (D. 26.7.7.5, 35 *ad ed.*).

ment to the pupil: "Tutoribus concessum est a debitoribus pupilli pecuniam exigere, ut ipso iure liberentur, non etiam donare vel etiam deminuendi causa cum iis transigere: et ideo eum, qui minus tutori solvit, a pupillo in reliquum conveniri posse" (Paul. D. 26.7.46.7, 9 *resp.*). The danger, of course, was that the tutor might grant special favors in renegotiating a loan, so that the legal authorities adopted a universal policy that was safer if sometimes disadvantageous to the individual pupil's financial interests.

The jurists clearly took the volatile nature of the Roman credit market into account as they regulated the lending of the pupil's money. As I already stated, Paul allowed the tutor six months to invest the pupil's funds before the tutor's liability to pay interest began (D. 26.7.15, 2 *quaest.*). This same jurist also exempted the tutor from responsibility if he was not able to find anyone to whom he might lend the pupil's money: "Si tutor pecuniam pupillarem credere non potuit, quod non erat cui crederet, pupillo vacabit" (D. 26.7.12.4, 38 *ad ed.*). The emperor Alexander Severus, as we have seen, released the tutor from the obligation to pay *usurae pupillares* if he was not able to find suitable borrowers or to invest in property (CJ 5.56.3, A.D. 228). These rules served to answer the problem of determining the extent of the tutor's responsibility to invest money in an economy in which the opportunities for investment were few and often highly risky. A characteristic area of dispute between a pupil and tutor at the *iudicium tutelae* would be whether the tutor had displayed sufficient energy in trying to invest funds belonging to the pupil. Gaius addressed this issue by ruling that a tutor who had managed to invest his own money had no defense if he claimed that he was unable to invest the money belonging to his pupil: "Non est audiendus tutor, cum dicat ideo cessasse pupillarem pecuniam, quod idonea nomina non inveniret, si arguatur eo tempore suam pecuniam bene collocasse" (D. 26.7.13.1, 12 *ad ed. prov.*). Clearly, the Roman legal authorities envisioned the tutor as seeking to invest the pupil's money in a haphazard credit market in which there were limited opportunities for investments yielding regular income.[60]

---

60. In extending the Severan legislation restricting the alienation of landed property to include urban and movable property as well (discussed later in this chapter), Constantine emphasized the difficulties involved in investing the cash derived from the sale of such property: "huic accedit, quod ipsius pecuniae, in qua robur omne patrimoniorum veteres posuerunt, fenerandi usus vix diuturnus, vix continuus et stabilis est: quo facto saepe intercidente pecunia ad

Finding suitable creditors presented the tutor with one difficulty, but another important issue for the Roman legal authorities centered around the tutor's responsibility for loans once they were contracted. As we have seen, the tutor was obligated to enforce the repayment of loans, being financially liable if he failed to carry out this duty, and the tutor was clearly responsible for the suitability of loans that he had contracted himself. The rules about the tutor's responsibility in this regard seem to have been a matter of dispute. Marcellus held that at the *iudicium tutelae* the pupil only had to acknowledge creditworthy loans, or "nomina integra," contracted by the tutor, whereas the risk for bad or poorly contracted loans fell to the tutor (apud Paul. *D.* 26.7.16, 6 *ad Sab.*). But Paul, correcting this view, held that the former pupil at this judgment had either to acknowledge all of the loans that the tutor made with his money or else to refuse them all. In this latter case, the tutor became responsible to pay the normal interest, the *usurae pupillares*.[61] This responsibility may, in many cases, have made tutors hesitate to make risky loans, in which case they would have resorted to the defense to which Gaius was reacting in the passage previously discussed (*D.* 26.7.13.1). The Roman legal authorities must have had to make difficult judgments to determine whether the tutor had in fact made a good-faith effort to invest the pupil's money. After all, assessing the risk of a loan depended on business judgment and was often only possible after the fact, that is, after a loan went bad. But a related legal issue concerned the responsibility on the part of tutors for loans whose administration and enforcement they took over from their predecessors as tutors. In general, tutors assumed the risk for loans made by their predecessors (Paul. *D.* 26.7.44 pr., 13 *quaest.*), but they could not be forced to take on the risk for loans that they considered bad (Papin. *D.* 26.7.35, 2 *quaest.*).

This discussion of the tutor's responsibilities with regard to loans of the pupil's money is meant not to be exhaustive but rather to indicate

---

nihilum minorum patrimonia deducuntur" (*CJ* 5.37 22.5a, A.D. 326). This passage is quoted by G. Cervenca, "Studi sulla Cura Minorum," pt. 3, "L'estensione ai minori del regime dell'*Oratio Severi*," *BIDR*, 3d ser., 21 (1979): 41–94 at 85 n. 154.

61. The tutor was responsible for loans that the father had made and that the pupil inherited; cf. Papin. *D.* 26.7.39 pr. (5 *resp.*), which states that the tutor carrying out his office by mistake after the pupil reached majority no longer bore this responsibility, since he could no longer sue the father's debtors.

some of the considerations that the jurists had to take into account in their effort to protect the interests of the pupil. Lending money posed a range of legal questions, all centered around an assumption that it was exceedingly difficult to invest the pupil's money safely and appropriately in loans. Clearly this type of investment carried substantial risk even under the best of conditions, and the income that a pupil or anyone else might realize from loans was subject to large variations. Some lenders might make high profits from their investments, but it was also possible to lose everything. As I have emphasized earlier, it is noteworthy that the jurists developed no rules concerning either investing in agricultural property or administering it. It seems warranted to conclude that the horizons for achieving an income from agricultural property were much more limited and that the range of investments that a typical landowner might make in an estate to bring it to full productivity or to keep it in that condition was not envisioned as being subject to the same type of variation inherent in investing in loans. Investment in land, then, unless it was undertaken under fraudulent circumstances, represented a form of security not at all achievable in any other type of investment available to the pupil.

The *Oratio Severi* and Restrictions on Alienating the Pupil's Property

This conception of land as the only form of investment capable of providing stable long-term income was so basic to Roman legal thinking that the imperial government viewed the alienation of landed property as the most serious compromise of a pupil's economic interests. The Roman state had long sought to protect pupils and minors against fraudulent efforts of unscrupulous individuals to deprive them of property, as the very antiquity of the institutions of tutorship and curatorship attest, going back to the Twelve Tables. In 192/1 B.C., moreover, the *lex Plaetoria* strengthened the legal protection for pupils and minors against fraud. The emperor Marcus Aurelius sought to protect the financial interests of pupils when he created the post of the *praetor tutelarius*; this official was assigned the duty of overseeing the administration of the pupil's property.[62]

As a general rule, the tutor was reckoned as serving *in loco domini* for the pupil, but such authority in no way authorized the tutor to plunder

---

62. The *praetor tutelarius* was authorized to use the institution of *cognitio extraordinaria* to compel the tutor to perform his duties: see Ulp. D. 26.7.1 pr. f.

the pupil's property (Julian. *D.* 41.4.7.3, 44 *dig.*).[63] Under the principate, the imperial government viewed the alienation of landed property as the most serious threat to the financial interests of pupils and minors, and this economic concern went hand in hand with a growing body of imperial legislation and juristic opinion establishing guidelines for all aspects of the institution of tutorship.[64] The imperial government's concern about the economic interests of pupils and minors reached its culmination in A.D. 195, when the emperor Septimius Severus introduced legislation, embodied in the *Oratio Severi*, establishing new guidelines on two key aspects of tutorship, namely, the selection of tutors and the ability of the tutor to authorize alienation of the pupil's property. This legislation is attested in several quotations in the fragments of Ulpian, as well as in a long quotation by Ulpian preserved at the beginning of *Digesta* 27.9 (Ulp. *D.* 27.9.1 pr.–2, 35 *ad ed.*).[65] To deal with the first issue addressed by the *Oratio*, the imperial government sought to ensure that pupils were represented by suitable tutors, so it refined the criteria prescribed for selecting tutors and restricted the right of a duly appointed tutor to name a substitute.[66]

(35 *ad ed.*) and Kaser, *RP* I, 357 (citing SHA *Marc.* 10.11), 361. On *cognitio extraordinaria*, see I. Buti, "La '*cognitio extra ordinem*': Da Augusto a Diocleziano," *ANRW* II 14 (1982): 29–59, especially 38, 40–41, for the steps taken to afford greater legal protections to pupils. Cf., for the added legal protection in enforcing trusts afforded by the *cognitio extraordinaria*, D. Johnston, *The Roman Law of Trusts* (Oxford, 1988), 222–55.

63. On restrictions on the ability of pupils and minors to alienate property and on the general responsibilities of tutors and curators, see Kaser, *RP* I, 83–91, 352–72. For a preliminary discussion of the issues raised here, see my "Investment in Estates," 221–23, as well as "Approaches to Profit and Management in Roman Agriculture," 47–49.

64. As Ulpian emphasized (*D.* 1.21.2.1, 3 *de omn. trib.*), the supervision of the pupil's property was considered such an important matter that the praetor or provincial governor responsible for overseeing this supervision could not delegate his authority in this matter.

65. On the *Oratio Severi* and the duties of tutors, see Sachers, *RE* 7 (1948): 1550, and Krause, *Witwen und Waisen*, 101–3. Further sources, in addition to *D.* 27.9, and Ulp. *Frag. Vat.* 158, 212–14 (*lib. de excusat.*), include [Paul.] *Sent.* 2.30 and *CJ* 5.70.2, 5.71–74; the sources are listed in Talbert, *The Senate of Imperial Rome*, 449 no. 134.

66. Ulp. *Frag.Vat.* 158, quoting the *Oratio:* "Promiscua facultas potioris nominandi nisi intra certos fines cohibeatur, ipso tractatu temporis pupillos fortunis suis privabit." See also Ulp., *Frag.Vat.* 212–14. On the application of this legislation to curators in charge of the property belonging to minors, see later in this section.

For our purposes, the important aspect of the Severan legislation was the restriction that it imposed on the authority of the tutor to alienate property belonging to a pupil. Seeking to minimize the chances that a pupil or minor might lose the only means that he or she had available to gain a safe income, the Severan legislation made it illegal for tutors and curators to alienate landed property, or "praedia rustica vel suburbana," belonging to pupils or minors, respectively, except under specific and tightly regulated circumstances (Ulp. D. 27.9.1.1–2). The only circumstance under which alienation of the pupil's property was to be allowed, apart from specific provision in the testator's will or codicil, was to pay off debt, in which case a decree from the praetor first had to be obtained. To quote Ulpian's version of the *Oratio Severi* :

> Praeterea, patres conscripti, interdicam tutoribus et curatoribus, ne praedia rustica vel suburbana distrahant, nisi ut id fieret, parentes testamento vel codicillis caverint. quod si forte aes alienum tantum erit, ut ex rebus ceteris non possit exsolvi, tunc praetor urbanus vir clarissimus adeatur, qui pro sua religione aestimet, quae possunt [quae possessiones *Lenel*][67] alienari obligarive debeant, manente pupillo actione, si postea potuerit probari obreptum esse praetori. (Ulp. D. 27.9.1.1–2)

> [Moreover, conscript fathers, I shall forbid tutors and curators to alienate rural or suburban properties, unless the parents have provided that this should happen in the will or in codicils. But if by chance there will be a debt so great that it cannot be paid off from the other property, then the urban praetor, a senator, should be approached, who is to judge, in accordance with his sacred duty, what [which possessions] should be alienated or obligated, with the action [sc. for guardianship] remaining open to the pupil, if afterwards it can be proved that the praetor was deceived.]

This text raises a major question about the application of the Severan law. In the juristic sources, the restrictions imposed on tutors are also

---

67. For defense of the text in the manuscripts against Lenel's emendation, see A. Biscardi, "L'*Oratio Severi* e il divieto di *obligare*," in *Studi in onore di Giuseppe Grosso*, vol. III (Turin, 1970), 245–66 at 254–55; Biscardi suggests *possint* for *possunt*.

applied to curators overseeing the management of property belonging to minors (males who had reached puberty but were not yet twenty-five). It has long been a scholarly controversy whether the application of the Severan legislation to *curatores minorum* represented classical Roman law or whether this was the work of Justinian's compilers. If the latter is the case, the compilers would have substantially rewritten the classical juristic sources to suit the requirements of sixth-century Byzantine law, in which the institutions of tutorship and curatorship were assimilated. Solazzi long ago argued for this position, viewing the application of the Severan legislation to *curatores minorum* as a post-classical development.[68] In his view, the Severan legislation, like other enactments by the emperors and rulings by the classical jurists, affected only pupils, since it was only much later that *curatores minorum* gained the authority to administer the property belonging to minors. The only curators affected by this legislation, then, were *curatores impuberum*, that is, curators who were appointed under very specific circumstances to assist or to replace the tutor administering the pupil's property. In this interpretation, accordingly, references to *curatores minorum* in the sources dealing with the Severan legislation are to be considered interpolated.

Much more convincing is the view, originally argued by Lenel, that the Severan legislation did in fact apply at a very early date to the *curator minoris* as well as to the *tutor impuberis*.[69] In Lenel's interpretation, the *curator minoris* did have the authority, in classical Roman law, to administer the property of a minor, and even, under certain circumstances, to alienate it. Although the *Oratio Severi* itself was only concerned with protecting the financial interests of pupils, jurists of the Severan Age, apparently recognizing the similarity between the interests of minors and those of pupils, soon applied to curators administering the property of minors the restrictions imposed on tutors by the Severan legislation. Lenel's view has been defended convincingly by Cervenca.[70]

Indeed, as Lenel argues, we should expect the imperial government to have been concerned to regulate the duties of *curatores minorum*

---

68. See S. Solazzi, *La minore età nel diritto romano* (Rome, 1912), especially 110–23, and idem, *Curator impuberis* (Rome, 1917), especially 118–21, 166–77.

69. See O. Lenel, "Die cura minorum in der klassischen Zeit," ZRG 35 (1914): 129–213, at 166–75.

70. Cervenca, *BIDR*, 3d ser., 21 (1979): 41–94.

much as it did the duties of tutors. In contrast to pupils, minors were not legally required to be represented by curators when they transacted business. Yet minors were protected against transactions that were contrary to their interests; under certain circumstances, such transactions could be declared null and void, with the minor receiving full restitution of his property, that is, *restitutio in integrum*. In view of this situation, many people must have balked at doing business with a minor not represented by a curator. Even if the law did not explicitly compel minors to seek out curators to represent them, the protections afforded the minor in effect compelled them to do so.[71]

To return to the restrictions imposed on tutors, the jurists and the Roman imperial government were stringent in their interpretation and application of the Severan legislation. The restrictions on alienating the pupil's property, it must be emphasized, were limited to agricultural property. Seeking to protect against any danger to the income of the pupil or minor, the imperial legislation, in the jurists' interpretation, made it illegal not only to alienate the property belonging to the pupil or minor but also to change in any way the legal status of productive land to the disadvantage of the pupil or minor, such as by pledging or hypothecating it. The legal principles established by the Severan legislation receive a clear expression in a later constitution of Valerian and Gallienus.[72]

---

71. Lenel, *ZRG* 35 (1914): 132–35.

72. On the authenticity of *vel adulescentes*, see Cervenca, *BIDR*, 3d ser., 21 (1979): 74–76; cf. P. Krueger ad loc. for problems with the text in this passage. As Lenel, *ZRG* 35 (1914): 167 argues, following Solazzi, *La minore età*, 114–17, the generalizing references to both pupils and minors that often appear in imperial rescripts are signs of interpolation, since the emperors, in issuing rescripts, would have addressed their remarks to the specific individual cases that they were judging. Their interpolation does not mean, however, that the generalizing references changed the law, since, in Lenel's view at least (175–76), the same law did in fact apply to both pupils and minors. Moreover, as Lenel argues (137–38), that the classical juristic sources would not always refer to the curator next to the tutor does not mean that the mentioning of a curator in the Digest or Code is a sign of interpolation that changed the law. The compilers of the Digest would have taken care to mention curators along with tutors to emphasize that the same law applied to both, but in doing so they were faithfully reproducing classical Roman law.

Cf., earlier, Maximinus Thrax, *CJ* 5.70.2 pr. (A.D. 238): "orationis divi Severi beneficium, quo possessiones rusticas sine decreto praesidis pupillorum seu

Non solum per venditionem rustica praedia vel suburbana pupilli vel adulescentes alienare prohibentur, sed neque transactionis ratione neque permutatione et multo magis donatione nec alio quoquo modo ea transferre e dominio suo possunt. (*CJ* 5.71.4 pr., A.D. 260)

[Not only are pupils and minors prohibited from alienating rural or suburban properties through sale, but they cannot transfer them from their possession either by any consideration of transaction or by any change in status, let alone by an act of donation or by any other means.]

The tutor or curator, accordingly, could not create a usufruct, renounce a servitude on another's property from which the pupil or minor benefited, or impose a servitude on the pupil or minor's own property to his or her disadvantage (Ulp. *D.* 27.9.3.5). Similarly forbidden, at least in subsequent application of this legislation, was the alienation of emphyteutic rights, that is, the perpetual rights granted by emperors to agricultural property in exchange for the payment of an annual charge (Ulp. *D.* 27.9.3.4). The strictness of the application of this legislation on the part of the jurists is indicated in a decision by Ulpian concerning the sale of a *fundus* to a minor. In this case, Ulpian ruled invalid the pledging of that property until such time as the purchase

---

adulescentium distrahi vel obligari prohibitum est...." Cf. also Paul. *Frag.Vat.* 45 (*lib.* 2 *manualium*), where a woman cannot alienate a usufruct or a servitude without the authority of a tutor. The prohibitions in the classical interpretation of the Severan legislation did not encompass urban properties; see Diocletian, *CJ* 5.71.16 pr. (A.D. 294): "Si praedium rusticum vel suburbanum, quod ab urbanis non loco, sed qualitate secernitur . . . sine decreto praesidis provinciae . . . venumdedisti...." For the text of this passage, see Krueger, ad loc. On the question of urban properties, see Cervenca, loc. cit., 56–57 and n. 60. Constantine extended the Severan legislation to urban and movable property; see *CJ* 5.37.22 (A.D. 326) and Cervenca, 84–88. In this same constitution, Constantine repealed an existing law requiring tutors and curators to convert into cash movable property belonging to pupils or minors, *praeter praedia et mancipia rustica* (ibid., pr.). The prohibition against pledging or imposing other liens on the land has been seen by many scholars as postclassical; for a convincing argument that this prohibition did in fact form part of the original Severan legislation, see Biscardi, "L'*Oratio Severi*," 245–66.

price was paid in full (*D.* 27.9.1.4).⁷³ Even the pupil's property that had been for sale before the death of the testator could not be sold unless the testator specifically provided for that in his will (Ulp. *D.* 27.9.1.3). The jurists, moreover, took a broad approach in interpreting what categories of property were affected by the Severan legislation. Thus Ulpian prohibited the alienation of land not strictly agricultural but still containing productive resources, such as quarries, mines, or salt beds (Ulp. *D.* 27.9.3.6, Paul. *D.* 27.9.4, *lib. sing. ad orat. div. Sev.,* Ulp. *D.* 27.9.5 pr.–1). The disposition of property bequeathed to a pupil might of course become a complicated matter, because the imperial government did allow the pledging of property to pay off an inherited debt, and it also allowed the sale of inherited property to pay off inherited debts if no other measures existed (see, e.g., Ulp. *D.* 27.9.1.2). But aside from these circumstances, commonplace though they might have been, the imperial government was very rigorous in protecting the pupil's undisturbed ownership of property passed on to him or her by the testator.⁷⁴

The Severan legislation provided a degree of protection to the pupil not available previously. Under the Severan legislation, the pupil had a direct claim on any property that was alienated without proper authorization, even when the purchaser acquired the property in good faith. Before the Severan legislation, the tutor had of course been liable for his administration of the pupil's property, and he could be forced to account for this in the *actio tutelae,* a remedy open to the pupil at the end of the tutor's service. The tutor who was sued under this action had to account for his management of the pupils' property. The pupil did have an action in rem to recover any property that had been fraudulently taken from him, but this action would be defeated if the acquirer of the property had gained good title over it by *usucapio,* that is, by acquiring it in good faith and holding it for two years.⁷⁵ In general, then, the pupil

---

73. The authenticity of this passage is disputed, based on the view that the application of the Severan legislation to minors is postclassical. If it is interpolated, it will originally have referred to a pupil rather than to a minor. For a defense of the authenticity of this passage, see Cervenca, *BIDR,* 3d ser., 21 (1979): 54–55.

74. Neither the *Oratio Severi* itself nor the juristic commentary on it makes any mention of the equipment, or *instrumentum,* but it seems likely that the alienation of this was also prohibited. Constantine, at any rate, considered slaves and livestock to be covered by the legislation; see Steinwenter, *Fundus cum instrumento,* 63–64.

75. On *usucapio,* see Kaser, *RP* I, 418–25.

would have to rely on the *actio tutelae*. Any judgment against the tutor would be subject to some of the same disadvantages as any other judgment in personam, in that the pupil was rewarded a monetary judgment and so bore some risk for the insolvency of the tutor. The pupil did hold a lien over the property of the tutor and so had better protection than other creditors, and there were also severe penalties that must have deterred many tutors from acts of fraud. In particular, a dishonest tutor might also be found *suspectus* and removed from his duties. A tutor convicted in an *actio suspecti* suffered *infamia,* and a senatus consultum from Trajan's reign provided for legal action against a magistrate or other official for appointing a tutor who had gone bankrupt and consequently was unable to pay off a judgment against him.[76] In the *actio rationibus distrahendis,* the tutor could be forced to pay the pupil double the value of property fraudulently managed. But all of these measures offered cold comfort if the pupil had no way to regain ownership of the property providing his or her livelihood. The Severan legislation provided immediate relief to the pupil, since it nullified ahead of time any transaction damaging the pupil's financial interests. The pupil had a direct claim on land that a third party might purchase in good faith, and we should expect that prospective purchasers of agricultural land would think twice before they acquired property belonging to a pupil.[77]

The only circumstance under which alienation of the pupil's property was permissible was to pay off debts, but tutors were to have recourse to this measure only as a last resort, and they required authorization from the praetor or the provincial governor to implement it (Ulp. *D.* 27.9.3.1). Otherwise, any transaction was illegal, and the praetor was enjoined to grant the pupil *restitutio in integrum.*[78] So strict was

---

76. See Ulp. *D.* 27.8.2 (3 *disputat.*) and Diocletian, Maximian, *CJ* 5.75.5 (A.D. 294?), with Talbert, *The Senate of Imperial Rome,* 444 no. 77.

77. For discussion of the measures taken to protect the interests of pupils, see Kaser, *RP* I, 363–67.

78. According to a rescript of Gordian (*CJ* 5.73.2, A.D. 241), the pupil or minor was to receive back not only possession of the property in question but all the profits, *fructus universi,* from a purchaser not acquiring the property in good faith. A later constitution provided for the restoration of the *fructus* in connection with the sale of property to pay off a creditor contracted without a decree of the governor and resulting in too low a sale price: ". . . ita ut fenebris pecunia cum competentibus usuris restituatur" (Diocl., Max., *CJ* 5.74.1, A.D. 290). According to Solazzi, *Curator impuberis,* 168–70, *restitutio in itegrum* was

the Severan legislation that the praetor or governor was not to authorize selling property belonging to a pupil, even when it could be alleged that such measures would serve the financial interests of the pupil, allowing him or her, for example, to purchase other, presumably more advantageous properties. Ulpian insisted on this principle with great emphasis, apparently in response to an opinion to the contrary (Ulp. *D.* 27.9.5.14).[79] Whereas the tutor was granted no latitude in the management of agricultural property, he had more freedom in managing other property belonging to the pupil. Ulpian, for example, allowed for a more rational management of the pupil's other types of property by granting the tutor the authority to ignore the wishes of the father by alienating the pupil's slaves, clothing, and houses (*D.* 26.7.5.9, 35 *ad ed.*).

The Roman legal authorities were similarly strict in carefully defining the circumstances under which problems with debt might justify alienating property. Before he could issue a decree allowing the alienation of property, the praetor or provincial governor was charged to ascertain that no other resource existed for paying off the pupil's or minor's debts, such as reserves of cash or crops or expectations of future income (Ulp. *D.* 27.9.5.9). Having determined that obligating or alienating property was necessary, the praetor was to minimize damage to the interests of the pupil or minor by making sure, first, that no property was unnecessarily alienated and, second, that, if possible, only the least advantageous properties be alienated (Ulp. 27.9.5.10).[80] Finally, the praetor was

---

not explicitly provided for in the Severan legislation but only implied as a remedy; see *CJ* 5.71.16 pr. (Diocl., Max., A.D. 294), concerning the unauthorized alienation of a *praedium rusticum vel suburbanum:* ". . . secundum sententiam senatus consulti dominium eius sive ius a te discedere non potuit, sed vindicationem eius et fructuum, vel his non existentibus condicionem competere constitit." But cf., e.g., Scaev. *D.* 4.4.47 pr. (1 *resp.*), according to which the pupil might already be entitled to *restitutio in integrum* even when the tutor sold off his or her property in good faith to pay off debts.

79. Ulp. *D.* 27.9.5.14: "Si aes alienum non interveniat, tutores tamen allegent expedire haec praedia vendere vel alia comparare vel certe istis carere, videndum est, an praetor eis debeat permittere. et magis est, ne possit: praetori enim non liberum arbitrium datum est distrahendi res pupillares, sed ita demum, si aes alienum immineat. proinde et si permiserit aere alieno non allegato, consequenter dicemus nullam esse venditionem nullumque decretum: non enim passim distrahi iubere praetori tributum est, sed ita demum, si urgueat aes alienum."

80. Ulp. *D.* 27.9.5.10: "Idem praetor aestimare debebit, utrum vendere potius an obligare permittat, nec non illud vigilanter observare, ne plus accip-

charged with making sure that the tutor actually paid the creditors with the money raised by alienating the pupil's property (Ulp. *D.* 27.9.5.13). A general conception concerning the unqualified economic value of agricultural property is underscored by a rescript of Septimius Severus and Caracalla, quoted by the jurist Paul. Those emperors refused a petition that unproductive property might be alienated (Paul. *D.* 27.9.13 pr., *lib. sing. ad or. div. Sev.*). In the emperors' view, the price of the land would match the income gained from it. There was nothing to be gained by selling totally worthless land, but the converse is that land with any value at all could provide some income.[81] The imperial government in this case applied an abstract conception of land as a resource representing security; any loss of this resource severely compromised the pupil's or minor's financial stability.[82]

In devising these strict rules, the jurists allowed no exceptions to the general principle that property be alienated only when debt made no other measure possible. To be sure, the Roman legal authorities were concerned with minimizing the effects of fraud or incompetence on the part of the tutor, who might present a variety of plausible reasons why it might be advantageous to the pupil to sell off property, but who at the same time, dishonestly or not, would still leave the pupil with no means of support. One protection against fraud on the part of the tutor was the prohibition against the tutor purchasing property belonging to the pupil (Paul. *D.* 18.1.34.7, 33 *ad ed.*). This principle was expressed in judgment rendered by Caracalla against a curator who had taken money set aside for the purchase of agricultural property for the minor

---

iatur sub obligatione praediorum faenoris, quam quod opus sit ad solvendum aes alienum: aut distrahendum arbitrabitur, ne propter modicum aes alienum magna possessio distrahatur, sed [*del. Mo.*] si sit alia possessio minor vel minus utilior pupillo, magis eam iubere distrahi quam maiorem et utiliorem."

81. Paul. *D.* 27.9.13 pr.: "Si fundus sit sterilis vel saxosus vel pestilens, videndum est, an alienare eum non possit. et imperator Antoninus et divus pater eius in haec verba rescripserunt: 'Quod allegastis infructuosum esse fundum, quem vendere vultis, movere nos non potest, cum utique pro fructuum modo pretium inventurus sit.'"

82. Similarly, Septimius Severus and Caracalla allowed tutors and curators, even those in debt in connection with their management of their office, to alienate their own property. The woman to whom this rescript was addressed apparently claimed that her curator, by pledging property (to the imperial Fiscus), compromised the liability that he had to bear from his own property for his administration of her affairs (*CJ* 4.51.1, A.D. 205).

whose interests he was supposed to be representing and instead had used it to purchase property for himself. The minor had the choice of taking over this property himself (as if the curator had been acting on his behalf) or exacting from the curator the interest set by law (*CJ* 5.51.3, A.D. 215). The financial interests of the minor in this question could be served if any agricultural property were purchased for him; the minor was not entitled to damages for the curator's failure to purchase a specific piece of property that might have offered the minor particular advantages.[83] In this connection, the fourth-century author of the *Opiniones* attributed to Ulpian urged harsh punishment for the tutor who obtained by fraudulent means the authorization from a provincial governor allowing him to sell his pupil's property and then bought it himself (Ulp. *D.* 27.9.9, 5 *opinionum*).[84] But we should note what sort of fraud the dishonest tutor is envisioned as perpetrating. The imperial authorities, in writing this legislation, like the jurists in implementing it, operated from an assumption that land was a preeminently valuable investment. It represented a form of economic security obtainable from no other source, and once lost it could not easily be regained. Land was more valuable than money, because only land could be counted on to provide an income, and the opportunities to acquire productive land were sufficiently rare that the selling of land, even at a good price, meant a loss of economic security. The dishonest tutor, then, would try to take advantage of his office by purchasing land from his pupil. The dishonesty did not just result from his paying too low a price. Rather, in purchasing the pupil's land he would be depriving the pupil of a source of income and financial security that could only with difficulty be replaced.[85]

---

83. As Johnston, *The Roman Law of Trusts*, 223–49, points out, the governor in the *cognitio extraordinaria* could issue a judgment for performance, but this provision did not guarantee that the plaintiff would receive a specific property.

84. On this work, see T. Honoré, *Ulpian* (Oxford, 1982), 120–28, who views it as the work of a third-century jurist. Cf. D. Liebs, "Römische Provinzialjurisprudenz," *ANRW* II 15 (1976): 288–362, at 332–38, who views it as a fourth-century work suggesting a provincial setting in the problems that it addresses.

85. The concern on the part of the imperial government to protect against this type of fraud on the part of the tutor is also evident in a constitution of Constantine, who ruled that a tutor could be prevented from selling off an *aedificium deformatum* belonging to the pupil and could instead be forced to undertake repairs on it (*CJ* 5.37.22.3a, A.D. 326). The annual income that the building

In implementing this policy, the imperial government was no doubt seeking to prevent the kind of situation that had arisen at an earlier date in connection with a certain Ptolema, an underage heir in Roman Egypt; we learn about this affair from a court protocol (*SB* I 5761, A.D. 81–96). The father of Ptolema had bequeathed to her a house (or a share in a house) and some land, and her guardians had sold this property to pay off debts. One of these debts had arisen in connection with a prodomatic lease (a lease with payment in advance; the lessor often granted the lessee the use of lands as a means of paying off an existing debt). The present document records a lawsuit brought on behalf of Ptolema against a certain Apia, apparently the wife of the creditor in the prodomatic lease; the purpose was to claim back the funds paid to her that had been collected from the sale of Ptolema's property. Such a lawsuit involved difficult and subjective judgments about the good faith of the guardians, and it must have been very difficult after the fact to restore property to a pupil whose guardians were judged to have acted against his or her interests. The point of the Severan legislation was to compel the tutor to seek permission before alienating the pupil's property; in this way, the legitimate interests of creditors could be balanced against the need to protect the financial security of pupils.[86]

The rules discussed here restricting the alienation of property belonging to pupils applied generally, regardless of class, but it seems

---

might provide was envisioned as offering the pupil a greater benefit than the proceeds from the sale, which the imperial chancellery simply assumed would be reduced as a result of the tutor's fraud: "ita enim annui reditus plus minoribus conferent quam per fraudes pretia deminuta."

86. For complaints against guardians in Roman Egypt, see Krause, *Witwen und Waisen,* 108–12, and, on this case, see 108–9. Ptolema won her case. We might also consider that the Severan legislation, if it was originally conceived as applying to minors, sought to prevent the kind of situation described by a minor to the royal scribe functioning as the *strategos* of the Oxyrhynchite nome (*P.Oxy.* VI 898, A.D. 123). In this petition, a minor named Didymos was the son of the deceased Dionysios, and he alleged fraud on the part of his mother Matreina in her guise as his guardian, or *epitropos.* According to the petition, the unscrupulous mother beguiled her son into going to the Small Oasis with her confidant Dioskouros, where some kind of agreement was drawn up involving the payment of one and one-half talents. The minor Didymos was, in his own description, tricked into mortgaging his vineyards in the Small Oasis in exchange for a deed indemnifying him from certain debts, and later his mother took this deed and refused to hand it back until he had formally absolved her from any responsibilities arising from her service as guardian.

beyond dispute that they were developed particularly with the upper classes in mind. Indeed, as we have seen, the jurists were particularly concerned with circumstances affecting the interests of upper-class Romans as they developed rules about tutorship. That this principle also holds true in the application of the Severan legislation seems inherently likely, and this hypothesis draws some confirmation from a rescript of the emperor Aurelian. The case in question concerned a senator named Saturninus, who approached the emperor instead of the provincial governor to obtain a decree allowing him to sell property belonging to a pupil (*CJ* 5.72.2, no date). In this constitution, what seems implicit in many other cases is made explicit: the parties whose interests the imperial government and the jurists had to address belonged to the highest classes of Roman society.

We must, however, recognize that in developing rules applying the Severan legislation to real-life situations, the jurists to some extent proceeded from what they perceived to be the internal logic of the legislation, rather than from their own considerations about the economic interests of the parties involved. This approach can be most readily seen in the ruling by Ulpian on whether tutors could sell a *fundus* pledged as security to the pupil. Ulpian held that such a *fundus* could be sold, since it in some sense still belonged to the debtor, but that once the *fundus* had passed into the ownership of the pupil, it could not be sold (*D.* 27.9.5.3).[87] Ulpian's method here was not to apply the general logic and spirit of the Severan legislation to considerations of what might best serve the economic interests of the pupil. Rather, he formulated his decision in terms of a strict interpretation of the letter of the law, allowing alienation when the property in question could be viewed as not, properly speaking, belonging to the pupil, and of course not allowing alienation when the pupil formally owned the property in question.[88] We might imagine that, when considered in terms of the

---

87. Ulp. *D.* 27.9.5.3: "Si fundus pupillo pigneratus sit, an vendere [possunt *ins. dett.*] tutores? hunc enim quasi debitoris, hoc est alienum vendunt. si tamen impetraverat pupillus vel pater eius, ut iure dominii possideant [possideat *Hal.*], consequens erit dicere non posse distrahi quasi praedium pupillare. idemque et [est *Mo.*] si fuerit ex causa damni infecti iussus possidere."

88. But see Ulp. *D.* 27.9.8.1 (2 *de omn. trib.*), where Ulpian extended the application of the *Oratio* to various types of curators, including to the *curator furiosi*. On the authenticity of this passage, see Cervenca, *BIDR*, 3d ser., 21 (1979): 65–70.

apparent intention of the Severan legislation, selling a property pledged as security to the pupil might be just as disadvantageous as selling any other property belonging to the pupil.[89]

Restrictions on Alienating Property in Dotal Law

To appreciate better the protection that the imperial government was seeking to provide for the financial interests of pupils, we might compare the Severan legislation with the somewhat less stringent restrictions on the alienation of land in the *lex Iulia de fundo dotali*. This measure, part of Augustus' moral reforms, prohibited the alienation, without the wife's consent, of Italian land included in a dowry.[90] One purpose of this law was to provide the wife a degree of financial security by protecting her ability to regain intact at the time of her husband's death or divorce whatever landed property she had included in her dowry. To this extent, the Julian dotal law is similar in purpose to the Severan legislation. Accordingly, as in the Severan legislation, the husband not only was prohibited from alienating the property or pledging it as security without the wife's consent (Gaius *D*. 23.5.4, 11 *ad ed. prov.*), but he was also enjoined not to impose servitudes on the dotal property or to renounce servitudes owed to it (Julian. *D*. 23.5.7 pr.–1, 16

---

89. The principle that Ulpian developed in this response is also found in a rescript of Caracalla referring to the *Oratio Severi*: "Venditio quidem praedii, quod iure pignoris vel in causa iudicati captum et distractum est, ad senatus consultum, quod de alienandis praediis pupillorum vel adulescentium auctore praetore vel praeside provinciae factum est, non pertinet" (*CJ* 5.71.1 pr., A.D. 212).

90. The *lex Iulia de fundo dotali* was apparently part of the *lex Iulia de adulteriis coercendis*: see A. Mette-Dittmann, *Die Ehegesetze des Augustus* (Stuttgart, 1991), 39, and Kaser, *RP* I, 334, 406. See also Berger, *Encyclopedic Dictionary of Roman Law*, s.v.; J.A. Crook, "Women in Roman Succession," in B. Rawson, ed., *The Family in Ancient Rome* (London and Ithaca, 1986), 58–82, at 68; S. Dixon, "Family Finances: Terentia and Tullia," in *The Family in Ancient Rome*, 93–120, at 94 and n. 3; J.F. Gardner, *Women in Roman Law and Society* (London, 1986), 103; and Treggiari, *Roman Marriage*, 348–49 and n. 141. Other references include Gaius *Inst*. 2.63, Iust. *Inst*. 2.8 pr., Pap. *D*. 31.77.5 (8 *resp.*), and Iust. *CJ* 5.13.1.15 (A.D. 530). For general discussion of the significance of dowries for the security of upper-class Roman women, see Gardner, 98–116, and Treggiari, 323–64, as well as R.P. Saller, "Roman Dowry and the Devolution of Property in the Principate," *CQ* 34, no. 1 (1984): 195–205, and idem, *Patriarchy, Property and Death*, 204–24.

dig., Ulp. *D.* 23.5.5, 2 *de omn. trib.*). The classical jurists extended the provisions of the Julian law to cover urban as well as agricultural properties (Ulp. *D.* 23.5.13 pr., 5 *de adult.*).[91] But the Julian prohibitions against alienation were not as strict as those in the Severan legislation for pupils. The wife could agree to the alienation of the dotal property, and Africanus addressed the possibility that the wife might provide a dowry on *aestimatio,* with her or her husband, respectively, having the choice of restoring the equivalent value of the *fundus dotalis,* rather than the dotal property itself (Afric. *D.* 23.5.11, 8 *quaest.*).[92]

It seems clear that the legislator of the Julian dotal law shared a similar conception of the preeminent value of agricultural property as a form of economic security. The Julian law, however, allowed greater flexibility in managing the dotal property, because the dowry served the financial interests of both husband and wife. To some extent, the jurists interpreted the Julian law without regard for economic consequences. Thus, according to Ulpian, the legacy of the *fundus dotalis* to a third party was valid if a wife instituted as heir received the equivalent value of her dowry in her share of the inheritance (*D.* 23.5.13.4, 5 *de adult.*). This same jurist cites the authority of Scaevola in allowing the wife to reclaim part of the *fundus dotalis* bequeathed to a third party through legacy, if such a step was needed to bring the wife's share of the inheritance up to the value of the dowry (ibid.). Breaking up of the *fundus dotalis,* one possible result of such a claim, could have the effect of making the property less valuable as an agricultural enterprise, but the point of the legislation was to protect the wife against any financial loss caused by the alienation of the dotal property, not to guarantee that the wife would reacquire a specific piece of land.[93] However, the jurists did debate whether certain types of expenditures that the husband might undertake on the dotal property were to be considered beneficial to the wife's interests, in which case the wife might have to compensate the husband. The alternative was that the wife was to be compensated for the changed (and therefore worsened) condition of the property.

---

91. Ulp. *D.* 23.5.13 pr.: "Dotale praedium accipere debemus tam urbanum quam rusticum: ad omne enim aedificium lex Iulia pertinebit."

92. Afric. *D.* 23.5.11: "Quod si fundus in dotem aestimatus datus sit, ut electio esset mulieris, negavit alienari fundum posse: quod si arbitrio mariti sit, contra esse."

93. Vindicating part of a *fundus* could also entail a cash compensation for the value of the property claimed.

Accordingly, Iavolenus took issue with Labeo's ruling that opening up a marble quarry on the *fundus dotalis* worsened the condition of the property (Iav. D. 23.5.18 pr., 6 *ex poster. Lab.*).

The perception underlying both the Julian law and the Severan legislation that the alienation of agricultural property represented the most damaging compromise to an upper-class Roman's financial security can be seen in a case discussed by Papinian (apud Ulp. D. 16.1.8.1, 29 *ad ed.*). This case concerned the application of the *Senatus Consultum Velleianum*, which was designed to protect women against the financial dangers involved in assuming obligations on behalf of third parties. The tutors of an underage son planned to alienate estates belonging to him, presumably to pay off inherited debts. The child's mother wished to prevent her son from losing his source of income in this way, so she undertook to indemnify the tutors against any potential lawsuit arising from their failure to pay off the debts. In effect, she took on herself the responsibility to pay off any obligations owed by the son, but in so doing she did not gain the protection of the *Senatus Consultum Velleianum*.[94] The woman in question apparently had the same concern for her son that the Roman legal authorities in general had for pupils, and she was willing to take drastic steps to avoid the possibility of her son losing his property.[95]

---

94. Pap. apud Ulp. D. 16.1.8.1: "Si mulier intervenerit apud tutores filii sui, ne hi praedia eius distraherent, et indemnitatem eis repromiserit, Papinianus libro nono quaestionum non putat eam intercesisse: nullam enim obligationem alienam recepisse neque veterem neque novam, sed ipsam fecisse hanc obligationem." For discussion of the *SC Velleianum* (or *Vellaeanum*), see Talbert, *The Senate of Imperial Rome*, 442 no. 53, who dates it to the reign of either Claudius or Nero, and Kaser, *RP* I, 667. My interpretation of the legal situation in the Ulpian text is based on Kaser, 353 n. 8. See also Gardner, *Women in Roman Law and Society*, 75–76, and J.A. Crook, "Feminine Inadequacy and the *Senatusconsultum Velleianum*," in B. Rawson, ed., *The Family in Ancient Rome: New Perspectives* (London and Ithaca, 1986), 83–92.

95. A similar principle was applied in Roman municipal law, since cities were discouraged from alienating lands that they owned. Thus Papirius Iustus cited a rescript of Marcus Aurelius and Lucius Verus requiring a *curator civitatis* to reclaim (*retrahere*) *agri* formerly belonging to a res publica but subsequently purchased in good faith by private individuals (D. 50.8.11.2, 2 *de constit.*). On the responsibility of the *curator rei publicae* to oversee the administration of property belonging to a city, see W. Eck, *Die staatliche Organisation Italiens in der hohen Kaiserzeit* (Munich, 1979), 219, 227.

Testamentary Prohibitions against Alienating Property

The economic assumptions implicit in the imperial government's regulation of property belonging to pupils also characterized the thinking of private upper-class Romans, to judge at least by private testamentary strictures against the alienation of agricultural property. A good example is a case ruled on by Scaevola, in which the testator prohibited his son and heir from either alienating or pledging the agricultural property left him. The testator's wish was that this property should be preserved intact for the heir's children from legitimate marriages and for other relatives as well (D. 32.38 pr., 19 dig.).[96] The concern of this testator was financial; he clearly viewed the imposition of a trust on his heir to prevent the alienation of the property that he was bequeathing as the best way to provide for the financial security of future generations of his family. Underlying the testator's method in drawing up his will was an assumption that the alienation of property could have no positive financial consequences. It was also common for testators to impose trusts to prevent the alienation of property outside of their family or extended circle of freedmen and freedwomen; in many cases, the motivation may have been less financial and more to preserve the prestige associated with the ownership of particular properties.[97]

The intention of the testator in the case discussed by Scaevola (D. 32.28 pr.) and of those in many comparable cases was to preserve an estate intact for the heir while at the same time providing for connec-

---

96. Scaev. D. 32.38 pr.: "Pater filium heredem praedia alienare seu pignori ponere prohibuerat, sed conservari liberis ex iustis nuptiis et ceteris cognatis fideicommiserat. . . ."

97. See, for example, Scaev. D. 32.38.1 (19 dig.), where two freedmen are to share a *fundus Cornelianus* and are not to allow it to leave the testator's name, *de nomine meorum*; 32.38.2, concerning a building; and 32.38.3, concerning a *fundus*. See also Johnston, *The Roman Law of Trusts*, 76–116, especially 88–97, with discussion of further examples. For detailed discussion of the legal issues surrounding such prohibitions, see idem, "Prohibitions and Perpetuities: Family Settlements in Roman Law," ZRG 102 (1985): 220–90. As Johnston argues, some such prohibitions were set up to provide *alimenta* for freedmen and freedwomen and other members of the *familia*. Very often, the purpose of these prohibitions was not financial but rather simply to perpetuate the nomen of the testator. On legacies to freedmen and freedwomen and prohibitions against alienation, see Champlin, *Final Judgments*, 133–36.

tions whose claims for support could not be ignored. In another case, treated by Papinian, a man bequeathed to his wife the profits from his properties, *fructus praediorum,* with the (mistakenly formulated) proviso that the properties and their revenues were to revert to the heirs after the wife's death (*D.* 33.2.25, 8 *resp.*). The testator was concerned to keep within his family not just a certain amount of wealth but rather a specific set of properties. Finally, in another type of arrangement, a testator bequeathed to a woman the income from a *fundus* for as long as she should live; the heir was given the choice of selling the property and offering the woman, as a substitute for the annual income, the rent that the estate had provided during the testator's lifetime (Scaev. *D.* 33.2.38, 3 *resp.*).[98]

Literary Evidence for Investing in Estates

To return to the legislation protecting the financial interests of pupils, the Roman government adopted an approach toward investment that even by Roman standards must have been extremely conservative. Yet protections of agricultural property in other areas of the law and the dispositions made by testators that I have already examined in this chapter suggest that the approach taken by the jurists to agricultural property as investment was not the product of the internal logic of the law alone. Rather, the Roman legal authorities applied a principle that informed upper-class economic thinking in general. According to this principle, agricultural land provided the preeminent form of security, so the acquisition and undisturbed ownership of land was essential to an individual's financial stability. To be sure, some upper-class Romans might take a very different approach, speculating in agriculture. For example, L. Tarius Rufus, the former officer of Augustus, squandered a fortune by making excessively risky investments in agriculture (Plin. *Nat.* 18.37). Later, Q. Remmius Palaemon, the wealthy freedman and grammarian who flourished in Julio-Claudian Rome,

---

98. Prohibitions against alienating agricultural land also turn up in the Roman wills preserved on papyri, as in the will of the veteran C. Longinus Castor (*P.Test.Roma.*[2] 12, lines 9–10, A.D. 189–94, also discussed in chap. 2) and possibly in the will of a woman named Sabinia, probably the wife of a Roman soldier (*P.Testa.Roma.*[2] 9, *PSI* XIII 1325, *SB* V 7630, A.D. 169–77). For the date of this document, see P.J. Sijpesteijn, "Further Remarks on Some Imperial Titles in the Papyri," *ZPE* 45 (1982): 177–96, at 182.

invested heavily in viticulture in the *ager Nomentanus* near Rome and after eight years sold his spectacularly productive vineyard to Seneca for four times its original purchase price of six hundred thousand sesterces (Plin. *Nat.* 14.49–51).[99] Rufus and Palaemon were led to engage in such heavy investment because of the particular circumstance that each had recently come into substantial fortunes.[100] These unusual circumstances aside, other Roman landowners could not have viewed their estates as investments whose value would grow; nor did they tend to acquire them to improve them and sell them for a profit.[101]

Much more characteristic no doubt was the approach taken by Seneca when he purchased Palaemon's estate. Pliny the Elder emphasizes that Seneca was captivated by a love for the property (*Nat.* 14.51). Seneca apparently was not interested in making a profit by finding an agricultural property that he might develop with a view toward achieving some level of wealth or income.[102] Seneca's approach was comparable to that exhibited by Pliny when he considered purchasing an estate adjacent to property that he already owned at Tifernum Tiber-

---

99. On Q. Remmius Palaemon, see J. Christes, *Sklaven und Freigelassene als Grammatiker und Philologen im antiken Rom* (Wiesbaden, 1979), 98–102, and R.A. Kaster, *Guardians of Language* (Berkeley and Los Angeles, 1988), 55–57. Palaemon's ability to purchase land and invest in its development may have been exceptional as well. He received a substantial income of four hundred thousand sesterces from his schools in Rome (Suet. *Gram.* 23); this income, with his other presumed sources of income outside of agriculture, may have provided him with a substantial surplus of cash to invest. We might also compare Palaemon's vintner, Acilius Sthelenus, who also sold a vineyard of sixty *iugera* in the *ager Nomentanus* for four hundred thousand sesterces (Plin. *Nat.* 14.48). For the involvement of wealthy freedmen in agriculture, see A. Łoś, "Les intérêts des affranchis dans l'agriculture itaienne," *MEFRA* 104 (1992): 709–53.

100. On the buying up of land for development, see also Duncan-Jones, *Economy*, 324.

101. Columella viewed agriculture as the safest way of increasing one's patrimony—"amplificandi reliquendique patrimonii" (1 pref. 7)—but it must be kept in mind that he was seeking to convince what he perceived to be a skeptical audience to take a greater interest in agriculture.

102. On Seneca's involvement in agriculture, see M. Griffin, *Seneca: A Philosopher in Politics* (Oxford, 1976), 289–91. Seneca certainly exploited this estate for income, since, as Columella reports, the estate's highest yield of eight *cullei* per *iugerum* was achieved after Seneca had bought it (3.3.3). The purchasing strategy adopted by a landowner like Seneca does not imply that such a landowner would not be involved in complex financial arrangements in order to keep the estate productive.

inum (*Ep.* 3.19).[103] Pliny balanced the attraction of extending his holdings against financial concerns centered around the condition of the tenants. What is striking about Pliny's deliberations is the implication that, had there been no problems with the tenants, there would have been no debate, and the purchase of the estate would have justified itself.[104] The approach to investing in land exhibited by Seneca and Pliny underlies the satirical portrait of the figure Trimalchio in the *Satyricon*. At the beginning of his career, of course, Trimalchio took ridiculous risks investing his inheritance to make a killing in commercial ventures (Petr. 76.2–10). Once having made his fortune, Trimalchio, in self-conscious imitation of the practices of upper-class Romans, began investing his wealth in land, limiting his involvement in commerce to lending money through freedmen (Petr. 76.8–9).[105] He absurdly hoped to acquire Sicily (Petr. 48.2–3) and to join his estates to Apulia (Petr. 77.3). Trimalchio's wishes are fantastic, but they would have little point if Petronius did not share with his upper-class audience certain assumptions about the way in which upper-class Romans invested their wealth. Petronius assumes is that any spare capital was to be invested in agricultural land. Thus when reciting the report for 26 July from the daily gazette from Trimalchio's property, the *actuarius*

---

103. The following discussion summarizes the conclusions that I have drawn in "Investment in Estates." For the physical characteristics of Pliny's Tuscan estates, see De Neeve, *Athenaeum* 78 (1990): 367–68, 373–75.

104. See Love, *Chiron* 16 (1986): 99–146, especially 143 f.; Capogrossi Colognesi, "Grandi proprietari, contadini e coloni," 355; idem, *Ai margini della proprietà fondiaria* (Rome, 1995), 264–68 and n. 127, and "Il regime degli affitti agrari," *Scienze dell'antichità, Storia, Archeologia, Antropologia* 6–7 (1992–93): 232–33 and n. 127; and De Neeve, *Athenaeum* 78 (1990): 364. Love argues on the basis of the evidence of Cato and Varro that Roman landowners, when purchasing property, sought estates that would fulfill their needs as members of a particular class in Roman society while at the same time allowing them to achieve a required level of income.

105. On Trimalchio's emulation of upper-class mores, see P. Veyne, "Vie de Trimalcion," *Annales ESC* 16 (1961): 213–47, and on his investment of available cash, see idem, *REA* 81 (1979): 272 n. 3. See also J.H. D'Arms, *Commerce and Social Standing* (Cambridge, Mass., 1981), 97–120. Trimalchio's investment strategy finds parallels in early modern Europe, when individuals amassing great fortunes from commerce would invest their wealth in land. On this see Pleket, "Wirtschaft," 41–42, 44–45. For the social importance of acquiring estates, see also J. Kolendo, "Ostentation sociale et grande propriété," in *Du Latifundium au Latifondo* (Talence, 1995), 425–36.

remarks that ten million sesterces that could not be invested were stored in the cash box, or *arca* (Petr. 53.4). Otherwise these funds were to be invested in property, such as the *horti Pompeiani,* which Trimalchio was not even aware he had purchased (Petr. 53.5–8). Here Trimalchio is probably to be understood as affecting the same aristocratic insouciance for financial matters that characterizes much of Pliny's correspondence, but again it seems that agricultural lands are understood as purchased for their own sake, for the prestige and security that they afford, not to achieve any more precisely articulated investment goal. The bottom line, from a financial standpoint, for Seneca, Pliny, and Trimalchio, was that any investment in a productive estate justified itself, and none of these landowners had alternative ways to invest securely the funds used to purchase agricultural property.[106]

The security represented by the long-term income from land meant that the acquisition of agricultural property needed little justification under ordinary circumstances, and that the loss of land was a serious compromise to a person's financial well-being. This conception of the value of agricultural property as an investment had to have a significant effect on the way in which Roman landowners managed their agricultural property, and it surely also had a profound effect on the market for buying and selling agricultural land.[107] Given the value of land as a resource providing financial and social security, there was very little to be gained for upper-class Romans in selling estates. Land sales might be spurred by special circumstances, such as Trajan's enactment that candidates for senatorial offices were required to have one-third of their property invested in Italian land (*Ep.* 6.19.4).[108] But in

---

106. But see the discussion of the profitability of viticulture by Columella (3.3.9), where Columella assumes an expected 6 percent return otherwise achievable from the money to be invested in viticulture. Love, *Chiron* 16 (1986): 127–28, argues that this assumption by Columella implies that the agronomist viewed the funds to be invested in viticulture as capital with alternative possibilities for investment; cf. Love, *Antiquity and Capitalism,* 91–93.

107. On this point, see Saller, *Patriarchy, Property and Death,* 202–3. For the process by which upper-class Romans acquired land and formed large estates, see D. Vera, "Dalla 'villa perfecta' alla villa di Palladio," *Athenaeum* 83, no. 2 (1995): 331–56, at 331–46.

108. On this enactment, see Sherwin-White, *Commentary,* 377. Similar legislation was enacted by Marcus Aurelius, who required non-Italian senators to have one-fourth of their property invested in Italian land (SHA *Marc.* 11.8). On this legislation, see Talbert, *The Senate of Imperial Rome,* 142. See also V. Sirago, *L'Italia agraria sotto Traiano* (Louvain, 1958), 271–74. As governor of Bithynia,

general, the actions of Pliny and the fictitious Trimalchio suggest that upper-class Romans rarely liquidated estates to reinvest the proceeds in some other undertaking. The restrictions of the Severan legislation concerning the alienation of a pupil's agricultural property were grounded in the realities of upper-class financial thinking.[109]

Conclusion

In this chapter I have begun to examine the implications of what living in a largely agrarian society had for upper-class Romans in their efforts to invest their wealth. The point of this examination is to define the general considerations that guided upper-class Romans as they went about the business of managing their wealth. It might be objected that we cannot draw general conclusions about upper-class Roman attitudes toward risk and investment from the juristic writings about the financial responsibilities of tutors toward their pupils' property, since in almost all societies with an institution like tutorship, the legal system imposes restrictions on the tutor's freedom in administering the pupil's property. Indeed, the tutor (or the corresponding official) in modern legal systems generally has a responsibility similar to that of the Roman tutor, to safeguard the pupil's uninterrupted use of his or her property.[110] Thus it should be no surprise that the imperial administration

---

Pliny also anticipated that it would be difficult to use surplus funds from Bithynian cities to purchase land, so he requested from Trajan guidance on how to lend this money out at interest (*Ep.* 10.54).

109. P.K. Bardhan, *Land, Labor, and Rural Poverty* (New York, 1984), 95, argues that in peasant societies peasants will rarely sell land. Not only does selling land entail the high transaction costs in investing in alternative business activities. In addition, the proceeds from selling land do not compensate the peasant for the loss of economic security represented by land. In this case a comparison between peasant agriculture and Roman estate owning is justified to the extent that Roman landowners would face similar transaction costs and risks if they sold off their chief means of livelihood. On the basis of Cato's recommendations concerning the purchase of an estate (*Agr.* 1, discussed in n. 43 in this chapter), Love, *Chiron* 16 (1986): 114, 143, argues that a market did exist in ancient Rome for estates bought and sold with a view toward their "profitability as enterprises" (114). I do not disagree with this conclusion, but the market for agricultural land would have been severely restricted by the limited ways for a landowner to invest capital to realize income.

110. In Church courts in medieval England, by contrast, the guardian had few responsibilities in managing the property of the orphan. See R.H. Helmholz, "The Roman Law of Guardianship in England, 1300–1600," *Tulane Law Review* 52, no. 2 (1978): 223–57.

required the Roman tutor to be conservative in administering the pupil's property. But what is striking is the form in which this expected conservatism found expression. The jurists clearly viewed any form of investment other than agriculture as too risky to serve the needs of a pupil, and their insistence that the tutor invest the pupil's money in agriculture whenever this was possible suggests an approach toward investment quite different from what we might understand today.

In the jurists' conception of the Roman economy, agriculture represented not only the safest investment but, for all practical purposes, the only investment capable of providing a reasonably reliable income for the long term.[111] An examination of other types of evidence for investing in land suggests not that the jurists' approach toward investment was developed simply as a juristic doctrine but that the jurists shared their conception of the Roman economy with the upper classes in general. Certainly some landowners also engaged in commerce on a large scale and accumulated large fortunes from such activity, but the imperial authorities chose to ignore these other possibilities for income. Their reasons for doing so in all likelihood derived from their understanding of the circumstances under which upper-class Romans achieved their incomes. To serve the needs of a crucially important social class in the empire, the Roman imperial government simplified reality by creating idealized pictures both of the economic interests of the pupil and minor and of the general circumstances surrounding the Roman economy. These idealized concepts provided the basis on which regulations with general applicability could be formulated.

---

111. For the importance of risk and the uncertainty of income from estates as a factor affecting the planning of upper-class Roman landowners, see the discussion of H. Pleket, "Agriculture in the Roman Empire in Comparative Perspective," in H. Sancisi-Weerdenburg et al., eds., *De Agricultura* (Amsterdam, 1993), 317–42, at 332–42. Veyne, *REA* 81 (1979): 261–80, argues that upper-class Romans saw investment in land as their greatest protection against risk.

2

Profit, Security, and the Law of Legacy

In this chapter, I investigate upper-class Roman concepts of investment and profit in agriculture by considering how upper-class Romans used agricultural property as a means of providing funds for civic and private benefactions, including permanent charitable foundations, and for supporting private individuals. This investigation builds on my conclusions in chapter 1, where I argued that the jurists' treatment of the law of tutorship indicates a broadly held view among upper-class Romans that investment in agriculture represented the preeminent form of financial security and that any loss of land was a serious compromise to the ability of upper-class Romans to maintain a standard of living appropriate to their class. We can better understand what agriculture meant to the financial interests of upper-class Romans by analyzing how they conceived of profit from an estate. For this purpose, I will examine epigraphical and papyrological evidence for the dispositions that upper-class Romans made in their wills to fund their benefactions, and I will compare this evidence with the jurists' treatment of legal problems involving the bequest of agricultural property. As I will argue, the jurists shared with the benefactors attested in the inscriptions and papyri a basic understanding of what constituted profit from an investment in agriculture. In the eyes of the jurists and upper-class testators, the overall constraints imposed by the Roman economy severely circumscribed a landowner's ability to profit from an estate, so for many landowners the overriding concern was to keep the yearly income from an estate as stable as possible. This conception of profit in agriculture certainly had important consequences for how Roman landowners managed their property and was a key factor in shaping the relationships between the groups involved in the Roman estate economy, including large landowners and small-scale tenants cultivating the land.

Providing for Permanent Foundations

Let us begin our investigation by considering the planning exhibited by upper-class Romans in numerous charitable and civic foundations. In establishing their foundations, benefactors faced the problem of securing long-term financial stability from an initial investment of wealth. We can gain some insight about the financial expectations of upper-class Romans by examining Pliny's efforts to establish an annual fund of thirty thousand sesterces to provide sustenance for children in his hometown of Comum.[1] Pliny thought that helping to increase the number of free citizens in one's town was one of the most significant contributions that a Roman aristocrat could make. We should imagine, then, that Pliny wanted to make every effort to ensure that his alimentary foundation would be securely provided with the needed resources long after he was dead and no longer in a position to oversee its management. On the surface, the goals of Pliny's foundation were clear-cut, and, as was the case with numerous other charitable foundations in the Roman Empire, the donor set aside a capital fund, the interest from which was to provide revenue sufficient to meet the foundation's yearly expenses. Thus an inscription commemorating his public acts of generosity toward Comum records simply that Pliny gave the city five hundred thousand sesterces for feeding freeborn children (*ILS* 2927, *CIL* V 5262).[2] At the conventional annual return on capital of 6 percent, Pliny's foundation could spend thirty thousand sesterces each year. The capital fund was to cover costs that would remain for the most part fixed on a yearly basis; in this respect, Pliny's alimentary program was similar to the contemporary imperial alimentary foundations established by Trajan and possibly earlier by Domitian and Nerva.[3] In the

---

1. For preliminary discussion of topics treated in this section, see my "Investment in Estates by Upper-Class Landowners in Early Imperial Italy: The Case of Pliny the Younger," in H. Sancisi-Weerdenburg et al., eds., *De Agricultura* (Amsterdam, 1993), 214–37, at 228–29.

2. The amount of five hundred thousand sesterces is restored from Pliny *Ep.* 7.18.2 For the details surrounding Pliny's alimentary program, see Sherwin-White, *Commentary*, 104–5, on *Ep.* 1.8.10; see especially *Ep.* 1.8.9–10, where Pliny comments on his wisdom in choosing an alimentary program as the target of his generosity.

3. On the purposes of the imperial *alimenta*, to facilitate the raising of children and therefore to contribute to increasing the free population of Italy, see especially R. Duncan-Jones, *The Economy of the Roman Empire*, 2d ed. (Cam-

later of the two imperial programs recorded in the well-known inscription from Veleia (*CIL* XI 1147, *ILS* 6675), for example, a fixed level of income determined the number of children who could be supported at any one time. The fifty-two thousand sesterces of annual income from this program provided for the monthly purchase of food for 281 children of various classifications. Presumably the composition of the group of children receiving support might change over time, but the annual fund was to remain constant, at least under ideal conditions.

Fixed yearly costs were also a feature of purely private alimentary foundations. The best example is the second-century private foundation from Sicca Veneria in Africa established by the equestrian P. Licinius Papirianus (*ILS* 6818, *CIL* VIII 1641). In this program, an annual income of sixty-five thousand sesterces, representing a 5 percent return on a capital sum of 1,300,000 sesterces, fed three hundred boys at ten sesterces per month and three hundred girls at eight sesterces per month.[4] It seems clear that the benefactor at Sicca Veneria established his foundation after balancing a suitable number of children who might be supported and a reasonable level of support against the resources available to him. Accordingly, Papirianus made careful provision for a method to select new children to replace those who died or

---

bridge, 1982), 288–319, and earlier P. Veyne, "La table des Ligures Baebiani et l'institution alimentaire de Trajan," pts. 1 and 2, *MEFRA* 69 (1957): 81–135; 70 (1958): 177–241. See also N. Criniti, *La Tabula Alimentaria di Veleia* (Parma, 1991), 245–74, who offers a new text and a full study of the Veleian inscription. For the date at which the *alimenta* were originally instituted, see Duncan-Jones, 291–93. G. Woolf, "Food, Poverty and Patronage: The Significance of the Epigraphy of the Roman Alimentary Schemes in Early Imperial Italy," *PBSR* 58 (1990): 197–228, argues that the imperial *alimenta* demonstrated the emperor's generosity to Italy and were not primarily aimed at providing relief to the poor. See also C. Bossu, "L'objectif de l'institution alimentaire: essai d'évaluation," *Latomus* 48 (1989): 372–82. But Pliny certainly envisioned his program as enabling indigent people to raise children (*Ep.* 1.8.10–11). For discussion of private alimentary foundations, see S. Mrozek, "Die privaten Alimentarstiftungen in der römischen Kaiserzeit," in H. Kloft, ed., *Sozialmassnahmen und Fürsorge* (Graz and Horn, 1988), 155–66.

4. The number of girl recipients was three hundred rather than two hundred as on the inscription; see Duncan-Jones, *Economy*, 102. On this inscription, see also M. Amelotti, *Il testamento romano attraverso la prassi documentale* (Florence, 1966), 23. Papirianus was *procurator a rationibus* under M. Aurelius and L. Verus, but his career, attested only in this inscription, is otherwise not known: see *PIR*² L 229.

grew up: his goal was to use a constant level of income to provide a consistent level of support for a number of boys and girls that would remain constant in perpetuity.⁵ A comparable approach explains the organization of an alimentary program established at Tarracina by a wealthy woman named Caelia Macrina (*ILS* 6278, *CIL* X 6328).⁶ She bequeathed one million sesterces to the town in memory of her son, and the income from this sum was to be distributed to one hundred boys and one hundred girls at the rates of twenty sesterces per month and sixteen sesterces per month, respectively. The yearly expenditure of this program would be 43,200 sesterces, or a 4 1/3 percent return on the original principal. Again, the donor had in mind the amount of capital that she wished to make as a gift and a number of recipients that she wished to support, and she set the level of support in accordance with a return that she could confidently expect to be achieved from her principal. Such alimentary foundations as these, then, represent extreme versions of the economic goal that I ascribed to upper-class landowners in the previous chapter: the investment of capital served the sole purpose of providing a set level of income to realize a social goal.

The goals of such alimentary foundations may have been straightforward, but achieving them was surely a difficult problem that required Roman benefactors to exercise all their ingenuity so that they could accommodate their financial needs to the realities of the Roman economy. Thus the simple record in the inscription commemorating his benefactions masks the difficulties that Pliny faced in assuring his program a steady income. We see the solution that Pliny adopted when he discusses with his fellow townsman Caninius Rufus the latter's efforts to provide Comum with funds to pay for a yearly banquet (*Ep.* 7.18). Pliny had nothing but scorn for the two conventional methods of funding such foundations, that is, giving the town either a lump sum of money or an amount of land of equal value (7.18.1).⁷ Pliny's later experience with the troubled finances of Bithynian cities must have confirmed his mistrust of his own city's ability to manage funds (*Ep.*

---

5. *ILS* 6818, *CIL* VIII 1641: ". . . curari autem oportet, ut in locum adulti vel demortui cuiusque statim substituatur, ut semper plenus numerus alatur."

6. See also *FIRA* III 55d, and Amelotti, *Il testamento romano*, 23.

7. Plin. *Ep.* 7.18.1: "Deliberas mecum quemadmodum pecunia, quam municipibus nostris in epulum obtulisti, post te quoque salva sit. Honesta consultatio, non expedita sententia. Numeres rei publicae summam: verendum est ne dilabatur. Des agros: ut publici neglegentur."

10.54).⁸ But the possibility of incompetent management and corruption were not the only factors making it difficult to assure a municipality a steady income over a long period of time. Even when attentively managed, agricultural estates might fail, putting in jeopardy the annual income that the owner needed to achieve from them. Pliny himself experienced continuing problems of tenant indebtedness on his own estates, and these problems compromised his own ability to achieve an income. He viewed the uncertainty of the income from his estates as the most serious compromise to his financial well-being.

Pliny's solution to the problem of funding his own alimentary program proved to be a variation on a method characteristically adopted by civic benefactors to fund bequests, and it indicates how difficult he perceived it was for a private individual to achieve a steady income on invested capital. We should expect Pliny to have exercised one of the options that he scorns; at least the legacy of property to benefit a city is regarded as completely conventional by the jurists. For example, Scaevola discussed a bequest of a set of estates, or *praedia*, to a city, in which case the income from the *praedia* was to pay for gladiatorial games (*D*. 33.2.17, 3 *resp.*), and this same jurist likewise addressed legal issues surrounding the will of a wealthy landowner who left all his

---

8. Pliny displayed a similar lack of confidence in the municipal government at Comum when he provided in a challenge grant one-third of the funds necessary to support a teacher of rhetoric (*Ep*. 4.13.5–8). Pliny hoped to encourage better administration and especially better care in choosing a teacher by leaving it up to Comum to provide the balance of the salary (cf. Sherwin-White, *Commentary*, 288). Also revealing is the inability of Claudius Polyaenus of Prusa to maintain, through a trust in his will, a temple dedicated to the emperor Claudius (*Ep*. 10.70.2). This temple was to be supported from the revenues provided by a building that Claudius Polyaenus had left to the city, but the city only derived its income for a short period of time and allowed the building and the temple to fall into ruin. Sherwin-White (422) links such fears of civic corruption with the establishment of the office of the *curator rei publicae*. On the duties of the *curatores rei publicae*, see W. Eck, *Die staatliche Organisation Italiens* (Munich, 1979): 205–28. The legal questions concerning bequests to towns in Roman law are analyzed in detail by D. Johnston, "Munificence and *Municipia*: Bequests to Towns in Classical Roman Law," *JRS* 75 (1985): 105–25, who discusses (112–21) the difficulty that a *testator* might have in seeing that the town actually fulfilled his requests; for Pliny *Ep*. 10.70, see Johnston, 116. But the long-term problem that restricted Pliny's options went beyond the possibility of corruption or incompetence; he was also hampered by a limited range of options for investing capital.

estates, or *choria,* in Syria to the (unnamed) city of his birth (*D.* 32.101 pr., 16 *dig.*).⁹ The alternative method for funding a civic benefaction, namely, earmarking a cash sum for a specified purpose, might function as in another case considered by Scaevola, in which a testator left a sum of money to his town to cover the costs of games to be held every second year (*D.* 33.1.21.3, 22 *dig.*). In this arrangement, the provincial governor was to select the most appropriate outstanding loans that had been made in the testator's name; the income from these loans would then be used to pay for the games.

Mistrustful of such conventional methods of funding a benefaction, Pliny undertook to solve the problem by imposing the costs of his foundation on a private landowner, first on himself and then on the future owner of one of his estates. Pliny set aside a portion of land worth substantially more than the five hundred thousand sesterces that he was nominally donating, and he mancipated this land to the town of Comum. The town in turn imposed an annual *vectigal* of thirty thousand sesterces, representing a 6 percent return on the capital value of Pliny's donation, and then returned the land to Pliny's private ownership.¹⁰ Clearly corruption could interfere with the collection of the annual charge just as much as it could interfere with the management of lands or cash. But the more conventional methods of funding foundations were fraught with so many problems that Pliny resigned himself to paying a substantial premium for the assurance of a steady income. The *vectigal* apparently represented only a small part of the

---

9. An alimentary program founded by Hadrian at Antinoopolis in Egypt was apparently at least in part funded by land set aside for this purpose. *P.Fam.Tebt.* 33 (A.D. 151), an application to register a child in the program, describes the funding as follows (lines 5 f.): . . . ἀπὸ τῶν προσερ/[χομένω]ν εἰς τοῦτο συνχωρ[η]θέντων ὑπ' αὐτοῦ [sc. Hadrian] χρημάτων καὶ ἄλλων προσόδων. Cf. also *P.Lond.* II 383 (p. 117, III c.), which may also refer to the setting aside of land for this program: ἐδάφων κατακληρουχηθέν/των Ἀντινοεῦσι (lines 2–3). On this program, see M. Zahrnt, "Antinoopolis in Ägypten: Die hadrianische Gründung und ihre Privilegien in der neueren Forschung," *ANRW* II 10, no.1 (1988): 669–706, at 696–97 and n. 106: the exact details surrounding the program are not known.

10. Plin. *Ep.* 7.18.2–3: "Nam pro quingentis milibus nummum, quae in alimenta ingenuorum ingenuarumque promiseram, agrum ex meis longe pluris actori publico mancipavi; eundem vectigali imposito recepi, tricena milia annua daturus." For the use of land to fund private foundations, see Duncan-Jones, *Economy,* 132–33.

income from the land in question, but the annual fixed charge permanently lowered the land's value (*Ep.* 7.18.4). It is not clear that other upper-class Romans establishing foundations would have followed Pliny's example, since his method involved compromising the value of the property that a landowner might bequeath to his or her heirs. Pliny, at any rate, considered his solution to be novel.[11] But in Pliny's view this solution provided perhaps the only means for a private individual, even a wealthy and influential senator, to be certain that his donation would always provide a desired level of income: "per hoc enim et rei publicae sors in tuto nec reditus incertus . . ." (*Ep.* 7.18.3).

The planning to which Pliny resorted in establishing his alimentary program resulted from the same type of reasoning displayed by the imperial government when it established its own alimentary programs. The imperial government, without a reliable alternative method of guaranteeing the funding of its programs, imposed the task of providing for the annual budget on third parties, in this case, on private landowners in the towns concerned. In the two best-known programs, at Ligures Baebiani and at Veleia (*CIL* IX 1455, *ILS* 6509; *CIL* XI 1147, *ILS* 6675), participating landowners pledged property for which they received loans from the imperial government; the loans represented approximately 8 percent of the stated value of the property. These landowners took on the responsibility of financing the monthly support of children by paying in perpetuity a 5 percent annual interest on their loans.[12] The interest payments, from this perspective, represented a tax on the land pledged.[13] Participation in the imperial foundations should therefore be viewed, to some extent at least, as an obligatory liturgy. As was the case with Pliny's private foundation, the imperial alimentary programs surely diminished the value of the land pledged,

---

11. The imposition of charges on privately held land, however, is assumed by Julian, apud Ulp. *D.* 19.1.13.6 (32 *ad ed.*), to represent a conventional way of providing funds to benefit municipalities. The passage from Julian concerns the sale of a *fundus* owing *legata* (*vectigalia, Cuiacius*) to several *municipia*.

12. On the sizes of the pledges and interest payments in the imperial alimentary programs, see also F.G. de Pachtère, *La table hypothécaire de Veleia* (Paris, 1920), 98–115.

13. See Duncan-Jones, *Economy*, 306–10. The alimentary loans could be paid off in twenty years with payments of 5 percent of the principal each year, if no interest was charged. But if the 5 percent payment is viewed as interest, the loans could never be paid off, since the participant would never begin to pay back the principal.

since all future owners would be required to pay the same interest.[14] The imperial government resorted to compulsory loans because there were few other options that could provide as reliable and steady an income in the long term. This method of funding the alimentary foundations was of course not foolproof. The land pledged would have to remain productive in the future, and there is evidence that even the imperial authorities experienced difficulties in collecting the interest payments.[15]

Several inscriptions from Italy attest landowners setting aside properties for their towns as a means of funding benefactions in much the same way that Pliny did.[16] The closest parallel is provided by the case of a local notable of Ferentinum, A. Quinctilius Priscus (*ILS* 6271, *CIL* X 5853). To provide for a yearly distribution of food, this individual purchased from the town three *fundi* as well as a tract of pastureland for seventy thousand sesterces. He then restored these properties to the town with a perpetual charge imposed on them of 4,200 sesterces, which represented a 6 percent return on the properties' value.[17] To cite another example, a certain L. Domitius Phaon donated four *fundi* in Lucania to maintain a college celebrating the cult of Silvanus and rites in honor of Domitian and Domitia (*ILS* 3456, *CIL* X 444, near Caposele). The revenue (*reditus*) from the estates was to cover in perpetuity the costs of annual rites established by the donor. At Auximum, a member

---

14. For the argument that the loans were designed to increase the production of food for Rome, see E. Lo Cascio, "Gli *Alimenta,* l'agricoltura italica e l'approvvigionamento di Roma," *RAL* 33 (1978): 311–52. Cf., however, Bossu's arguments against this view in *Latomus* 48 (1989): 376–82.

15. The difficulty of guaranteeing revenues even through the alimentary foundations is demonstrated by the arrears into which many subscribers to the imperial programs in Italy fell during the later second century. According to a report in the SHA (*Pert.* 9.3), the emperor Pertinax undertook to relieve a growing problem by assuming responsibility for nine years' worth of arrears on obligations to the *alimenta.* See E. Lo Cascio, "Gli *Alimenta* e la 'politica economica' di Pertinace," *RFIC* 108 (1980): 264–88.

16. These inscriptions are cited by Duncan-Jones in *Economy,* 296 and n. 3.

17. *ILS* 6271, lines 7 ff.: "H(ic) ex s(enatus) c(onsulto) fundos Ceponian(um) / et Roianum et Macrianum et pratum Exosco ab r(e) p(ublica) redem(it) / HS LXX m(ilibus?) n(ummum) et in avit(um) r(ei) p(ublicae) reddid(it), ex quor(um) reditu de HS IV (milibus) CC / quo<t>annis. . . ." For an explanation of the term *avitum* and the arrangements made in this text, see Mommsen (*CIL* ad loc.), who suggests that Priscus himself retained a perpetual lease to the land.

of the equestrian order left behind a sum of fifty thousand sesterces and several *fundi* as a means of supporting annual rites there (*ILS* 3775, *CIL* IX 5845). We might also compare how L. Septimius Liberalis, a *sevir Augustalis* at Ariminum, sought to cover the costs of annual distributions of money there (*ILS* 6663, *CIL* XI 419). To fund this program, Liberalis pledged twenty-one *fundi*, but with the stipulation that the ownership of the *fundi* pass to his heir.[18] In the same town, an equestrian named C. Faesellius Rufio, among other benefactions, gave twenty thousand sesterces for the purchase of property whose income would pay for the annual distribution of *sportulae* and cash on his birthday (*ILS* 6664, *CIL* XI 379). A woman from Brixia named Valeria Ursa mancipated her one-half share of an *agellus* to her town so that the revenue from it might be used to celebrate funeral rites in honor of herself and her husband. This woman mancipated another farm to fund celebrations for the *collegium Pharmacopolarum* (*ILS* 8370, *CIL* V 4489). Finally, in the later empire, P. Aelius Apollinaris Arlenius, the deceased young son of a like named equestrian, was commemorated for covering the costs of biannual *convivia* at Praeneste by handing over to the town a nearby estate. The father also provided a sum that would provide income to purchase gardens for the town. For its part, the town was prohibited from alienating both the estates and the gardens (*ILS* 8376).[19]

This practice of setting aside lands for towns was also observed in the provinces. In Greece, the wealthy first-century jurist Caninius Rebilus, who committed suicide in A.D. 56 to avoid the diseases of old age (Tac. *Ann.* 13.30.2), left to Thasos and Philippi certain agricultural

---

18. See also *FIRA* III 118, and Amelotti, *Il testamento romano*, 24. Apparently the value of the property pledged exceeded the limits for legacies and trusts allowed by the *lex Falcidia*. The tutors of the heir, Septimia Prisca, accordingly removed one-sixth of the property pledged to bring the trust into accordance with the *lex Falcidia*, but Septimia Prisca's heir, Lupidia Septima, restored this property out of her own free will. For discussion of the legal problems involved, see Arangio-Ruiz in *FIRA* III 383–84 n. 4, and Amelotti, *Il testamento romano*, 24 n. 2, with references to other literature. According to the *lex Falcidia* (40 B.C.), the values of legacies in a will could not exceed three-fourths of the value of the inheritance as a whole; the *SC Pegasianum* (Vespasianic date) applied the same restrictions to trusts. See Kaser, *RP* I, 756–57, 762–63.

19. Cf. the manner in which L. Laecanius Primitivus set aside *praedia* for the Augustales at Puteoli to fund annual yearly celebrations on his birthday (*ILS* 6328a, *CIL* X 1880).

lands, the income from which was to be used to buy grain for the towns in question. At Thasos, a decree established the inalienability of these lands.[20] In Egypt, the equestrian Aurelius Horion wished to promote the status of the city of Oxyrhynchus, where he owned estates. We learn of this from a papyrus containing two decrees about this individual's generosity issued by Septimius Severus and Caracalla (*P.Oxy.* IV 705, A.D. 202).[21] One of his aims was to provide some sort of financial support for the individuals performing the annual liturgies. To accomplish this goal, he purchased an estate for the city, the revenues from which were to be set aside to support liturgists. This act of generosity was clearly the one on which Horion placed great priority. Wanting to secure funding for his program for the long term, the benefactor set aside a private estate—in whose continued productivity he apparently had confidence—rather than simply setting aside a sum of money.[22]

The previously mentioned methods for funding benefactions are to be contrasted with the arrangements made by the imperial freedman Aelius Onesimus for the town of Nakoleia, in Phrygia (*ILS* 7196, *CIL* III

---

20. The inscriptions attesting these dispositions are reported in J. Robert and L. Robert, "Bulletin épigraphique," *REG* 61 (1948): 168 no. 106; see Amelotti, *Il testamento romano*, 31, and W. Kunkel, *Die Herkunft und soziale Stellung der römischen Juristen*, 2d ed. (Graz, Vienna, and Cologne, 1967), 120–21.

21. This text is also included in the collection of J.H. Oliver, *Greek Constitutions of Early Roman Emperors from Inscriptions and Papyri* (Philadelphia, 1989), nos. 246–47; the second of the two decrees recorded in this papyrus is also printed in *W.Chr.* 407.

22. Oliver, *Greek Constitutions*, no. 247 (*W.Chr.* 407), ii.75 f.: ἐγὼ [ο]ὖν καὶ τοῦ φιλανθρώπου καὶ τοῦ χρησίμου στοχα/ζ[όμε]νος βούλομαι εἰς ἀνάκτησιν αὐτῶν ἐπίδοσίν / τ[ινα] βραχεῖαν ἑκάστῃ ποιήσασθαι εἰς συνωνὴν / χ[ωρί]ου οὗ ἡ πρόσοδος καταθήσεται εἰς τροφὰς καὶ / δ[απά]νας τῶν κατ᾽ ἔτος λειτουργησόντων ἐπὶ τῷ [-]. My interpretation is based on Wilcken's restoration of χ[ωρί]ου in *W.Chr.* 407 instead of χ[όρτ]ου in *P.Oxy.* IV 705; for a different interpretation, see J. Rowlandson, *Landowners and Tenants in Roman Egypt* (Oxford, 1996), 192 and n. 55. Horion also provided a cash fund of at least ten thousand Attic drachmas to provide prizes for ephebic games celebrating the suppression of the Jewish rebellion in Egypt in A.D. 117 (ii.46–53). *P.Oxy.* XXXVIII 2848 (A.D. 225) records the loan of one talent and twelve hundred drachmas from the funds set at the disposal of Oxyrhynchus by Horion, described as acting in common with his sons and C. Calpurnius Firmus, an Alexandrian councillor discussed in chap. 1 in connection with the purchase of a vineyard for his son. See also *C.Pap.Jud.* II 450 and D. Delia, *Alexandrian Citizenship during the Roman Principate* (Atlanta, 1991), 86 and n. 71 for the nature of ephebic games.

6998, 13652).²³ Like Rebilus, Onesimus desired to assure that his hometown would have a sufficient supply of grain every year. Unlike Rebilus, however, Onesimus set aside a sum of cash, two hundred thousand sesterces, which was to be loaned out under the watchful eyes of two individuals, both Roman citizens, and both presumably trusted associates of the testator. The income from this sum was to be used to purchase grain for three years, and after the end of the three-year period, the interest was to be divided each year among the town's citizens to celebrate Hadrian's birthday. We might speculate that Onesimus was willing to establish his benefaction in this way because he had associates on whom he could rely to oversee the investment of the town's money. The most critical aspect of his program, the provision of grain, was only to last three years, so Onesimus clearly did not require the same long-term assurances that Pliny did.

Pliny's concerns suggest that maintaining the productivity of land donated to a municipality was no easy task, and this impression is confirmed by the dispositions that one M. Megonius Leo made in a bequest to support the Augustales in the town of Petelia in Bruttium. This individual's benefactions toward Petelia are revealed in a series of inscriptions dated to the reign of Antoninus Pius (*ILS* 6468–71).²⁴ Among his many benefactions, Leo bequeathed to the town a vineyard, which he hoped would furnish high-quality Aminean wine for their feasts (*ILS* 6469, *CIL* X 114).²⁵ The vineyard was to be leased out by the town; apparently the yearly rental was to include a specific amount of Aminean wine, although the bulk of the yearly rental may have consisted of cash.²⁶ The bequest of this vineyard, then, offered the town a

---

23. See Amelotti, *Il testamento romano*, 36, and cf. *FIRA* III 53; the inscription has a Hadrianic date.

24. See Amelotti, *Il testamento romano*, nos. 2a–b, discussed on pp. 20 f.; cf. D. Johnston, *The Roman Law of Trusts* (Oxford, 1988), 211.

25. *ILS* 6469, lines 3–5: ". . . item vineam Caedicianam cum / parte{m} fundi Pompeiani ita uti optima maxi/maq(ue) sunt finibus suis qua mea fuerunt." Leo apparently bequeathed the land in question to the town, adding the trust that the produce be dedicated to the use of the Augustales; see Amelotti, *Il testamento romano*, 21. On the testator's confusion between trust and legacy in this document, see Johnston, *The Roman Law of Trusts*, 260 n. 12

26. *ILS* 6469, lines 26 f. seems to preserve the beginning and end of a lease contract for the vineyard: "Locatio vineae partis Pompeiani vin[e]/am colere poterint"; see Dessau, *ILS* ad loc., referring to Mommsen, *CIL* ad loc.

steady source of income, as well as a constant supply of the wine needed for the Augustales' public feasts.

The difficulty of the bequest, in the benefactor's eyes, lay in making sure that the vineyard remained in such a condition that it could accomplish these goals. Vineyards can only be productive with continual investment in replanting and maintaining the vines, and Leo clearly did not trust either the town itself or the potential tenant of the vineyard to carry out these tasks. Instead, he made it possible for some of the income from another fund to be used for the expensive task of trenching the vines, or *pastinatio* (ILS 6469.13–16).[27] One of the most expensive aspects of viticulture was to provide for the training of the vines; this consideration was especially important in agreements regarding vineyards producing high-quality wine. In Roman Egypt, landowners would often contract with their tenants to share the costs involved in training the vines.[28] Such a flexible approach to this problem might in general benefit both landowner and tenant, but it was one that a testator bequeathing a property to a town could hardly adopt. Instead, Leo imposed a trust on his heirs to provide the materials used in the training of the vines from other property that they inherited.[29] There were clearly disadvantages to this type of solution—the town, over the course of years, might have difficulty enforcing the trust—but it was the best way that the benefactor found to deal with the difficult problem of ensuring the high productivity of the vineyards.[30]

---

27. On *pastinatio,* a very time-consuming method of planting vines, see Col. 3.13.6–13, and K.D. White, *Roman Farming* (Ithaca, 1970), 236–37, 357.

28. Two contracts in particular, from the archive of the peasant farmer Soterichos, indicate how owners of vineyards of modest size might divide the costs of training vines with a small-scale tenant; see *P.Soter.* 1–2, with S. Omar, *Das Archiv des Soterichos* (Opladen, 1979), 59–61. For a detailed study of labor contracts for viticulture in Roman Egypt, see A. Jördens, *Vertragliche Regelungen von Arbeiten im späten griechischsprachigen Ägypten* (Heidelberg, 1990), especially 222–38, 249–59, and Rowlandson, *Landowners and Tenants in Roman Egypt,* 228–36.

29. *ILS* 6469, lines 28 ff.: "Hoc amplius ab heredibus meis volo praestar[i] / rei p(ublicae) Petelinorum et a re p(ublica) Petelinorum corpori Augustalium ex praedi<i>s ceteris meis ridica[m] / omnibus annis sufficiens pedaturae vineae / quam Augustalibus legavi."

30. On difficulties encountered by towns in enforcing trusts, see n. 8 in this chapter.

Providing for Dependents

Pliny viewed his solution to the problem of funding a civic foundation as unique, but the method that he used, to impose the costs on a third party, was characteristic of the economic planning practiced by upper-class Romans in using bequests as a means of providing for the financial needs of their connections. As we will see, this method of funding benefactions was a logical response to an economy with few possibilities for making stable long-term investments. It was a widespread custom among upper-class Romans to use their wills as a means to provide for the financial needs of dependents in addition to their heirs; these people might include wives, freedmen and freedwomen, and foster children. The upper-class Roman writing a will would accomplish this purpose by subtracting money and individual properties from the inheritance passing to the heirs; the testator would then bequeath these properties through legacy or trust to individual beneficiaries or sometimes even to groups of beneficiaries.[31]

A trust described by Scaevola in very general terms seems to be representative of the common pattern (*D.* 33.1.21.1, 22 *dig.*). In this case, a testator required a freedman, to whom he had bequeathed a *fundus* carrying a fixed notional annual income (of sixty), to provide an annual income of ten to a freedwoman.[32] Here, one-sixth of the fixed annual income derived from an estate was earmarked to provide a beneficiary

---

31. For the legal norms for writing Roman wills, see Kaser, *RP* I, 91–112, 668–765, as well as Amelotti, *Il testamento romano*. On the social background, see E. Champlin, *Final Judgments* (Berkeley and Los Angeles, 1991), especially 103—54; on trusts, see Johnston, *The Roman Law of Trusts*. For discussion of the legal institutions of legacies and trusts, see Kaser, 109–12, 740–63. In general on the strategies of upper-class Romans to pass on property, see R.P. Saller, *Patriarchy, Property and Death in the Roman Family* (Cambridge, 1994), 161–80.

32. Scaev. *D.* 33.1.21.1: "A liberto, cui fundum legaverat ferentem annua sexaginta, per fideicommissum dederat Pamphilae annua dena." Here Scaevola asserted that, if the application of the *lex Falcidia* should reduce the freedman's legacy, the income to the beneficiary of the trust nevertheless should not be reduced. Scaevola also addressed this question in *D.* 35.2.25.1 (4 *resp.*). For discussion of bequests of subsistence annuities, in particular of the cases addressed by Scaevola, see B.W. Frier, "Subsistence Annuities and Per Capita Income in the Early Roman Empire," *CP* 88 (1993): 222–30. Scaevola's responses in this area of the law were apparently based on real-life cases (Frier, 224–25).

of the testator's generosity with a fixed annual allowance. In another case, also discussed by Scaevola, a testator bequeathed a *fundus* to a group of freedmen and included in his bequest a prohibition against their alienating the farm (*D.* 33.1.18 pr., 14 *dig.*).[33] The testator imposed an additional requirement on these freedmen, that they provide the testator's heir with an annual income (of ten) until he reached the age of thirty-five. In this case, the testator wanted to make sure that his heir would receive a needed level of income, one presumably set at a level to maintain his current standard of living. The testator's solution was to provide his *legatarii* with ample means and then to impose a trust requiring them to furnish that income. This same method might also be used to finance *alimenta* left to the dependents of a testator, as is seen in two cases discussed by Scaevola. In the first, a woman bequeathed the usufruct over a *fundus* and then required the trustee to provide an income of one hundred *nummi* to two individuals for as long as they should live (*D.* 34.1.20.2, 3 *resp.*). In the other, a woman bequeathed the usufruct over a *fundus* and established a trust according to which the trustee was to provide two freedmen with an annual income of six hundred sesterces for the remainder of their lives (*D.* 33.1.19 pr., 17 *dig.*). Similarly, Scaevola discussed a trust whereby a town to which estates had been bequeathed was to provide *alimenta* for the testator's freedmen and freedwomen out of these estates' revenues (*D.* 34.1.20.1, 3 *resp.*). Papinian also considered a case in which the *fructus* from a group of estates was bequeathed so as to provide *alimenta* for the testator's freedmen and freedwomen (*D.* 7.1.57.1, 7 *resp.*). We might also compare a legacy, also discussed by Scaevola, of one-fiftieth of the revenue from *praedia;* this income was to support a freedman for the rest of his life (*D.* 33.1.21 pr., 22 *dig.*).

As an alternative to providing a group of estates, the testator might set up a trust, putting a sum of money at the disposal of a third party and requiring that individual to provide money to support a dependent or other connection of the testator. This method of funding individual *alimenta* was probably seen as a safer alternative to the more straightforward method of simply setting aside a sum of money for that pur-

---

33. Scaev. *D.* 33.1.18 pr.: "Codicillis testamento confirmatis fundum libertis legavit eumque alienari vetuit, sed pertinere voluit et ad filios libertorum vel ex his natos: deinde haec verba adiecit: 'a quibus praestari volo heredi ex reditu eius fundi decem per annos singulos usque ad annos triginta quinque a die mortis meae.'"

pose (see, e.g., Papin. *D.* 34.1.8, 7 *resp.*). Thus, to reexamine a case discussed in chapter 1, one wealthy individual, to provide *alimenta* for a foster child, or an *alumnus,* required a trustee to take a sum of four hundred thousand sesterces left to the *alumnus* and then to pay an annual 5 percent interest to cover the costs of the *alimenta* (Scaev. *D.* 34.1.15 pr., 17 *dig.*). The legal issue in this case arose from the possibility that the trustee might refuse to take up the sum of money set aside for the foster child. In this case, the testator did not set aside land to support the foster child but rather set a sum of money at the disposal of a third-party trustee. The trustee was to invest it as she saw fit, and the 5 percent interest that she was to provide each year must have represented a level of income that could have been realized on a regular basis. That the trustee might refuse this money suggests that there may not have been enough profit for her in this sort of business to offset her liability.[34] We might compare another case, discussed by Papinian, in which the testator set aside the sum of ten aurei (ten thousand sesterces) to provide for the support of each of his foster children (*D.* 34.1.9 pr., 8 *resp.*). The testator set up a trust for an individual (who was not his heir) to take this money and invest it as he deemed appropriate, with the proviso that he use the interest to pay for the *alimenta.* Imposing the costs of providing an income on a third party was also the method used by a prefect of a legion who offered his son's tutors the option of using themselves the money that he was leaving his son, on the condition that they pay the son interest of 1 percent each year (Scaev. *D.* 26.7.47.4, 2 *resp.*)[35]

In a somewhat more complicated arrangement, a father who had instituted his two sons as heirs sought to provide for the education and tax liability of the younger, underage son (Scaev. *D.* 33.1.21.5, 22 *dig.*). To the younger son he bequeathed a set of estates as well as a sum of money, both of which were to be given to him on his fourteenth birthday. The father required the older brother to furnish a yearly fund, to be given to the mother, to pay for the younger brother's education from his twelfth until his fourteenth birthday and to pay his *tributum.* For his

---

34. For a similar case see Scaev. *D.* 34.1.16.2 (18 *dig.*), where a testator, in seeking to provide *alimenta* for a freedwoman, bequeathed to her a sum of money but then required two freedmen to take this money and invest it, maintaining her from an annual interest of 5 percent. The principal was to be restored to the woman at her twenty-fifth birthday.

35. This case is also discussed in chap. 1.

part, the older brother was to be able to keep the income from the aforesaid properties until the younger son's fourteenth birthday.

In wills from Roman Egypt, it was customary for testators simply to impose on the heirs the obligation to provide for a surviving spouse or other dependents.[36] The testator might bequeath to a dependent a right to live in a specified house but would impose on the heirs the obligation to provide for food and clothing, without setting aside any property or funds to pay for this. For example, the veteran C. Iulius Diogenes provided legacies of land for his two children, but he imposed on them the obligation to feed and clothe his wife, Iulia Primilla, apparently for the remainder of her life (*P.Test.Roma.*² 7).[37] Similarly, a wealthy landowner from third-century Oxyrhynchus, who also made substantial bequests of land to his children, imposed on his children the obligation to provide his surviving wife a generous allowance of grain, wine, and other foodstuffs (*P.Oxy.* XXVII 2474, lines 15–20, *P.Test.Roma.*² 26). In both of these cases, we should imagine that the testator considered his children to have received an amount of land adequate to support themselves and at the same time to pay the allowance. An important factor affecting the form of such bequests was the confidence that the testator had in the ability of his or her heirs to manage their land wisely and in their willingness to fulfill their duties under the trust. It does not seem likely, then, that the testators in these examples had a different conception of the profit that agricultural investments might provide than the conception exhibited by the testators envisioned in the legal sources. Indeed, wealthy landowners in Roman Egypt might also provide for the funding of their benefactions by setting aside land. Thus Aurelius Hermogenes, an officeholder and councillor at Oxyrhynchus, designated lands at a particular village to provide a yearly allowance of wine and wheat for a friend named Aurelius Dionysammon (*P.Oxy.* VI 907, lines 23–24, *P.Test.Roma.*² 24, A.D. 276). Landowners in Egypt writing Greek wills also used this method to provide for their beneficiaries.[38] One

---

36. For Roman wills in Egypt, see L. Migliardi Zingale, *I testamenti romani nei papiri e nelle tavolette d'Egitto*, 2d ed. (Turin, 1991 [= *P.Test.Roma.*²]), who provides texts with commentary of many of the most important documents.

37. For additional discussion of this document, dated ca. A.D. 127–48, see later in this chapter under "Bequests of Farmland in Egypt."

38. On Greco-Egyptian wills, see R. Taubenschlag, *The Law of Greco-Roman Egypt in Light of the Papyri, 332 B.C.–640 A.D.*, 2d ed. (Warsaw, 1955), 181–222; see also E. Seidl, *Rechtsgeschichte Ägyptens als römischer Provinz* (St. Augustin, 1973), 226–32. For discussion of strategies of transmitting property in Egypt, see Rowlandson, *Landowners and Tenants in Roman Egypt*, 139–75.

example is the will of Hermodoros son of Herakleides, who bequeathed land to his maternal granddaughter, Chenepeis daughter of Hermias, but in doing so set aside three *arourae* of land, the usufruct to half of which was reserved for the woman's mother Hermis, daughter of Hermodoros (*P.Amh.* II 71, A.D. 178–79, Hermopolis).[39]

The characteristic practice of upper-class Romans to impose the costs of providing for permanent foundations on third parties seems to have been a response to a very uncertain climate for investments. In the Roman economy, not only was it difficult to find a suitably safe investment for substantial amounts of capital, but it was also, in all likelihood, difficult for the interested parties to establish sufficient oversight so that they could be assured that the investment would be managed wisely and yield the desired income. The testators whose planning I have described in this chapter sought to sidestep these difficulties by establishing a level of income as the fixed point to be reached. They could be relatively certain that this target would be achieved year after year, because by providing the third party sufficient property or resources, they obviated the need to supervise how these resources were invested or managed. The income accruing to the foundation would not change, unless it were set as a portion of a larger income. What would change was the much larger income made available to the party charged with the task of funding the foundation.

Another method of responding to these conditions for investment was simply to set aside land as a means of providing for a dependent. This method was similar to the one previously discussed, in that the testator would normally set aside a greater amount of property than would be needed in an average year to achieve the desired level of income, but the testator did not impose the risk for providing this income on a third party. For example, Scaevola considered a trust involving the bequest of a usufruct to a wife with the specific purpose of providing her with a set level of income (*D.* 33.2.32.4, 15 *dig.*). The wife was to hold the usufruct for fifteen years, and her annual income was to be set at a very generous four hundred thousand sesterces; any

---

39. The document is a report about the will's provisions; cf. *P.Amh.* II 71, lines 13 f.: ὧν ἡ [κ]αρπεία τοῦ (ἡμίσους) μέρους / τέτηρηται τῇ προγεγρ(αμένῃ) μου μητρὶ Ἕρμιος / ἐφ' ὃν περίεστι χρόνον. According to H.-A. Rupprecht, "Zum Ehegattenerbrecht nach den Papyri," *BASP* 22 (1985): 291–95, wives were generally not the chief beneficiaries of husbands writing Greco-Egyptian wills, but these wills did tend to provide for their needs by assuring them a place to live, furniture, and other items needed for the home.

revenue accruing from the usufruct over that amount was to revert to the testator's heirs. Apparently the testator estimated somewhat generously the amount of property needed to provide his wife with her income in any given year; the goal was to eliminate the risk that her income might fall short. In another case, a testator bequeathed an annual allowance for food and clothing (*cibaria* and *vestiaria*) to a group of freedmen, and he assigned a specific cash value for this bequest (Paul. *D.* 34.1.12, 14 *resp.*). To pay for this annual allowance, the testator set aside a group of *fundi*, which were to be "obligated" for this purpose.[40] Clearly the testator set aside these farms as a means of assuring adequate funding for his bequest; nevertheless, as Paul ruled, the heirs would be obliged to provide the full amount of the bequest in the event that the income from the *fundi* proved insufficient.

We see comparable planning in a case treated by the third-century jurist Modestinus in which a wealthy woman sought to make sure that her freedmen and freedwomen would receive the same level of support for food and clothing as she had been accustomed to granting them while she was alive (*D.* 34.1.4 pr., 10 *resp.*). As in other cases that we have examined, her need to achieve a set level of income was absolute. To realize this intent, she set aside a discreet set of estates for her freedmen and freedwomen.[41] This arrangement raised a number of legal questions: whether the people concerned were to receive their *alimenta* from the estates or from the heirs; whether the property itself or only a usufruct was bequeathed; whether any income over and above what was needed for the *alimenta* was to revert to the testator's heir; whether the share of an already deceased beneficiary should pass to the survivors; and finally, whether the shares of beneficiaries dying before the

---

40. Paul. *D.* 34.1.12: "Lucius Titius libertis suis cibaria et vestiaria annua certorum nummorum reliquit et posteriore parte testamenti ita cavit: 'obligatos eis [esse volo *ins. Mo.*] ob causam fideicommissi fundos meos illum et illum, ut ex reditu eorum alimenta supra scripta percipiant'."

41. Mod. *D.* 34.1.4 pr.: τοῖς τε ἀπελευθέροις ταῖς τε ἀπελευθέραις μου, οὓς ζῶσα ἐν τε τῇ διαθήκῃ ἐν τε τῷ κωδικίλλῳ ἠλευθέρωσα ἢ ἐλευθερώσω, δοθῆναι βούλομαι τὰ ἐν Χίοις μου χωρία, ἐπὶ τῷ καὶ ὅσα ζώσης μου ἐλάμβανον στοιχεῖσθαι αὐτοῖς κιβαρίου καὶ βεστιαρίου ὀνόματι. We might compare the so-called *testamentum Dasumii* (*FIRA* III 48, *CIL* VI 10229, A.D. 108, Amelotti, *Il testamento romano*, no. 16), in which the testator provided for a number of legacies and trusts in cash, including annual grants to *liberti* for clothing, *vestiarii nomine*. The testator appointed three men to serve as procurators with the duty to oversee these annual payments (lines 60 ff.).

trust became valid should pass to their heirs or to the heirs of the woman originally making the will. Modestinus ruled that the freedmen and freedwoman received full ownership over these estates, as if in a conventional legacy. The point of interest for our discussion, though, is that the aristocratic woman undertook to provide for a fixed level of income by setting aside a group of estates whose income we might expect to have varied considerably from year to year. Even third parties assigned the trust of providing *alimenta* and similar benefits might themselves set aside land to meet their financial obligations. Thus in a case discussed by the second-century jurist Valens, an individual who was obliged by a trust to provide *alimenta* for the freedmen of his brother bequeathed to them vineyards as a means of covering this obligation (*D.* 34.1.22.1, 1 *fideicommiss.*).[42] This type of arrangement raised questions about the status of the vineyards and the obligations of the brother's heirs under the trust. In this case, the brother would have to free his heirs from the obligation to provide the *alimenta* in order for the freedmen to gain ownership of the vineyards in question.

We might compare this arrangement to provide for *alimenta* with a will of a member of the provincial elite in third-century Roman Egypt. We know of this will through a partially preserved document from Oxyrhynchus (*PSI* XII 1258). An Antinoite citizen, Dionystheon, alias Theon, son of Theon, had died, leaving behind a widow and a daughter, Aurelia Diogenis, who was his sole heir. This daughter was underage, and the father provided a yearly living allowance for her, which was to include payments of grain. In the present document, the widow, whose name is lost but who had the *ius trium liberorum*, made a claim to the daughter's guardian, or *epitropos*, Aurelius Apollonios, for payment of the allowance. It is noteworthy for our purposes that the father set aside land as a means of covering the expenses of the allowance. The land was to provide either the stipulated foodstuffs in kind or at least the income with which the foodstuffs might be purchased.[43]

Pliny and the testators discussed here shared an understanding of the difficulty of maintaining a stable income for the long term. The characteristic solution for upper-class Romans was to set aside land in generous portions, so that they could be assured that the parts of the land's

---

42. Valens *D.* 34.1.22.1: ". . . vineas cum hac adiectione reliquerat 'ut habeant, unde se pascant.'"

43. *PSI* XII 1258, lines 18 f.: . . . ἅμα καὶ ὑπό σου ἀφ' ὧν κατέ/λειψεν ὑπαρχό[ντων-].

income allocated to their charitable purpose could be realized year after year. This type of approach does not mean that Pliny and comparable landowners considered land a completely safe investment or that agriculture was absolutely lucrative. In fact, we should imagine that the chronic vicissitudes of Mediterranean agriculture caused the production of an estate to vary considerably from year to year, and so the income must have varied to some extent at least. The question to be considered was where these chronic swings were to be felt. As I will argue, leasing provided a reliable solution to the problem of keeping revenues from estates as stable as possible. We might expect landowners leasing out their estates to have had modest expectations about what kind of income their estates might furnish. With few alternatives for safely investing large amounts of wealth, upper-class Romans would have sought to manage their land in such a way as to minimize their risk and assure themselves of an income whose chief virtue was its reliability. The problem that lessors faced, as we will see, was to manage their tenants in such a way as to keep income stable. We should expect that people making dispositions through their wills were especially cautious in planning their investments, while landowners able to supervise their finances personally were probably willing to take more risks. Even so, the dispositions that testators made to cover the costs of their benefactions suggest to us how Roman landowners in general understood the constraints on their ability to profit from their investments in agriculture. In pursuit of stable incomes, Roman landowners often contented themselves with what frequently amounted to limited incomes from their estates, renouncing the higher incomes that might be achieved through greater investment or more intensive management.

Income from Bequests of Agricultural Property

Now let us consider the nature of the income derived from investments in agriculture by considering the jurists' treatment of the bequest of agricultural property through legacy. In developing regulations for the legacy of estates, the jurists had to consider the interests of the testator, who characteristically sought to ensure that the legatee would be able to derive the same benefits from an estate as he or she had. This intention on the part of the testator is given explicit expression in several wills discussed by the jurists. One common formulation was to specify that the legatee was to receive a property in the same condition as it was on the day of the testator's death. In the interpretation of the jurists, this meant that the legatee was to receive all of the equipment needed to cultivate the estate as well as any other furnishings contributing to the owner's enjoyment of the estate.[44] Other testators might specify at great length the equipment and furnishings that they wanted included in the legacy, but with the same general intention in mind.[45] The testator bequeathing an estate, then, sought to provide the legatee first and foremost with a source of income, and in some cases with a property that would provide chiefly aesthetic benefits to the legatee.

To be sure, the lines between the economic and noneconomic benefits of an estate were often blurred for the testators and legatees envisioned by the jurists. Indeed, the jurists accounted for the aesthetic aspects of an estate in their regulation of the bequest of a *fundus cum instrumento*, that is, an estate with its equipment (on this subject, see the next section in this chapter). Thus Pomponius emphasized that the villa, or the farmhouse in which the estate owner resided, was to be considered an integral part of the *fundus* (D. 33.7.15.2, 6 *ad Sab.*), while Ulpian discussed the conditions under which slaves serving the villa might be included in a legacy of *fundus cum instrumento* (D. 33.7.8.1, 20 *ad Sab.*).[46] In recognition of the testator's desire to pass on an estate from which the legatee might derive not only economic but also aes-

---

44. See, e.g., Papin. D. 32.91.1 (7 *resp.*), concerning the legacy of *praedia* with a *praetorium*, "'sicut a me in diem mortis meae possessa sunt, do' instrumentum rusticum et omnia, quae ibi fuerunt, quo dominus fuisset instructior, deberi convenit: colonorum reliqua non debentur." Cf. Scaev. D. 33.7.20.6 (3 *resp.*), a legacy to a freedwoman of a *fundus*, "cum instrumento et his quae in eodem erunt cum moriar."

45. See, e.g., Paul. D. 32.92 pr. (13 *resp.*), concerning a *praelegatio* of *fundi* to one of the testator's female heirs: "Si mihi Maevia et Negidia filiae meae heredes erunt, tunc Maevia e medio sumito praecipito sibique habeto fundos meos illum et illum cum casulis et custodibus omnium horum fundorum et cum his omnibus agris, qui ad coniunctionem cuiusque eorum fundorum emptione vel quolibet alio casu optigerint, item cum omnibus mancipiis pecoribus iumentis ceterisque universis speciebus, quae in isdem fundis quove eorum cum moriar erunt, uti optimi maximique sunt utique eos in diem mortis meae possedi et, ut plenius dicam, ita uti cluduntur." Cf. Scaev. D. 33.7.20 pr. (3 *resp.*), where the testator again bequeathed *fundi* to a female heir; the *fundi* were described in the will as "instructos cum suis vilicis et reliquis colonorum . . ."; in a codicil the testator described the properties further: "ita ut sunt instructi rustico instrumento suppellectile pecore et vilicis cum reliquis colonorum et apotheca habere volo."

46. Ulp. D. 33.7.8.1: "Quibusdam in regionibus accedunt instrumento [sc. servi], si villa cultior est, veluti atrienses scoparii, si etiam viridiaria sint, topiarii. . . ."

thetic enjoyment, the jurists developed the category of the *fundus instructus*. The *fundus instructus* was to be distinguished from the *fundus cum instrumento*, in that the former included more, not just the equipment, slaves, and livestock needed to cultivate the estate, but also any furnishing or slaves that served the owner of the estate during his visits there: ". . . omnia quae eo collocata sunt, ut instructior esset pater familias . . ." (Sab. and Cassius apud Ulp. D. 33.7.12.27, 20 *ad Sab.*). This type of legacy would include, for example, furniture and cloth furnishings; glass, gold, and silver; wine; utensils; various types of slaves serving the household, including learned slaves who entertained dining guests (*paedagogia*); libraries; paintings and sculpture; and even medicine (D. 33.7.12.28–41). The development of this category of legacy, which dates back at least to the first-century jurists Sabinus and Cassius (apud. Ulp. D. 33.7.12.27), responded to the social needs of upper-class Roman landowners, for whom estates provided a tangible sign of social prestige.[47]

In regulating the bequest of agricultural property, the jurists were faced with the difficulty of drawing up a sufficiently general definition of what constituted the productive resources needed to cultivate an estate, so that they could determine on a case-by-case basis what might be included in a legacy involving an agricultural property. To accomplish this purpose, the jurists concentrated on determining the intentions of the testator in a given will on the basis of general assumptions about the conventional meanings of the terms that testators might

---

47. For discussion of the ancient controversy over whether the legacy of a *fundus instructus* was to be distinguished from the legacy of a *fundus cum instrumento*, see A. Steinwenter, *Fundus cum instrumento* (Vienna and Leipzig, 1942), 73–74. By Diocletian's time, the bequest of a *fundus instructus* was considered to include more than the *fundus cum instrumento;* see Diocl., Max., *CJ* 6.38.2 pr.–1 (A.D. 293): "Fundo 'sicut instructus est' legato sive per fideicommissum relicto vilicum hominesque et omnia, quae vel, ut ipse pater familias, cum ibi ageret, vel fundus esset instructus, non temporis causa habuit in eo, relicta esse iuris auctoritate definitium est: ea etiam, quae tam fructuum colligendorum quam servandorum. Item pecora stercorandi vel pascendi causa ibi constituta, ut fructus de his capiantur vel ut fundus sit instructior, fideicommisso cedere certi iuris est." The "social" aspect of an estate was also recognized in the law of usufruct. For example, Ulpian ruled that a legatee receiving a usufruct of a *praedium voluptarium* could not remove amenities contributing to the owner's entertainment to replace these with vegetable gardens or other types of cultivation that would produce a greater income (D. 7.1.13.4, 18 *ad Sab.*).

use.⁴⁸ Since their overarching purpose was to settle potential disputes between heirs and the beneficiaries of legacies, the jurists' were not chiefly concerned with influencing how estates might be organized and cultivated. Instead, they interpreted the intentions of the testator without necessarily considering the economic consequences of the testator's dispositions. Thus the jurists might determine that a testator did not intend that the equipment of an estate be included in the legacy of the estate in question, and in making such a determination they would not be influenced by the likelihood that removing its equipment would make an estate impossible to cultivate and therefore economically valueless.

But even if the jurists did not prescribe how estates should be organized and cultivated, they were still concerned with responding to the financial needs of the parties involved in a legacy. The expressions of this concern that are found in the jurists' interpretation of what was to be included in a legacy of an estate provide us with a set of assumptions about the Roman economy that can be used to investigate Roman economic planning. To cite one example of how the jurists approached this problem, Ulpian cited the authority of the Republican jurist P. Mucius Scaevola in defining what slaves might be included in a legacy (*D.* 28.5.35.3, 4 *disputat.*). In Ulpian's view, the mere physical presence on an estate was not the crucial criterion for determining whether a given slave counted as part of the legacy. The presence of certain slaves on the estate at the time of the testator's death might be purely fortuitous, while using the presence of the slaves on the estate at the time of death as the criterion could exclude other slaves on whose labor the cultivation of the estate depended. Consequently, the deciding factor for Ulpian was whether or not the slave was regularly involved in the estate's cultivation.⁴⁹ Clearly Ulpian's purpose was to develop a work-

---

48. For a thorough discussion of the extent to which the jurists took into account economic considerations in formulating rules for the legacy of estates, see Steinwenter, *Fundus cum instrumento,* especially 9–10, 30, 34–35, 82–84, 87–88, 100–101.

49. Mucius apud Ulp. *D.* 28.5.35.3 (4 *disputat.*): "ut est apud Mucium relatum, cum fundus erat legatus vel cum instrumento vel cum his quae ibi sunt: agasonem [i.e., "groom"] enim missum in villam a patre familias non pertinere ad fundi legatum Mucius ait, quia non idcirco illo erat missus, ut ibi esset. proinde si servus fuerit missus in villam interim illic futurus, quia dominum offenderat, quasi ad tempus relegatus, responsum est eum ad villae legatum non pertinere. quare ne servi quidem, qui operari in agro consuerunt, qui in

able definition that would make it possible to interpret the dispositions of individual testators.

In more general terms, the jurists must have followed the principles outlined by Celsus when discussing the difficulty of determining the testator's intentions in a legacy of *suppellex,* or furnishings (D. 33.10.7.2, 19 *dig.*). Celsus approved the opinion of Servius, who had recommended interpreting the wording of a will in accordance with the common usage of terms: "non enim ex opinionibus singulorum, sed ex communi usu nomina exaudiri debere."[50] The jurists treated the legacy of an estate according to a comparable principle, relying on accepted and conventional notions of the economic value of an estate to determine the testator's intentions. Accordingly, Ulpian assumed that the "typical" testator made sensible arrangements for cultivating an estate, and, more important, that these arrangements were conventional. In this connection, as Steinwenter points out, the jurists displayed technical knowledge of the workings of an estate, a knowledge no doubt influenced by the writings of the Roman agronomists, as well as by the jurists' own firsthand experience. The assumptions that the jurists made about the Roman estate economy, then, merit a great deal of scrutiny. Consequently, in this area of law, as in others impinging on the Roman estate economy, the economic assumptions made by the jurists serve as the basis for drawing broader conclusions about the considerations affecting upper-class Romans in managing their wealth.

To generalize, the treatment of legacy by the jurists envisions the estate as a resource from which the beneficiary of the legacy was expected to draw an income with little involvement in the management of the estate and little investment in improving the property. In other words, an agricultural estate, as treated in the Roman regulations concerning legacy, was not a potential source of profit in the sense of an investment toward which the beneficiary of a will might dedicate resources to achieve some level of return. Instead, the revenue that the estate was to furnish in any given year was more or less fixed, and the

---

alios agros revertebantur, et quasi ab alio commodati [cum in alios agros revertantur, quasi ab alio commodati *Mo.*] in ea sunt condicione, ut ad legatum pertineant, quia non ita in agro fuerant, ut ei agro viderentur destinati."

50. Servius' view was hardly without controversy, since, as Celsus relates in the same passage, it was attacked by Tubero, who stressed the importance of ascertaining the *voluntas* of the testator. In Servius' view, the testator's *voluntas* could only be determined by the common meaning of the words chosen, not in the more subjective fashion that Tubero seems to have called for.

beneficiary of the estate was envisioned as making the minimum possible investment needed to maintain this level of income. I am not of course suggesting that, for the jurists or for Roman landowners in real life, the task of keeping an estate productive year after year was simple. Rather, the horizon for investing in such a way as to change substantially the revenues that an estate might be relied on to provide was severely restricted.

Let us first consider the notion that estates produced a fixed level of income. Scaevola made this assumption quite explicit when he discussed the case of a legacy to a freedman of a *fundus* carrying an annual income of "sixty" (D. 33.1.21.1, 22 *dig.*). The association of a fixed annual income with an estate underlies another case addressed by Scaevola (D. 32.37 pr., 18 *dig.*), in which a man bequeathed to his mother through a legacy a *fundus* (that in fact belonged to her), with a request that the mother should restore it in her own will to the testator's wife. The mother, willing to follow her son's wishes although the *fundus* was rightfully hers, made an agreement in the presence of a magistrate to hand the *fundus* over to the wife if the wife gave her the equivalent of two years' income. Neither of these transfers happened, however.[51] Apparently the estate had provided a fixed yield, and the mother sought to keep this fixed income for two years. It is not possible to prove this point, however, by means of the details of this case that are preserved in the Digest.

The assumption that an estate provided a set level of income also seems to stand behind a ruling made by Scaevola in connection with a bequest to a freedman of one-fiftieth of the income from a set of properties for as long as he should live (D. 33.1.21 pr., 22 *dig.*).[52] Scaevola addressed the question of the heirs' obligations to the freedman in the event that they sold the estates; the freedman apparently was to receive one-fiftieth of the customary interest realized from the sale price.[53] The

---

51. Scaev. D. 32.37 pr.: "paratamque se fundum Flaviae Albinae tradere, si sibi annua bina praestarentur redituum nomine: sed neque possessionem tradidit neque annua bina accepit."

52. Scaev. D. 33.1.21 pr.: " 'praestari volo Philoni, usque dum vivet, quinquagesimam [quinquagesimas *Mo.*] omnis reditus, quae [de *ins. Mo.*] praediis a colonis vel emptoribus fructus ex consuetudine domus meae praestantur'."

53. This conclusion depends on the additions to the text made by Mommsen (D. 33.1.21 pr.): "quaesitum est, an pretii <quinquagesima debeatur> usurae<que>, quae ex consuetudine in provincia praestarentur, quinquagesima debeatur, <licet praedia vendita sunt>. respondit reditus dumtaxat quinquagesimas legatas, [licet praedia vendita sunt]."

phrasing of the legal question in this case presupposes a close relationship between the income realizable from an investment of a given amount of capital in agriculture and the interest that same amount of capital might yield on the open market.[54] An expectation of a fixed level of income from an estate explains the dispositions made by a woman for her son in another case discussed by Scaevola (*D*. 33.2.32.7, 15 *dig*.). In this case, a mother set up a trust whereby her heir was to provide a set annual income for her (the testator's) son or, alternatively, to purchase property and assign the usufruct to the son, as long as the usufruct provided the previously established level of income. The son, on receiving the estates, then leased them out.[55] The son's goal was to achieve the level of income prescribed for him by his mother in her will; this orientation almost precluded any investment to improve the property or to take advantage of changing market conditions. That the son held only a usufruct would also work against his investing in the estates assigned to him so as to take advantage of putative market conditions by cultivating riskier but more lucrative crops. The point is that the annual rent did not change, and the son, to benefit from this income, made no financial commitment in order to exploit the estates. Nonetheless, the dispositions in this case strongly suggest an attitude whereby an individual piece of property constituted an asset providing a given level of income, rather than an investment, in a modern sense, that could be liquidated with the proceeds being invested elsewhere.

In several cases an estate's fixed income is envisioned to be the same as the annual rent. For example, Scaevola considered the legacy of the income (*reditus*) from a *fundus* to the wife of a defunct landowner; the woman was to receive this income for as long as she lived (*D*. 33.2.38, 3 *resp*.). Scaevola ruled that the heir had the right to sell the *fundus* and to offer the legatee instead an annual sum equal to the estate's income.

---

54. In fact, the market interest rate was probably higher than the expected return on an agricultural investment, so the freedman's interests were being protected to the extent that he received a yearly income based on the market interest rate for his share of the property.

55. Scaev. *D*. 33.2.32.7: "Heredis instituti fidei commisit filio suo annua decem praestare aut ea praedia emere et adsignare, ut usum fructum haberet, reditum efficientia annua decem: filius fundos sibi ab herede secundum matris voluntatem traditos locavit. et [*del*. Mo.] quaesitum est, defuncto eo reliqua colonorum utrumne ad heredem filii fructuarii an vero ad heredem Seiae testatricis pertineant. respondit nihil proponi, cur ad heredem Seiae pertineant."

In this logic, the annual income was equated with the estate's annual rental.[56] Several observations can be made in connection with this case. In contrast to what I have been arguing thus far, the heir is envisioned as gaining financial advantages by selling off the estate, but we must keep in mind that he would be selling a property that was not providing him with any income. Presumably, the heir sought to rationalize his landholdings in some unspecified way. As was the case with Pliny's alimentary foundation, the testator clearly viewed providing land as the surest way to assure steady income for his wife. More important for the present argument, the question posed to the jurist assumes that a direct link existed between annual income from the estate and its annual rental. No one is envisioned as having any capacity for improving the estate and achieving a higher rent; it is simply assumed that the estate had a set level of income, that the estate was leased out, and that the estate's income was the same as the annual rental. Even the conception of what constituted the estate's income (*reditus*) is noteworthy. Scaevola took no account of the necessity on the part of the landowner to contribute resources on a continuing basis to keep the estate productive.

We can trace what must have often been the financial responsibility of a beneficiary of a will in yet another case considered by Scaevola, concerning the bequest of a *fundus* with a trust that the beneficiary provide another party with a usufruct of half of the land (*D.* 33.2.32.5, 15 *dig.*). The legal issue in this case concerns the claim that the trustee was to have compensation for the costs involved in repairing a villa needed

---

56. Scaev. *D.* 33.2.38: "quaero, an possit tutor heredis fundum vendere et legatario offerre quantitatem annuam, quam vivo patre familias ex locatione fundi redigere consueverat." There seems to be confusion in the text over whether this was a legacy or a trust. This blurring between legacy and trust seems especially characteristic of Scaevola: see Johnston, *The Roman Law of Trusts*, 256–71. For additional discussion of this text and in general for the legal questions surrounding trusts designed to preserve property for future generations, see D. Johnston, "Successive Rights and Successful Remedies: Life Interests in Roman Law," in P. Birks, ed., *New Perspectives in the Roman Law of Property* (Oxford, 1989), 153–67. Johnston argues that in classical Roman law a trust preserving "a residue of an estate" was conceived of as a fund rather than as a specific property or set of properties. The responsibility of the trustee, then, is to provide the beneficiary with a fund of a particular value. This state of affairs gave the property owner bound by a trust a great deal of freedom in disposing of his or her property.

for the production of crops but in disrepair because of age.[57] In the circumstances of this case, the trustee was entitled to compensation for his expenses if the repairs were necessary and carried out before the usufruct was passed on; the villa had to be maintained in order for the third party to benefit from the usufruct. This case provides an example of the type of investment in the infrastructure of an estate that a landowner would always have to make. The landowner had to maintain the fixed capital of his estate, and this responsibility corresponds broadly to that of the lessor in the normative farm-lease contract in Roman law. The lessor in the Roman farm lease was expected to maintain the fixed capital within the estate, such as wine or olive presses, storage buildings, and the like, whereas the tenant provided the movable capital, including tools, livestock, and slaves.[58] Scaevola envisioned a generally similar responsibility on the part of a landowner in the case, already discussed, concerning the legacy of the revenues from a *fundus* to the testator's widow for as long as she lived (*D.* 33.2.38). Scaevola required the heir to repair the property in question if the income had been reduced because of any action on his part.[59] In and of themselves these passages do not tell us much about how landowners or beneficiaries of legacies exploited their estates. But it is noteworthy that, in the second case discussed here, Scaevola envisioned that the responsibility of the landowner was to maintain an estate in such a condition that it could be leased out to a tenant.

Other cases also assume an estate providing a fixed income derived from leasing. For example, in a case considered by Julian (*D.* 23.4.22, 2 *ad Urs. Fer.*), a husband received as part of a dowry a *fundus*, and he agreed with his wife to provide her with the rent from the *fundus* as her annual allowance.[60] The husband leased the *fundus* to the mother of his wife, but legal complications arose when that mother died, leaving arrears: she had made her daughter the sole heir, and there was a

---

57. Scaev. *D.* 33.2.32.5: ". . . villam vetustate corruptam cogendis et conservandis fructibus necessariam. . . ." Cf., for the same case, Scaevola apud Paul. *D.* 7.1.50 (3 *ad Vitell.*).

58. For the division of capital in the Italian farm lease, see chap. 3.

59. Scaev. *D.* 33.2.38: "item quaero, an compellendus sit heres reficere praedium. respondit, si heredis facto minores reditus facti essent, legatarium recte desiderare, quod ob eam rem deminutum sit."

60. Julian. *D.* 23.4.22: ". . . interque eos convenerat, ut mercedes eius fundi vir uxori annui nomine daret. . . ."

divorce. The husband sought from the ex-wife the *mercedes* that the mother owed, and his suit was allowed, since the annual allowance that he had given was treated as a gift.[61] For our purposes the important assumption is that this *fundus* produced a fixed annual income that could suitably serve as an allowance; this fixed income was to be achieved from leasing.[62] We see this same assumption in a somewhat different form in a rescript from Diocletian and Maximian (*CJ* 5.12.18, A.D. 294).[63] The petitioner was a husband who was seeking to retain control of property that had been given to him as a part of his wife's dowry. The wife's mother had given her an estate, minus the usufruct. Apparently the mother-in-law then leased out the usufruct to her daughter, for a fixed annual payment. The wife in turn gave this property to her husband as the dowry, and the mother-in-law also bestowed on the husband her usufruct over the property. When the wife died, the husband sought to keep both the property and the usufruct, but the mother-in-law claimed the usufruct for herself. The emperors ruled in the husband's favor, pointing out that the death of the wife/daughter did not end the usufruct. The estate in this case served only as a source of a fixed level of income. It provided the mother-in-law with a fixed income as a means of support, even if this amount was only a fraction of the overall rent. It seems likely that the other parties in this case saw the estate as fulfilling the same function.

Several papyri attest that beneficiaries of wills in Roman Egypt also conceived of their property as providing a set level of income, which was equal to the annual rent. Let us first consider a legal dispute over what was probably inherited land from second-century Hermopolis in

---

61. Ibid.: "placuit exceptionem mulieri dari non debere . . ."; ". . . nam quod annui nomine datur, species est donationis."

62. A similar assumption may stand behind the dispositions made by a father for his daughter and quoted in Papin. *Frag.Vat.* 258 (12 *resp.*). A certain Pompeius Philadelphus, to provide his daughter with a dowry, handed over estates, the rents from which were to be paid to his son-in-law: "Pomponius Philadelphus dotis causa praedia filiae Pomponiae, quam habuit in potestate, tradidit et mercedes eorum genero solvi mandavit. . . ."

63. *CJ* 5.12.18: "Si socrus tua fundum deducto usu fructu uxori tuae donavit tibique in dotem uxor quidem proprietatem, socrus autem usum fructum dedit, uxore tua rebus humanis in matrimonio exempta fundum apud te mansisse secundum placiti inter vos fidem non ambigitur. nam si acceptura certum quid annuum filiae suae usum fructum locavit, mortua conductrice usus fructus extingui minime potuit."

Egypt (*P.Gen.* I 31, A.D. 145/46). In this document, a petition to the *epistrategos*, the petitioner accused his adversary, the husband of his daughter, of exploiting the property for himself, by leasing the property out to tenants. Leasing was apparently the way that either party intended to exploit the property, and the involvement of the lessor in such a lease arrangement would consist primarily of collecting the rent. Similarly, in a dispute over inherited property, twenty *arourae* of grain land in the Hermopolite nome, an Alexandrian woman accused her adversaries of evicting the tenants who were in place when the land in question was bequeathed. In this case, again, leasing appears to have been the customary method of exploiting inherited property (*P.Flor.* I 58, A.D. 226 or later). Significantly for our purposes, the plaintiff describes the tenants as having held their leases "for a long time" [πάλαι]; if the plaintiff had her way, she would inherit the right to become lessor to these tenants. Also, the weak-sighted Gemellus, alias Horion, grandson of the well-known veteran C. Iulius Niger, described in a petition how he and his sister derived the income from the land that they had inherited from their uncle C. Iulius Longinus; this land too apparently remained under lease as it changed hands. When Gemellus took over the property, he essentially gained the right to exact the "income" (*P.Mich.* VII 422, A.D. 197, Karanis).[64] Finally, when acknowledging to her brother Tesenouphis her receipt of her share of an inheritance, a certain Stotoetis, daughter of Tesenouphis, revealed that she and her brother had exploited their modest patrimony of two-thirds of an *aroura* of grain land by leasing it out (*SB* XVIII 13300, ca. A.D. 204/5, Soknopaiou Nesos).[65]

A consistent assumption underlies the dispositions that these testators made. The land that the they were setting aside to support their dependents produced a relatively restricted range of income. In all likelihood, furthermore, the testators had hopes that the income from their estates would remain stable over a long period of time. Certainly the

---

64. *P.Mich.* VII 422, lines 13 ff.: ὁμοίος / δὲ συνέβη καὶ τὸν θεῖον μου / Γάϊον Ἰούλιον Λογγεῖνον / τελευτῆσαι πρὸ ὀκταετίας / καὶ τούτου τὰ ὑπάρχοντα / ἐπεκράτησα καὶ συν(ε)κομισά/μην τὴν πρόσοδον μηδενὸ(ς) κωλύσαντος. See the discussion of the archive of the grandfather by H.C. Youtie and O.M. Pearl in *P.Mich.* VII, pp. 117–19, and R. Alston, *Soldier and Society in Roman Egypt* (London and New York, 1995), 129–32.

65. On this last document, see J. Bingen, "Documents de l'Egypte romaine," *BASP* 22 (1985): 14–21, at 15–17.

woman bequeathing a set of estates to provide for the support of her freedmen and freedwomen (Mod. D. 34.1.4 pr.) had a fairly exact idea about the income that her estates regularly produced, since she was making arrangements that a set number of individuals would continue to receive benefits that they had already been receiving during her lifetime. Similarly, when Pliny sought to provide for his former nurse by giving her a small estate as a means of support (*Ep.* 6.3), he operated under the assumption that the estate had a set level of revenue; this revenue formed the basis on which the selling price of the estate was set.[66] Pliny was dismayed to find, however, that the income from the estate had begun to fail, and for this reason he secured the services of a certain Verus to oversee the estate's management. For the most part, close and rigorous management was more of an ideal than a reality for upper-class Roman landowners, many of whom had no interest or expertise in financial matters.[67] But also important, substantial investment in higher-yielding crops, such as wine, carried risks. Even if it might raise the level of income that an estate might provide, investment in farm improvements rather than in land itself reduced the reliability of the return on the capital invested. The beneficiaries of bequests involving agricultural property might be expected to avoid any risk, contenting themselves with a stable, if modest, income.

The expectations that the jurists attributed to the beneficiaries of the bequests in the cases I have discussed find their analogue in what was desired by a certain Lollianus, alias Homoios, a disgruntled public grammarian from third-century Oxyrhynchus. We learn of Lollianus' situation from a fragmentary document containing two drafts of a peti-

---

66. Plin. *Ep.* 6.3.1: "Erat, cum donarem, centum milium nummum; postea decrescente reditu etiam pretium minuit, quod nunc te curante reparabit."

67. A petition to the prefect of Egypt in A.D. 303 records what must have been a common situation for people lacking the expertise to manage their land directly (*P.Oxy.* I 71 no. ii). The petitioner was a woman who, during the absence of her sons in military service, hired two overseers, only to find out that they were dishonest and mismanaged her property. We might compare the way in which Cicero compares Verres in his management of the *decuma* in Sicily with a *vilicus* who provides the landowner with a larger income for a short time by selling off all of the equipment of the estate (2 *Verr.* 3.119). In making this comparison, Cicero assumes that a *vilicus* might manage an estate with a great deal of independence from the landowner. On this point, see T.J. Chiusi, "Landwirtschaftliche Tätigkeit und actio institoria," *ZRG* 108 (1991): 155–86, at 164–65.

tion to the emperors Valerian and Gallienus as well as a letter to an influential friend who Lollianus hoped would approach the imperial court and present his petition (*P.Oxy.* XLVII 3366, A.D. 253–60).[68] Lollianus, according to the description of his situation that he offers to his influential friend, had been appointed by the council of Oxyrhynchus as a public grammarian, and this appointment carried with it a stipend of wine and grain (lines 28 f.). Unfortunately for Lollianus, the food provided for his maintenance was not to the standard he thought fitting for someone of his station: the wine was more like vinegar, and the grain was eaten by worms. As a solution, Lollianus proposed that he be given the income from a specific orchard within the city. In support of his request, Lollianus reminded the emperors of the traditional policy of affording generosity to teachers of grammar (lines 6–16, 50–54). What is interesting for our purposes is Lollianus' description of the income from this orchard. The orchard, which apparently belonged to the city of Oxyrhynchus, included water for irrigation, and it provided a regular annual income of six hundred *atticae* (or 2,400 drachmas). This sum would cover his annual salary, the value of which was apparently set at five hundred *atticae*.[69] In Lollianus' description, this orchard produced a fixed annual income, which was derived completely from the annual rent paid by a tenant. It is hard to imagine that Lollianus himself would play any role in managing the orchard or in maintaining its productivity. Rather, the orchard was to serve as a resource from which he might derive a fixed annual income, and Lollianus clearly viewed this method of providing for his maintenance as much more desirable than a direct payment of foodstuffs by the city.

Lollianus was hoping to be supported by the emperors in much the same way that many other philosophers and teachers were. A good example of the type of support that the emperors might afford a philosopher is provided by the case of Flavius Archippus of Prusa in

---

68. For discussion of this document, see the source in which it was originally published, P.J. Parsons, "Petitions and a Letter: The Grammarian's Complaint," in A.E. Hanson, ed., *Collectanea Papyrologica,* vol. II [= *P.Coll.Youtie* II 66] (Bonn, 1976), 409–446. See also R.A. Kaster, *Guardians of Language* (Berkeley and Los Angeles, 1988), 115–16.

69. *P.Oxy.* XLVII 3366, lines 61–63: . . . ὥστε κῆπον τῆς πόλεως ἔνδον τ(ε)[ι]χῶν ὄντα κα/λούμενον παράδ(ε)ισον Δικτύνου σὺν τοῖς [ἄλλοις] φυτοῖς καὶ / [τ]ῷ πρὸς ἀρδείαν ὕδ[α]τι, φέροντα [ἐ]ν μισθώσ(ε)[ι]] Χ ἀτ'τικάς, / δοθῆναί μοι κ[ελεῦ]σαι . . . .

Bithynia. This individual gained the good graces of the emperor Domitian, who ordered a proconsul of Bithynia, Terentius Maximus, to purchase for the philosopher land worth one hundred thousand sesterces, from which the philosopher was supposed to derive sufficient income to support himself and his family.[70]

It might be argued that the dependence of an intellectual like Lollianus on such an income represented a peculiar situation and that the nature of the arrangements struck to support such an individual cannot be considered to reveal very much about how Roman landowners generally conceived of profits from their holdings. But I think that the opposite is true. Although some landowners might take advantage of favorable market conditions and invest heavily in improving their lands, many others viewed their holdings as resources providing a more or less fixed level of income, with that income almost always being derived from the rent paid by a tenant. Such an assumption must explain an otherwise curious expression in an act of sale from fourth-century Hermopolis (*CPR* XVIIA 17a, b, ca. A.D. 321). This document records a sale of seven *arourae* of grain land to Aurelius Adelphios, a councillor and officeholder at Hermopolis, and one of the city's major landowners.[71] The significant phrase comes in the clause stipulating the tax liability for the parcel of land sold. The seller stipulated that the purchaser would be responsible for all taxes in the future, since he would own all the future rents: διὰ τὸ ⟨εἶ⟩ναι σοι ἐκφόρια καὶ φόρους (line 8). The seller clearly envisioned this land as having one primary use, and he also expected that the future owner would exploit it in exactly the same way that he had.

This same assumption about the value of land can also be found in an earlier period, in a second-century court case from Egypt (*P.Oxy.* VIII 1102, ca. A.D. 146). The case involved a dispute over an inheritance

---

70. Domitian's letter is quoted by Pliny in connection with a legal dispute that he was forced to address as governor of Bithynia (*Ep.* 10.58.5). For discussion of Flavius Archippus (*PIR*[2] F 216), known only from Pliny, see Sherwin-White, *Commentary*, 640–43. For the privileges accorded to philosophers, see J. Hahn, *Der Philosoph und die Gesellschaft* (Stuttgart, 1989), 100–108. Cf. the constitution of Constantine upholding the right of grammarians, professors, and physicians to salaries and immunity from public charges (*CJ* 10.53 (52) 6 pr.–1, A.D. 333).

71. An archive of some thirty documents attests Adelphios' activities as an officeholder and landowner in 312–322; this archive is published by K. Worp, as *Corpus Papyrorum Raineri,* vol. XVIIA (Vienna, 1991).

between the representatives of a city, apparently Oxyrhynchus, and an individual named Eudaimon, and we learn about this case from a report of a hearing in the court of the *hypomnematographos* Cerealis.⁷² At issue were claims made by the city on property that Eudaimon had inherited from his brother, since this latter person had apparently died leaving substantial debts to the city associated with his service as *gymnasiarch*. In an earlier hearing, the prefect Valerius Proculus (lines 6 f.) and the *iuridicus* Neocydes (lines 15 f., 24) had decided against Eudaimon, ruling that he had to hand over to the city one-fourth of the inherited property, with a dowry that the brother had set aside for his daughter subtracted. But there was still a dispute outstanding on the status of some land. Eudaimon claimed this land, while the city for its part also claimed title, insisting that the land was not included in the disputed inheritance. In the second hearing, the *hypomnematographos* ordered an inquiry by the *strategos* to see whether this land was in fact included in the inheritance, and, at the request of the city's representatives, he awarded the city the income from the land for one year (lines 16–18).⁷³ At this point Eudaimon, fearing the loss of his income if the entire property remained sequestered, requested that the city's lien be removed from the property, a request that the *hypomnematographos* was only willing to grant after Eudaimon had made good the outstanding debts to the city (lines 18 f.).⁷⁴

Important for our purposes is the assumption that the land in question produced an income with little if any direct involvement on the part of the owner. Otherwise, it seems hard to imagine how the city would be able to take up the management of the land for a single year or, for that matter, how, after the city had garnered the income for one year, Eudaimon could receive back a resource that could provide him with an income. The most likely explanation is that the land in question

---

72. The property in dispute was certainly in the Oxyrhynchite nome, but the *hypomnematographos* was an Alexandrian official assigned to hear cases involving Alexandrian citizens; see A.K. Bowman and D. Rathbone, "Cities and Administration in Roman Egypt," *JRS* 82 (1992): 107–27, at 117.

73. *P.Oxy.* VIII 1102, lines 16–18: τῶν πρέσ/βεων ἀξιωσάντων ἐπὶ τῆς προσόδου τῶν ἀρουρῶν στῆσαι ὁ ἱερεὺς καὶ ὑπομνηματογρά/φ[ο]ς· ἐνιαυ[τοῦ] τὰς προσόδους ἀπολήμψεται ἡ πόλις.

74. *P.Oxy.* VIII 1102, lines 18 ff.: Εὐδαίμονος διὰ τῶν παρεστώ/των λέγοντος κατεσχῆσθαι αὐτοῦ τὰς προσόδους καὶ ἀξιώσαντος ἀπολυ/θῆναι αὐτάς, ὁ ἱερεὺς καὶ ὑπομνηματογράφος· ἐπὰν τὰ ὑπ' ἐμοῦ κελευ/σθέν[τ]α γένηται, κ[α]ὶ ἡ πόλις τὸ προσῆκον μέρος κομίσηται, ἀπολυθή/[σο]ν[τα]ι.

was leased out to tenants, who were themselves responsible for managing its cultivation and maintaining its productivity.[75] We see a similar assumption about the nature of the income provided by agricultural land in a loan from third-century Hermopolis, which we learn about from a legal case involving the debtor and the heirs of the creditor (*M.Chr.* 189, A.D. 240).[76] The loan was for eight thousand drachmas and was secured against a one-half share of forty-one and one-fourth *arourae* of agricultural land (as *hypallagma*). This loan was a *misthokarpeia*, that is, a loan in which the creditor received the income from the property pledged in lieu of interest. In this case, again, the income from the land is envisioned as simply the rental payment made by the tenants: κ[αρ]πείαν καὶ διαμίσθωσιν καὶ πρόσοδον πᾶσαν (line 38).

The expectation that a given estate could furnish only a narrow range of possible income would help to explain why the income immediately available from an estate became the single most important factor in determining an estate's value to a Roman landowner and hence its selling price. This assumption underlies a case ruled on by Julian (*D.* 30.92 pr., 39 *dig.*), in which an heir, to pay off an inherited debt, sold off a *fundus* that had been bequeathed to another individual through a trust. The heir had to compensate the beneficiary for the *fundus*, and the price of this property was determined as a direct function of its annual income: "excusso pretio secundum reditum eius fundi." We can also see this principle illustrated in the case of Pliny the Younger. In *Epistulae* 3.19, for example, Pliny describes the loss of income occasioned by the poor condition of the tenants as the major factor in reducing the price of an estate from five to three million sesterces: "sestertio triciens, non quia non aliquando quinquagies fuerint, verum et hac penuria colonorum et communi temporis iniquitate ut reditus agrorum sic etiam pretium retro abiit" (3.19.7). It is not likely that the price of the slaves that Pliny would have had to buy accounted for all of the two million sesterces by which the purchase price had declined. Rather, we

---

75. See A.S. Hunt's discussion of the details in this fragmentary document, in *P.Oxy.* VIII 1102, introd., p. 169. If the city received a share of the income, the case that I have made is stronger, since it is hard to imagine how a private individual could divide the task of managing land with a city.

76. This document was first published as *P.Lips.* 10. On antichretic leases, in which the creditor received a lease for property of the debtor as fulfillment of the debt, see J. Herrmann, *Studien zur Bodenpacht im Recht der graeco-aegyptischen Papyri* (Munich, 1958), 236–44.

can presume that the tenants' farms had deteriorated and that the lowered purchase price reflected the income that would have been lost during the time that it would have taken to restore these farms to full productivity.[77] A similar situation developed on the estate that Pliny used to finance his alimentary foundation at Comum (*Ep.* 7.18). As we have seen, Pliny imposed on an estate to remain in his private possession a permanent yearly *vectigal* of thirty thousand sesterces, representing a 6 percent income from the five hundred thousand sesterces that he was nominally donating. Pliny realized that there was a cost associated with this action: the imposition of a permanent *vectigal* would lower the selling price of the estate by even more than the five hundred thousand sesterces represented by the annual payment of thirty thousand sesterces (7.18.2–4; see the discussion of Pliny's donation earlier in this chapter, under "Providing for Permanent Foundations"). The *vectigal*, representing a fixed loss of income every year, would have made the estate marginally less desirable than other estates, where the owner would have had a wider margin for achieving a desired level of income.[78]

---

77. The phrase *hac penuria colonorum* (3.19.7) seems to refer to a shortage of tenants, not to poverty on the part of tenants; on this question, see most recently De Neeve, *Athenaeum* 78 (1990): 381–82, 393–94. This passage has been interpreted as attesting a general agricultural slump; see Sherwin-White, *Commentary*, 259; cf. V. Sirago, *L'Italia agraria sotto Traiano* (Louvain, 1958), 103–125.

78. On the ratio between land prices and rents, see C. Clark and M. Haswell, *The Economics of Subsistence Agriculture*, 4th ed. (London, 1970), 157–71, and Duncan-Jones, *Economy*, 49 n. 2. For full discussion of how land prices in ancient Rome were determined, see P.W. de Neeve, "The Price of Agricultural Land in Roman Italy and the Problem of Economic Rationalism," *Opus* 4 (1985): 77–109: the price of agricultural land was a direct function of the income that it produced; see also idem, *Colonus* (Amsterdam, 1984), 171–73. For a contrasting interpretation of the evidence provided by Plin. *Ep.* 3.19 for land prices, see L. Capogrossi Colognesi, *Ai margini della proprietà fondiaria* (Rome, 1995), 285–86 n. 143, and idem, "Il regime degli affitti agrari," *Scienze dell'antichità, Storia, Archeologia, Antropologia* 6–7 (1992–93): 243–44 n. 143. For land prices in Italy, see Duncan-Jones, *Economy*, 48–52, 377–78, who views Columella's suggested price of one thousand sesterces for one *iugerum* of virgin land for the cultivation of vines as exaggeratedly high even for Italy. It may not be possible to offer any average price for Italian land, however; see De Neeve, loc. cit., and *Athenaeum* 78 (1990): 370–71.

Fundus cum Instrumento

We can appreciate how the jurists envisioned an estate as serving the financial interests of upper-class Romans by considering their treatment of the legacy of an estate with its equipment, that is, the legacy of the *fundus cum instrumento* (see especially D. 33.7).[79] To summarize before proceeding to the details, the jurists conceived of one primary benefit from an estate that might accrue to the legatee, namely, the achievement of an income, or *reditus,* derived from the production of a crop.[80] The testator, in the jurists' conception, desired to provide the beneficiary of the legacy with a means to secure an income, and this income depended solely on the ability of the estate to achieve a level of physical productivity. The jurists took little account of external market conditions that might affect the yearly revenues of an estate. Rather, they simply assumed that the legatee might draw an income from the sale of crops produced on the estate.[81] Clearly there were limits on how the jurists might have accounted for factors extrinsic to the condition of the estate bequeathed through legacy, and the profitability of the estate in comparison to other forms of investment was not a concern for the jurists. Changes, moreover, in the market conditions for various crops would not have altered what constituted an agricultural estate or the equipment assigned to it.

In this area of the law, the jurists' principal task was to define what equipment, livestock, or slaves on the *fundus* were to be included in the legacy of a *fundus cum instrumento.* The underlying principle was to answer the needs of the testator and the *legatarius;* the testator is

---

79. For the jurists' conservatism in applying economic considerations when developing rules for the legacy of a *fundus cum instrumento,* see Steinwenter, *Fundus cum instrumento.*

80. See Steinwenter, *Fundus cum instrumento,* 35.

81. As Steinwenter points out (*Fundus cum instrumento,* 70), the jurists concerned themselves neither with the organization of labor on the estate nor with the methods used to market the crops. The jurists envisioned the estate as producing crops to be sold on the market. This assumption is implicit not only in the jurists' treatment of the legacy of estates but in other areas of the law as well, including, for example, the *actio institoria.* On this area of the law and the commercial aspects of a Roman estate, see A. di Porto, "Impresa agricola ed attività collegate nell'economia della 'villa': Alcune tendenze organizzative," in *Sodalitas,* vol. VII (Naples, 1984), 3235–77.

assumed to have sought to provide the *legatarius* with an agricultural property intended to furnish a given level of income.⁸² In the jurists' definition, the economic value of an estate bequeathed with its *instrumentum* consisted exclusively in its ability to produce a crop, or *fructus*, and so they included in *instrumentum* only such equipment, livestock, or slaves as were directly used in the production of that crop: "In instrumento fundi ea esse, quae fructus quaerendi cogendi conservandi gratia parata sunt, Sabinus libris ad Vitellium evidenter enumerat" (Sab. apud Ulp. D. 33.7.8 pr., 20 *ad Sab.*). This *instrumentum* included all the equipment and livestock needed for the production, storage, and transport of crops; slaves employed in the production of the crops; and even the foodstuffs needed to maintain the slaves and livestock (Ulp. D. 33.7.12 pr.–14, 20 *ad Sab.*). To generalize, in Ulpian's definition, the *instrumentum* of a *fundus* included equipment permanently on the estate and without which the estate could not be cultivated: "quippe instrumentum est apparatus rerum diutius mansurarum, sine quibus exerceri nequiret possessio" (D. 33.7.12 pr.). They key was that whatever was included in the *instrumentum* had to have some direct bearing on the production of a crop. Excluded from the legacy of a *fundus cum instrumento*, accordingly, were slaves who were not employed directly in the production of the crop but were rather leasing the estate in question as if they were tenants. Thus Ulpian approved the opinions of Labeo and Pegasus for excluding from the *instrumentum* a slave leasing the estate as a *servus quasi colonus* (D. 33.7.12.3, 20 *ad Sab.*).⁸³ Such a slave was treated as analogous to a free tenant. The slave's presence on the estate was coincidental to the way in which the estate produced its crop, since he did not serve directly as a unit of labor attached to the

---

82. See Steinwenter, *Fundus cum instrumento*, 71–86. In a definition of a *fundus* for the purpose of distinguishing between the rights of seller and purchaser in a sale, Ulpian considered the *stercilinium* and the *stramentum*, both critical to the production of fertilizer, to be part of the estate, so he allotted them to the purchaser in a sale, or to the legatee in a legacy of an estate. Cut wood, by contrast, remained the property of the seller in a sale, or of the heir in a legacy (D. 19.1.17.2, 32 *ad ed.*). Ulpian cited Trebatius as distinguishing between whether the *stercilinium* served to fertilize the *fundus*, in which case it went to the purchaser, or whether it provided dung for sale, in which case it remained with the seller.

83. Ulp. D. 33.7.12.3: "Quaeritur, an servus, qui quasi colonus in agro erat, instrumento legato contineatur. et Labeo et Pegasus recte negaverunt, quia non pro instrumento in fundo fuerat, etiamsi solitus fuerat et familiae imperare."

estate in question but rather supervised the labor force producing the estate's crop.

Some jurists defined broadly the "crop" that an estate might produce, recognizing special categories of estates not producing grain, olive oil, or wine as their chief crops; such special estates might be devoted to raising livestock or to hunting. For example, Ulpian included flocks of sheep and slave shepherds in the *instrumentum* of a *fundus* if the estate encompassed pasturelands, *saltus* and *pastiones* (D. 33.7.8.1, 20 *ad Sab.*). This livestock might be considered as additional to the livestock used for plowing and fertilizing the fields, which by a more stringent definition alone counted as *instrumentum fundi* (D. 33.7.8 pr.). Ulpian similarly considered as belonging to a legacy of *instrumentum* slaves serving the villa, such as doorkeepers, sweepers, and gardeners (D. 33.7.8.1). In this jurist's view, beehives and honey might likewise be counted as part of a legacy, if income was derived from honey (D. 33.7.10, 20 *ad Sab.*). Finally, Ulpian included hunting and bird-catching equipment, as well as slaves and dogs employed in those activities, especially if income might be derived from such sources (D. 33.7.12.12–13). Iavolenus had considered birds as part of a legacy of *instrumentum* if the estates in question were located on maritime islands (D. 33.7.11, 2 *ex Cassio*).

Other jurists took a more restrictive approach, focusing exclusively on whether the livestock or equipment in question contributed to the production of a more conventional crop. In Paul's definition, a flock of sheep could only be included in the *instrumentum* if it was essential to the income from the *fundus:* "si vero ideo, quia non aliter ex saltu fructus percipi poterit . . ." (D. 33.7.9, 4 *ad Sab.*); otherwise, livestock were excluded, even if they provided some income. For Paul, direct use in the production of a crop was paramount. Accordingly Paul approved the position of Neratius, who had excluded from a legacy a mill and the ass used to drive it (D. 33.7.18.2, 2 *ad Vitell.*). Presumably the mill, although potentially an important source of income, did not directly contribute to the production of a crop. Paul's restrictive interpretation is to be contrasted with the more flexible position of Iavolenus, who approved the earlier opinions of Labeo, Cascellius, and Trebatius that millstones were to be considered part of the *instrumentum* (D. 33.7.26.1, 5 *ex poster. Lab.*). By contrast, Paul did count as part of the legacy the ox-driver in charge of plowing the fields or tending the plow oxen as they grazed (D. 33.7.18.6, 2 *ad Vitell.*). Earlier Alfenus had allowed the inclusion of sheep in a legacy of estates with their equipment only when their dung was used to fertilize the crops; likewise a shepherd might be included in a legacy if he tended sheep used for this purpose (*D.* 32.60.3, 22 *dig. a Paul. epit.*).[84]

Although there might be some latitude in the eyes of certain jurists and even substantial disagreement about what properly constituted a crop, an estate as treated in the legacy of *fundus cum instrumento* was generally conceived not as a piece of property with alternative potential uses but rather as an enterprise strictly oriented toward the production of a crop. For his part, Ulpian included items used in the transportation of crops off of the estate, such as draft animals, wagons, ships, storage jars (*cuppae*), and sacks (*cullei*). In so doing, he simply assumed that the income from the estate would be derived from the sale of crops produced there (*D.* 33.7.12.1, 20 *ad Sab.*). From this point of view, the estate bequeathed in a legacy had a value only insofar as it could provide an income derived from the sale of a crop; the desires of the testator and the needs of the legatee were fulfilled if the estate was sufficiently equipped to make possible the production and sale of this crop.

In formulating legal rules about legacies, of course, the jurists did not consider all the many possible uses to which a Roman estate owner might put his or her land. Indeed, in a manner analogous with their treatment of tutorship, the jurists simplified reality to facilitate the adjudication of cases in the real world. But behind the "ideal" type of a *fundus cum instrumento* lay a basic understanding of the economic interests of the parties involved in a legacy, and the principle underlying these economic interests was the very restricted range of ways in which a landowner might achieve an income from an estate. In the "idealized" form in which this principle is expressed in the Digest, the characteristic way for a beneficiary of a will to achieve an income was by producing and selling a crop, and the income that an estate produced was not likely to change.

The interests of the "typical" legatee envisioned by the jurists were probably not much different from those of Pliny's former nurse, to whom he had given a farm originally worth one hundred thousand sesterces as a means of support (*Ep.* 6.3). In all likelihood, this woman was

---

84. The jurist did not refer specifically to a legacy of a *fundus cum instrumento,* but rather to "praediis legatis et quae eorum praediorum colendorum causa empta parataque essent" (Alf. *D.* 32.60.3).

completely dependent on this farm for her livelihood, and her role in managing it seems to have been completely passive. Thus when, for unknown reasons, the small estate had lapsed into a poor state of repair and its income had gone down, Pliny had to come to the aid of his former nurse and seek out a new manager for it. Pliny's personal intervention and the employment of a new manager were the only factors that could restore the estate and assure the freedwoman of the income on which she had no doubt grown to depend. The interests of the legatees envisioned by the jurists must have been in large part similar; the estate bequeathed in legacy was not an investment to be exploited with a view toward obtaining a given level of profit but a resource that the individual concerned had no choice but to exploit to achieve a needed income. As we consider the economic purpose behind the legacy of the *fundus cum instrumento,* we might compare the description in the vita of Alexander Severus of the generosity that emperor displayed toward retiring officials (SHA *Alex. Sev.* 32.3). The biographer describes the emperor as bestowing land, oxen, horses, grain, iron, funds and labor for building a house, and marble for decorating a house. The point of this generosity was that the retiring official "could live as a private citizen in accordance with his station."[85]

These assumptions did not simply derive from the internal logic of the law on legacy. They can be traced in other related areas of the law as well, such as the jurists' regulation of the legacy of a usufruct.[86] When defining the mutual obligations of the owner of the property held under usufruct and the beneficiary of the usufruct, the jurists assumed that the parties to such an arrangement would benefit from an estate in the same way as the parties in a legacy, that is, by cultivating conventional crops. Accordingly, in Ulpian's view, the owner was not to prevent the usufructuary from using the property, as long as the latter party did not make the condition of the property worse. In this con-

---

85. SHA *Alex. Sev.* 32.3: ". . . eumque muneratus est, ita ut privatus pro suo loco posset honeste vivere, his quidem muneribus: agris, bubus, equis, frumento, ferro, impendiis ad faciendam domum, marmoribus ad ornandam, et operis quas ratio fabricae requirebat." Cf. the generosity that Alexander Severus is described as bestowing on former officeholders who had fallen into poverty. The emperor provided them with land, slaves, animals, flocks, and farm tools (SHA *Alex. Sev.* 40.2).

86. For the rationale behind the bequest of usufructs, see Saller, *Patriarchy, Property and Death in the Roman Family,* 173–74.

nection, the owner also had no right to prevent the usufructuary from using an estate's equipment, or *instrumentum* (Ulp. D. 7.1.15.6, 18 *ad Sab.*), even though, in the interpretation of some jurists, the usufructuary's use of this equipment worsened the condition of the property.[87] Similarly, the owner was not permitted to bar the usufructuary from entering the property held under usufruct or to bar access to light and water (Ulp. D. 7.4.29.2, 17 *ad Sab.*). The usufructuary, for his or her part, had the right to use the property as any pater familias would. Generally, this meant that the usufructuary could be compelled to cultivate the estate (Celsus apud Ulp. D. 7.1.9 pr., 17 *ad Sab.*).[88] The usufructuary was also permitted to gain income from the other sources of income associated with agricultural property, such as beekeeping (Ulp. D. 7.1.9.1), quarries, clay pits and sand pits (Sab. apud Ulp., sec. 2), mines (sec. 3), falconry, hunting and fishing (Cassius apud Ulp., ibid., sec. 5), and seedbeds (ibid., sec. 6). But the usufructuary was only allowed to alter the way in which the *fundus* was exploited, such as by digging new quarries, clay pits, sand pits, or mines, if he or she could do so without harming the property's original type of agricultural productivity (Ulp. D. 7.1.13.5, 18 *ad Sab.*).[89] The underlying principle for the

---

87. Ulp. D. 7.1.15.6: "Proprietatis dominus non debebit impedire fructuarium ita utentem, ne deteriorem eius condicionem faciat. de quibusdam plane dubitatur, si eum uti prohibeat, an iure id faciat: ut puta doleis, si forte fundi usus fructus sit legatus, et putant quidam, etsi defossa sint, uti prohibendum: idem et in seriis et in cuppis et in cadis et amphoris putant: idem et in specularibus, si domus usus fructus legetur. sed ego puto, nisi sit contraria voluntas, etiam instrumentum fundi vel domus contineri."

88. For the principle that the usufructuary was not to worsen the condition of the property and was in general to demonstrate the same standard of care as a diligent pater familias, see Sab. apud Ulp. D. 7.1.9.2 (17 *ad Sab.*), Ulp. D. 7.1.13.4, 6 (18 *ad Sab.*), Pomp. D. 7.1.10 (5 *ad Sab.*), and Pomp. D. 7.1.65 pr. (5 *ex Plaut.*).

89. Ulp. D. 7.1.13.5 : "Inde est quaesitum, an lapicidinas vel cretifodinas vel harenifodinas ipse instituere possit: et ego puto etiam ipsum instituere posse, si non agri partem necessariam huic rei occupaturus est. proinde venas quoque lapicidinarum et huiusmodi metallorum inquirere poterit: ergo et auri et argenti et sulpuris et aeris et ferri et ceterorum fodinas vel quas pater familias instituit exercere poterit vel ipse instituere, *si nihil agriculturae nocebit*" (emphasis added). Ulp. D. 7.1.13.5 envisions the possibility of eliminating agricultural operations to invest in a more profitable enterprise, although this last part of section 5 is considered a gloss by Noodt or Justinianic by Pampaloni (see the app. crit. of the Mommsen and Krueger text): "Et si forte in hoc quod instituit plus reditus sit quam in vineis vel arbustis vel olivetis quae fuerunt, forsitan etiam haec deicere poterit, si quidem ei permittitur meliorare proprietatem."

jurists was to address the interests of the owner of a property held under usufruct by preserving the property's economic value. This value consisted first and foremost in its agricultural productivity, and other forms of income, even mines, were secondary and temporary.

To return to their conception of profit from agriculture in the law of legacy, the jurists drew a static picture of the relationship between a *fundus* and its income, as if the range of investments that a landowner might make in an estate, like the range of potential revenues, was severely restricted. This principle is especially apparent in the treatment of cash and other financial assets associated with an estate. In the real world, of course, upper-class Romans achieved incomes from a variety of sources, including forms of business other than agriculture, such as lending money out at interest. The significance of lending money to the financial interests of upper-class Romans is suggested by the fact that the jurists had to deal with problems involving it in adjudicating the bequest of *fundi*. *Fundi* were commonly the locations at which landowners might not only keep assets and documents but also transact business.[90]

When they treated this problem, the jurists maintained a strict distinction between cash and other financial assets that were intrinsically linked with the cultivation of an estate, on the one hand, and cash and assets that had an essentially coincidental association with an estate, on the other. The jurists sharply distinguished between funds that formed part of the yearly income from the estate and funds physically present on the estate but not derived from income associated with the produc-

---

90. The jurists also had to address this situation in other areas of the law. In this connection, we might note the scholarly dispute about the competence of the *vilicus* of an estate to engage in transactions beyond those associated directly with the production of crops. In a discussion of the application of the *actio institoria* to *vilici* managing estates, Chiusi, ZRG 108 (1991): 155–86, argues that the *vilicus* was characteristically entrusted only with duties directly concerned with the cultivation of the crops. In her view this type of *vilicus*, with a restricted range of duties, was a readily recognized "type" in the Roman economy, so the jurists, when regulating the liability of the employer of a *vilicus* managing an estate, restricted the *actio institoria* to transactions directly connected with the production of crops. Chiusi is arguing against a broader view of the competence of a *vilicus* and of the availability of the *actio institoria*. According to this view, the regular duties of a *vilicus* might include not only marketing the crops produced on the estate but also financial activities not at all connected with agriculture, such as lending money out at interest. This view has recently been argued by Di Porto, "Impresa agricola."

tion of a crop. In so doing, the jurists remained consistent with the principle that the sole economic function of an estate was to produce a crop. We see this distinction made especially in connection with the *reliqua colonorum*, or arrears of the tenants, that might form part of a legacy.[91]

For example, Papinian, in defining what might be included in a legacy of *reliqua colonorum*, included only those arrears that when paid formed part of the yearly income from the estate (*D*. 32.91 pr., 7 *resp*.).[92] In this case the money that had been exacted from the tenants and assigned to the account book, or *calendarium*, for investment did not constitute part of the legacy; only the arrears of the tenants that could properly be viewed as forming part of the yearly income from the estate might be included. To judge by this decision, the jurists did not think that cash should be counted as an asset for reinvestment in the estate, since the profits gained from the tenants' leases, if they were to be reinvested at all, would be invested in loans and other financial arrangements that had nothing to do with the agricultural side of an estate.[93] We might compare a case discussed by Paul, in which a testator bequeathed to one of his heirs through *praelegatio* a set of *fundi* with all the equipment, livestock, and slaves that were on the estate at the time of the testator's death (*D*. 32.92 pr., 13 *resp*.). Paul excluded from this legacy documents recording purchases and loans that were found in a *tabularium*, or records office, on one of the estates. The documents in the *tabularium* had nothing intrinsically to do with the cultivation of the estate in question. Again, the value of a *fundus* consisted solely in its ability to produce a crop; the *fundus* was not to be considered a business engaging in various types of investment.

We see a distinction between cash used for expenses and cash that represented a profit from an estate in another case discussed by Paul, in which the jurist played an adjudicatory role (*D*. 32.97, 2 *decret*.).[94] In this

---

91. The *reliqua colonorum* did not, however, constitute part of the *instrumentum fundi*.

92. Papin. *D*. 32.91 pr.: "Praediis per praeceptionem filiae datis cum reliquis actorum et colonorum ea reliqua videntur legata, quae de reditu praediorum in eadem causa manserunt: alioquin pecuniam a colonis exactam et in kalendarium in eadem regione versam reliquis non contineri neque colonorum neque actorum facile constabit, tametsi nominatim actores ad filiam pertinere voluit."

93. See Scaev. *D*. 33.7.20.3 (3 *resp*., quoted in n. 102 in this chapter).

94. Paul's *Decreta* are reports of actual cases in which the jurist participated as a member of the emperor's council: see T. Honoré, *Emperors and Lawyers*, 2d ed. (Oxford, 1994), 20. On this case, see C. Sanfilippo, *Pauli Decretorum Libri Tres* (Milan, 1938), 81–84.

case, a landowner named Hosidius instituted his daughter (Valeriana) as heir to his ample estate. At the same time, Hosidius made generous provisions for a slave business manager named Antiochus. The testator manumitted Antiochus and bequeathed to him a group of estates, *praedia certa,* together with his *peculium,* his own debts to the testator as well as the arrears of the tenants on the estates in question. The freedman subsequently produced evidence that he had a substantial debt with the testator because he was responsible to render an account for the grain, wine, and other agricultural products stored on the estates under his charge. He then sued the heir for the agricultural products stored on the estate, and he won his case before the provincial governor.[95] Antiochus was apparently a large-scale manager of the estates belonging to Hosidius; he may have held these estates under some sort of lease arrangement, functioning as a *servus quasi colonus.*[96] At any rate, he was responsible to the testator both for the rents collected from the tenants and for the disposition of crops grown and stored on the estates under his charge. The heir disputed his claim on the products stored on the estate and appealed to the emperor; the point of the appeal was the distinction between the debts owed by the tenants and the commodities stored on the estate. The emperor granted the appeal, awarding the commodities to the heir. The emperor offered the analogy of a substantial sum of money used to meet daily expenses on the estate; what was left over from this sum of money could not be considered part of the legacy.[97]

The cash that the emperor, in deciding this case, hypothetically assigned to the estate was to be used solely to meet daily expenses. This cash, then, would fulfill a similar function to the cash maintained both

---

95. Paul. *D.* 32.97: "Hosidius quidam instituta filia Valeriana herede actori suo Antiocho data libertate praedia certa et peculium et reliqua relegaverat tam sua quam colonorum: legatarius proferebat manu patris familiae reliquatum et [se *ins. Mo.*] tam suo quam colonorum nomine: item in eadem scriptura adiectum in hunc modum: 'item quorum rationem reddere debeat,' scilicet quae in condito habuerat pater familias frumenti vini et ceterarum rerum: quae et ipsa libertus petebat et ex reliquis esse dicebat: et apud praesidem optinuerat."

96. See De Neeve, *Colonus,* 159 n. 192, 163 n. 213. For further discussion of *servi quasi coloni,* see chap. 3.

97. Paul. *D.* 32.97: "ex diverso cum diceretur reliqua colonorum ab eo non peti nec propria, diversam autem causam esse eorum, quae in condito essent, imperator interrogavit partem legatarii: 'quaerendi causa pone,' inquit, 'in condito centiens aureorum [sestertium *Sanfilippo*] esse, quae in usum sumi solerent: diceres totum, quod esset relictum in arca, deberi?' et placuit recte appellasse."

in the central account and in the individual divisions of the large estate of Aurelius Appianus in Egypt.[98] On this estate, a central account was maintained of all livestock and agricultural products, and the agricultural products were often accounted for in terms of a cash value. Every month, the manager of an individual division would have to account for all the commodities, including agricultural products, livestock, and cash, that he used in connection with the cultivation of land assigned to his supervision. But on both the estate of Appianus and on the estate conceived of by the emperor in the case just discussed, cash was an asset only to the extent that it was used to meet expenses involved in raising crops. It had no independent status as an asset for reinvestment in the estate. Likewise the agricultural products kept on the estate of Hosidius were not to be considered part of the equipment of the estate; the profits from their sale represented wealth for the owner, not a source of funds for maintaining the productivity of the estate.

The decision by the emperor in this case seems consistent with the general thinking of the jurists in maintaining a strict distinction between cash and other assets used to meet immediate expenses in producing a crop, on the one hand, and cash and assets simply present on the estate, on the other. This principle is expressed in most general terms by Pomponius, who did not include in a legacy of a *fundus* money that was kept there to be lent out at interest (*D.* 32.44, 2 *ad Sab.*).[99] Similarly, Paul excluded from a legacy of a *fundus* with its slave staff both the debts owed to the owner and the cash present on the estate (*D.* 32.78.1, 2 *ad Vitell.*). To bequeath cash and other financial assets, the testator would have to make specific provision. Thus in his discussion of a bequest by trust of *praedia* with a *calendarium* to a grandson of the testator, Africanus went against previous opinions and ruled that the grandson was entitled to both the debts recorded therein and any cash that might be designated for reinvestment in loans (*D.* 32.64, 6 *quaest.*).[100] Otherwise, the distinction between cash and other equip-

---

98. For the central accounts of the estate of Appianus, see D. Rathbone, *Economic Rationalism and Rural Society in Third-Century* A.D. *Egypt* (Cambridge, 1991), 331–87.

99. Pompon. *D.* 32.44: "Si fundus legatus sit cum his quae ibi erunt, quae ad tempus ibi sunt non videntur legata; et ideo pecuniae, quae faenerandi causa ibi fuerunt, non sunt legatae."

100. This interpretation is based on Mommsen's emendation of *cum certo* for *excepto* in the manuscripts: ". . . certa praedia quaeque in his mortis tempore sua essent nepoti per fideicommissum dederat <cum certo> kalendario. . . ."

ment was so strictly maintained that the *reliqua colonorum* would be excluded from a legacy of an estate described in very general and inclusive terms unless specific provisions were made to include them. Thus Paul allowed only the slaves to be included in a legacy to a foster daughter of a *fundus* worth two hundred aurei (Paul. D. 32.78.3, 2 *ad Vitell.*).[101] The revenue from the leases on the estate was considered not as an asset of the estate but only as the income of the owner or heir. Similarly Papinian, in awarding to a legatee of *praedia* both the *instrumentum rusticum* and the furnishings contributing to the owner's comfort, excluded the arrears of the tenant (D. 32.91.1, 7 *resp.*).

Even when the testator specifically bequeathed the arrears of the tenants, the legacy was restricted to the arrears of the sitting tenants. Thus Scaevola excluded from a legacy of estates that included the *reliqua colonorum* the arrears of those tenants who had left the property at the end of their leases and provided a *cautio* for their outstanding debts (D. 33.7.20.3, 3 *resp.*).[102] Significant here is that, since the concept of *reliqua colonorum* is restricted to the arrears of sitting tenants, by extension the productivity of the sitting tenants could be considered an integral component of the productive capacity of an agricultural property. These arrears were one step removed from the stored crops, which the testator might also bequeath. But arrears from previous tenants are treated like other debts that have merely a coincidental association with an agricultural property. The estate was envisioned not as an enterprise with assets and liabilities but only as a physical entity geared toward the production of a crop.

Bequests of Farmland in Egypt

Thus far I have sought to infer, on the basis of assumptions made by the jurists when formulating legal rules for the bequest of estates, what economic goals Roman landowners undertook to accomplish through their investments in agriculture. The point of inferring these goals is to determine, by examining what the jurists reported and assumed about

---

101. Paul. D. 32.78.3: "Peto, ut fundum meum Campanianum Genesiae alumnae meae adscribatis ducentorum aureorum ita uti est."

102. Scaev. D. 33.7.20.3: "Praedia ut instructa sunt cum dotibus et reliquis colonorum et vilicorum et mancipiis et pecore omni legavit et peculiis et cum actore: quaesitum est, an reliqua colonorum, qui finita conductione interposita cautione de colonia discesserant, ex verbis supra scriptis legato cedant. respondit non videri de his reliquis esse cogitatum."

the economic background to the cases that they considered, how Roman landowners envisioned themselves as deriving a "profit" from their estates. The next step is to see whether the relationships inferred from the legal evidence can be traced in the papyri from Roman Egypt. Even when allowances are made for the very substantial differences between Egypt and other provinces of the Roman Empire, the papyri offer an important source of evidence for examining the social background to the legal principles formulated by the jurists. The evidence provided by documentary papyri suggests that landowners in Roman Egypt had an understanding of the value of agricultural property consistent with the one that I have ascribed to the testators in the wills that I have already discussed. Land represented a resource providing economic security, not a source of potential profit realizable after a given level of investment.

We can detect such an approach in the case of one L. Ignatius Rufinus, who died in the early third century. Although this individual can hardly be counted among the privileged elite of the Roman Empire, he did enjoy the double status of being both a Roman citizen and a citizen of the city of Antinoopolis. This latter status was comparable to Alexandrian citizenship, in that it conferred on the holder certain privileges. Antinoites were exempt from the poll tax, and they could not be compelled to perform liturgies outside the territory of Antinoopolis, although they might own property elsewhere.[103] L. Ignatius Rufinus, then, at least belonged to the more privileged class of landowners in the Arsinoite nome (the Fayum), where he had property. In his will, Ignatius made his brother, L. Ignatius Nemesianus, his sole heir, but he also sought to provide for the security of his wife, Lucretia Octavia (*P.Diog.* 10, A.D. 211).[104] He did this by bequeathing to her through legacy a small farm, consisting of five and one-half *arourae* of grain land, as well as a one-half share of a house. Here the testator undertook to provide security in terms of a source of income and a place to live; this point remains valid even if, as is highly likely, the modest parcel of

---

103. For a discussion of the privileges of Antinoite citizenship, see P. Schubert, *Les archives de Marcus Lucretius Diogenes et textes apparentés* [= *P.Diog.*] (Bonn, 1990), 24–33. See also Bowman and Rathbone, *JRS* 82 (1992): 199–20 and n. 68, as well as Zahrnt, *ANRW* II 10, no.1 (1988): 669–706, especially 690 ff.

104. This text, a report about the opening of the will, has also recently been published as *P.Test.Roma.*² 20.

land was not her sole source of income.[105] We should imagine that this parcel filled the same function for Lucretia Octavia as the *fundus* that Pliny gave to his nurse did for that individual. Both women depended on property given them to maintain a decent standard of living; in exploiting their agricultural property, both will have been concerned first and foremost that their land furnished a stable yearly income.

A similar concern to provide security for a dependent must have influenced another testator, C. Longinus Castor, a veteran of the Misenum fleet (*P.Test.Roma.*[2] 12, A.D. 189–94).[106] In his will Castor manumitted two female slaves, named Marcella and Kleopatra, and he made each one heir to half of his estate. Castor further sought to provide for a daughter of Kleopatra named Sarapias. He manumitted her in his will, and he provided a legacy to assure that she would continue to have a source of income as well as lodging. This legacy included several parcels of farmland, including five *arourae* of grain land at the village of Karanis (in the Fayum), a one-third share of a date palm grove, and one and one-fourth *arourae* of a ravine, as well as a two-thirds share of a house (i.17 f.). Again, we cannot be certain that this land would have provided the woman in question her sole source of income; she may have had other sources, such as a trade or craft in which she was trained. But the purpose behind this bequest was clearly to provide the freedwoman economic security. The bequest of land was qualitatively much different from the other bequest that Castor made, namely, one

---

105. On the number of *arourae* needed for subsistence in Roman Egypt, see R.S. Bagnall, "Landholding in Late Roman Egypt: The Distribution of Wealth," *JRS* 82 (1992): 128–49, at 138, who posits that ten *arourae* of land could support an average household of five persons. Clearly, no exact figure can be offered for the minimum land needed to support a single individual, and the widow in the will discussed in the text was presumably to be maintained above the subsistence level. We should imagine that Lucretia Octavia owned some land in her own right.

106. This text, a Greek translation of a Latin original and therefore not the actual will, was first published in *BGU* I 326 and was reprinted in *M.Chr.* 316, *FIRA* III no. 50, *New Primer* 50, and *Sel.Pap.* I 85. There is also an English translation in N. Lewis and M. Reinhold, *Roman Civilization*, 3d ed., vol. II (New York, 1990), 191–93. For discussion of this will, see Amelotti, *Il testamento romano*, 49 f. For discussion of the social background to this will and the relationship between Castor and his beneficiaries, see J.G. Keenan, "The Will of Gaius Longinus Castor," *BASP* 31 (1994): 101–7, as well as Alston, *Soldier and Society in Roman Egypt*, 127–29.

of four thousand sesterces to a relative named Iulius Serenus (ii.17 f., a codicil). To judge by the size of these legacies, which in Roman law cannot have constituted more than three-fourths of the value of the estate, Castor was at least a moderately comfortable landowner. In all likelihood, his estate was much larger than a quarter of the wealth represented by his legacies. In any case, the legacy to the freedwoman was not simply a distribution of wealth, as a simple payment of cash would be. Instead, the testator probably devised a legacy of grain land with a share in a date palm grove, the latter of which might provide a much higher return for each unit of land than grain land, as a means of providing the beneficiary with the securest income possible for the long term.[107]

The testator might also undertake to provide this type of security for the heirs instituted in the will. In one example, a veteran named C. Iulius Diogenes allotted specific agricultural properties to each of his two children, C. Iulius Diogenes and Iulia Isarous, although they were apparently not his heirs (*P.Test.Roma.*² 7, ca. A.D. 127–48).[108] A fragmentary document preserves a Greek translation of Diogenes' will, but despite gaps in the text, it seems possible to establish what was included in each of the legacies. The son Iulius Diogenes apparently was to receive at least two important parcels of land, including one consisting of six *arourae* of grain land, and a second one also encompassing six *arourae*; this latter parcel was probably a vineyard, since the bequest also included a reed plantation of one-fourth *aroura* (lines 8–10).[109] The

---

107. Several wills of soldiers are preserved on papyri, and in such wills, the testators, who probably did not own significant amounts of land, commonly restricted their bequests to cash. See, e.g., the will of Antonius Silvanus, a cavalry officer at Alexandria (*P.Test.Roma.*² 5, *FIRA* III 47, A.D. 142). Silvanus named as heir his son M. Antonius Satrianus, but, among other dispositions, he left legacies in cash of 500 denarii to his mother, Antonia Thermoutha, and of 750 denarii to his brother Antonius R[-], heir in the second degree, to be paid to him as long as he should not be instituted as heir. On this will, see Amelotti, *Il testamento romano*, 38–39. Cf. *CPR* VI 76 (*P.Testa.Roma.*² 18, II–III c.), a Greek translation of a Roman soldier's will, in which the legacies are all in cash.

108. I have discussed this document earlier in this chapter under "Providing for Dependents." This document was originally published as *P.Select.* 14 (*Pap.Lugd.Bat.* XIII). The mother of the children, Iulia Primilla, was formerly the testator's slave.

109. *P.Test.Roma.*² 7, lines 8–10. Reed plantations were commonly sold or bequeathed in association with vineyards; the reeds were used to train the vines.

share of the daughter, Iulia Isarous, seems to have included ten *arourae* of grain land, a date palm orchard of two *arourae*, a one-third share of a barn or granary (*pyrgos*), and a share in a house (lines 16–20). If the son also received a share in a house, then the testator in this case again sought to provide each of his children with both shelter and a continuing source of income. The testator diversified the types of land bequeathed to each child, combining grain land with vineyard land, and in so doing he afforded each of his heirs the greatest security possible under the circumstances. The testator also provided for his wife (or more properly, his *contubernalis*), Iulia Primilla, who had earlier been his slave. The will included a provision whereby the two children were to feed and clothe her, presumably for her lifetime (lines 25 f.).[110]

We might compare the dispositions in the will of C. Iulius Diogenes with those made by an individual named Aurelius Hermogenes, alias Eudaimon (*P.Test.Roma.*² 24, A.D. 276).[111] Hermogenes was a member of the provincial elite in Roman Egypt, since when he wrote his will he was *exegetes* and *prytanis* at Oxyrhynchus, as well as a councillor there.[112] His wife, Aurelia Isidora, alias Prisca, had the rank of *matrona stolata*, designating her as an individual of equestrian rank.[113] Hermogenes was the father of a large family, leaving behind a wife and five children, including three sons, the Aurelii Hermeinos, Horion, and Herakleides, and two daughters, the Aureliae Ptolemais and Didyme.

---

110. The son may also have become a soldier. *SB* I 5217 (*FIRA* III 6, A.D. 148, Theadelphia) is a copy of an *epikrisis* protocol concerning C. Iulius Spurii f. Diogenes and his suitability for army service; this document includes (1) the *epikrisis* document of A.D. 104–5 of the patron of Iulia Primilla, C. Iulius Diogenes (the testator in the document discussed in the text); (2) her own *tabula manumissionis* (A.D. 127–28); (3) a sealed tablet attesting the birth of C. Iulius Spurii filius Diogenes and Iulia Isarous. The point was to prove that the son was eligible to become a soldier; see *P.Select.* 14, pp. 37–38. See also Amelotti, *Il testamento romano*, 41–42 and 41 n. 4.

111. This document was first published in *P.Oxy.* VI 907; cf. *M.Chr.* 317, *FIRA* III 51. I have discussed this document earlier in this chapter under "Providing for Dependents."

112. See A.K. Bowman, *The Town Councils of Roman Egypt* (Toronto, 1971), 132.

113. For the significance of the term *matrona stolata*, see B. Holtheide, "Matrona Stolata–Femina Stolata," *ZPE* 38 (1980): 127–34; cf. Rathbone, *Economic Rationalism*, 47–48, who argues that the title was accorded to women with the wealth appropriate to the equestrian order, not just to wives or daughters of equestrian procurators.

He wrote his will in such a way as to provide as much economic security as possible for every member of his family. Accordingly, he made all his children heirs, but, following a practice of Greco-Egyptian wills, he allotted his property specifically among the individual heirs.[114] In so doing, Hermogenes apparently followed the practice that we have traced of seeking to assure each heir access to a variety of different types of farmland, including both vineyards and grain lands. In this way Hermogenes minimized the financial risks for each heir to the extent that it was possible for him to do so.

Thus the three brothers received common ownership of a house in the city of Oxyrhynchus as well as of a vineyard and grain lands at several villages outside the metropolis (lines 7 ff.). Unfortunately, the fragmentary document does not preserve the amount of land bequeathed to the brothers, so it is difficult to determine what level of wealth the children would enjoy.[115] Hermogenes made an additional bequest of grain land and a slave to his son Hermeinos, no doubt because he was the eldest son (lines 10 f.). Hermogenes also provided for the financial security of his daughters and wife. To the daughters he likewise left common ownership of a vineyard and grain lands (lines 11–13).[116] The testator also provided his daughters with dowries, giving one daughter a dowry of four talents, and confirming the gift of a dowry (probably also four talents) that he had already bestowed on the other daughter. It is significant that two of the sons and one of the daughters were underage, so their affairs had to be managed by a guardian, that is, an *epitropos* (lines 18 f.). Having the grown children share their inherited property with their younger siblings provided one way to minimize the chance that the young children would lose their primary source of

---

114. For the interpretation of the specific bequests to heirs as a continuation in a Roman will of Greco-Egyptian testamentary practices, see Amelotti, *Il testamento romano*, 61–62; cf. Rowlandson, *Landowners and Tenants in Roman Egypt*, 169. Alternatively, the specific bequests to individual heirs could be described as legacies *per vindicationem*.

115. The vineyard and grain lands (?) are described as bequeathed with all equipment and appurtenances: . . . καὶ χρηστήρια καὶ συνκυροῦντα πάντα . . . (line 9); for the relationship of this formulation to the Roman legacy of *fundus cum instrumento*, see Steinwenter, *Fundus cum instrumento*, 40–69, especially 48–52.

116. Cf. the description of the bequest made to the daughters: . . . ὃ ἔχω κοινὸν πρὸς τὸν αὐτ[ὸν . . . ἀμπελικὸν] χωρίον καὶ σειτικὰς ἀρούρας πάσας καὶ προχρείας καὶ χρηστήρια καὶ συνκυροῦντα πάντα . . . (lines 12–13). For the meaning of προχρεία, see the translation in *P.Oxy.* VI 907.

income, since the elder heirs were presumably in a position to make sure that this property was managed suitably. The two older children, Hermeinos and Ptolemais, received individual slaves (lines 11, 14–15), while the other children were allotted four slaves to share among them (lines 15–16). Hermogenes provided for his wife by leaving her grain land in accordance with a previously existing agreement, as well as of course the dowry that she had brought into the marriage (lines 16 f.). Finally, Hermogenes sought to provide a yearly allowance of wine and grain for a friend, Aurelius Dionysammon. To accomplish this purpose, he imposed a trust whereby the allowance was to be paid every year from the lands situated at a particular village, Moa (lines 23–24). The fragmentary condition of the document does not allow us to determine to which heirs these lands were bequeathed.

We might compare this treatment of land with the dispositions made by a woman also belonging to the wealthier level of society in Roman Egypt, to the extent that these can be reconstructed on the basis of a fragmentary will (*P.Test.Roma.*[2] 23, *P.Princ.* II 38, A.D. 264). Aurelia Serenilla, alias Demetria, who wrote her will sometime after the middle of the third century A.D., was from a family in the bouleutic class at Hermopolis. Her father, Philippianos, alias Kopreos, was a member of the town council of Hermopolis, while her husband, Aurelius Herminos Achilleus, was *eutheniarch* and *kosmetes* there. One indication of the social circle in which Aurelia Serenilla lived is the fact that her curator assisting her in drawing up her will was a veteran, named Aurelius Valerius Longus. Her will is somewhat different from the other wills discussed, since she made no provision for her sons, whom she disinherited. Her concern instead was for her mother Aurelia Asklatarion, alias Koprilla, whom she made heir, and for her husband, to whom she left a legacy of land. This legacy consisted of several modest parcels, apparently of grain land, encompassing seven, five, and three *arourae* (lines 6 f.). Certainly the husband had far more substantial holdings in land, but the legacy of land that he received provided him with an additional source of income, and it was one that he could bequeath to the couple's children, who were presumably to be the principal heirs of the father's estate. Precisely how the surviving mother was to be maintained is unfortunately impossible to determine on the basis of the extant document.[117]

A division of a large inheritance of landed property in third-century

---

117. For discussion of this will, see Amelotti, *Il testamento romano*, 59.

Hermopolis provides the clearest evidence that upper-class landowners understood the economic value of an estate in much the same way as the jurists. A lengthy but somewhat fragmentary document records how three brothers and one sister from a wealthy family in Hermopolis— Claudius Eudaimon, Claudius Isidorianos, Claudius Theon, and Claudia Theodora—divided among themselves a considerable inheritance that they had received from their parents and a deceased brother (named Maikenas, *P.Flor.* I 50, A.D. 268).[118] The lofty status of the family is revealed by the fact that one brother, Claudius Eudaimon, was a member of the town council at Alexandria. The family estate at Hermopolis apparently consisted of at least several hundred *arourae,* and it included grain land, substantial vineyards, and orchards—all scattered among several villages—as well as a number of houses in the city of Hermopolis and in the villages. As was characteristic of Egyptian estates in this period, the family's holdings included numerous properties scattered among various villages in the nome, as well as a central facility. It is quite possible that this family, like the wealthy Alexandrian Aurelius Appianus, owned property in other nomes as well.

The four siblings apparently undertook to divide the property evenly, although the fragmentary state of the surviving document makes it impossible to prove this. It is noteworthy that the siblings in many cases agreed to own and cultivate in common various parcels of land, in particular vineyards, and also to maintain common ownership of the reed plantations and the irrigation facilities needed to support the vineyards. In this connection, it is significant that the family agreed to maintain common ownership of the estate's central facility, the *epoikion* of Nearchos, located at the village of Psobthos Chenarsiesein (*P.Flor.* I 50, lines 102 ff.). This place included facilities for storing crops, winepresses, jars, a drying area, a dovecote (for fertilizer), shops for producing storage jars, and extensive irrigation facilities. The irrigation facilities served what was apparently the prize property that the family owned, a vineyard at the same village with date palms and fruit trees, comprising some forty-two and one-half *arourae,* which the four siblings agreed to own and operate in common (cf. *P.Flor.* I 50, lines 9–10, 44–45, 86–87). The document includes specific provisions that each member of the family was to have access to this central facility, particularly to the irrigation installations.

---

118. On this document, see M.H. el Abbadi, "P.Flor. 50: Reconsidered," *Proceedings of the XIV International Congress of Papyrologists* (London, 1975), 91–96.

The dispositions that the family members made for dividing their inheritance demonstrate quite clearly how they conceived of the economic value of an estate. These people were concerned that each one of them be provided with the greatest level of financial security possible, so each individual received a share of all the types of land comprising the estate. The division of the estate served to guarantee each sibling's access to the productive resources needed to produce a marketable harvest of wheat, wine, and orchard products. There was no consideration of consolidating holdings or of one member of the family selling his or her share to another to invest the proceeds elsewhere. The family had worked long and hard to acquire and develop its estate, and when individual parcels could not be divided without compromising their productivity, they were to be owned and cultivated in common.

The various dispositions that I have described suggest that security was far more important to landowners in Roman Egypt than the hypothetical value of their land as a freely exchangeable commodity. To some extent bequests of agricultural land were designed to provide the same type of security represented by the bequests of yearly allowances of food and cash for maintenance so characteristic of wills from Roman Egypt. The point in bequeathing land was not principally to pass on a commodity that could be freely traded or improved to increase the recipient's wealth or even his or her capacity to make a profit, but rather to assure to the extent possible that the recipient would always have the income in foodstuffs and cash required to maintain his or her standard of living.[119] This principle, implicit, as I have hypothesized, in the wills I have already discussed, finds explicit expression in the will of a wealthy Oxyrhynchite from the mid–third century A.D. (*P.Test.Roma.*² 26, *P.Oxy.* XXVII 2474).[120] The beginning of the will is lost, so the name of the testator is not preserved, but enough of the specific provisions of the will survive to make clear how the testator intended his heirs to benefit from the land that he was bequeathing to them. The surviving part of the will begins in the specific division of the

---

119. Saller, *Patriarchy, Property and Death in the Roman Family*, 131, citing comparative evidence, argues that testators made detailed provisions for the support of surviving parents in order to protect against the danger of neglect on the part of their children.

120. See L. Koenen's general discussion of the organization of the will in *P.Oxy.* XXVII 2474, introd., pp. 155 f. This will is also published in Amelotti, *Il testamento romano*, no. 17, with discussion on pp. 65–67.

property among the heirs. The extant document preserves bequests to a daughter (lines 1 ff.); to the testator's wife, Aurelia Chaeremonis (lines 5–13, 15–20); and finally to his mother, Asklatarion (lines 13–15). These dispositions are followed by the appointment of a tutor for the underage sons and a series of *fideicommissa*.[121]

Although the name of the testator in this will is not preserved, it is clear that he belonged to the wealthier class of people resident at Oxyrhynchus. A certain indication of the social standing of the family derives from the identity of the individual appointed as guardian of the testator's underage heirs. This individual, Aurelius Achillion, the son-in-law of the testator, was a member of the bouleutic class at Oxyrhynchus, since he is described as a former *prytanis* of the city (lines 21–22).[122] The wealthy testator sought to provide for the welfare not only of his heirs but also of a variety of dependents. The conditions under which he manumitted two household slaves are of legal interest, since the form of manumission combines the Greek practice with the Roman one, with the slaves remaining under the power of the wife until her death (lines 28–31).[123] He bequeathed some farmland to an individual named Epimachos, described as the testator's *epitropos*, or steward. Epimachos was the general manager of the testator's property, and the testator urged him to remain on in his post and perform the same service for the testator's sons. As compensation for this ser-

---

121. Again, the division of property is probably in keeping with the Greco-Egyptian custom of allotting to the heirs specific properties (see Amelotti, *Il testamento romano*, 66 n. 1), rather than a legacy in the Roman pattern as Koenen (*P.Oxy.* XXVII 2474, p. 156) suggests. On the *fideicommissa*, see Amelotti, 66.

122. See Koenen's discussion of the possibilities in identifying the family in *P.Oxy.* XXVII 2474, pp. 159 f. Koenen suggests that the mother Asklatarion might be the Dionysia, alias Asklaratarion, who, along with an individual named Quintus Mareinos Klaudianos, delivered grain to the *sitologoi* of the village of Paomis in the Thmoisepho toparchy (*P.Oxy.* XII 1541, A.D. 192); this woman in turn was probably the mother of Aurelius Herakleides, alias Ptollas, son of Diogenes, himself *sitologos* in the village of Sepho in the same toparchy in A.D. 236/37 (*PSI* X 1121). This latter individual could be the testator in the present will, although this identification is far from certain. For the possibility that the testator's wife was a Christian (based on the testator's description of the woman's attitude toward God and humankind in line 6), see Koenen's discusion at *P.Oxy.* XXVII 2474, p. 156.

123. See Koenen's discussion of the legal issues connected with this manumission, in *P.Oxy.* XXVII 2474, line 29 n.

vice, the testator provided a yearly salary, or an *opsonion*, the details of which are not preserved (lines 36–42).[124] Also, the testator requested that his sister Theognoste remain with his wife, for which service she was to have a yearly allowance of twenty-four *keramia* of wine, to be paid by the heirs and their guardian (lines 31–36).

Significant for our purposes are the dispositions that the testator made for his wife and children. The wife was to receive lodging, in the form of half of two houses (lines 8–9),[125] a one-quarter share of some land (line 10), and a share of a hay barn and apparently also of a grain barn (line 11), as well as a yearly allowance of foodstuffs. She was also to keep a number of slaves that the testator had purchased and then assigned to her name (lines 18–20). In other words, the wife was to be able maintain the standard of living to which she had become accustomed and one that was appropriate to her social station. Accordingly, the yearly food allowance was quite generous, consisting as it did of far more food than a single individual could consume in one year; it included one hundred *artabae* of wheat, one hundred *keramia* of wine, twenty-five *keramia* of another commodity whose name is lost, and six *artabae* of vegetable seed (lines 15–18).[126] Presumably this allowance would help the wife to maintain her servants.

The provisions made for the testator's sons and heirs, who were at least three in number,[127] would have been in the lost beginning part of the document, but it seems certain that they received the bulk of the testator's land. The testator revealed how he conceived of these lands as sources of income when he described the duties of the guardian whom he appointed for his still underage sons. The guardian's role consisted of managing these lands until the sons came of age: "When they have come to the proper age, all that I have instructed shall be bestowed on them together with all future rents in money and in dry and liquid

---

124. Koenen (*P.Oxy.* XXVII 2474, lines 36–42 n.) argues that Epimachos was required to remain and work for the family under a sort of *paramone* contract to receive the land bequeathed to him. For the meaning of the term *epitropos* used here, see ibid.

125. One of these houses is called the *logisterion*, or "countinghouse" (line 9).

126. See Koenen's discussion at *P.Oxy.* XXVII 2474, lines 17–18 n.

127. That there were at least three heirs is indicated by the provision for the share of any son dying childless or intestate to pass on to the remaining sons (lines 25–28); see Koenen's discussion at *P.Oxy.* XXVII 2474, introd. p. 156.

measure which shall be produced from now on" (lines 23–25).[128] The testator had clearly envisioned the lands that he was bequeathing to his sons only as sources providing a fixed level of income, which consisted of rents in kind (*ekphoria*) and in cash (*phoroi*). The economic assumption underlying these instructions to the sons' guardian is that the value of this agricultural property was to be measured primarily in terms of the rents immediately available from them, and it is assumed that the income for the testator's sons would consist of rents, both while they remained minors and after they reached adulthood. These rents no doubt would provide the sons with more wealth than the allowances made to the mother, but the principle behind this and similar bequests of agricultural land is not completely different. The guardian was envisioned as exploiting these lands in but one way, by leasing them out to tenants, for rents that probably remained stable over a long period of time. The only "investment" of resources that the guardian was to make in these lands was that required to keep the rents from falling. The financial commitment of the owner of such lands, then, was to invest only what was needed to keep the lands in such a condition that they could be leased out for the same rents that they customarily provided. As far as the writer of this will is concerned, then, "wealth" consisted primarily of having access to resources providing a stable yearly income. The achievement of this income, moreover, depended on maintaining the capacity of agricultural land to furnish an adequate yearly harvest. In practical terms, the landowner in Roman Egypt was concerned with maintaining the irrigation facilities on the land so that cultivating a crop was possible. Beyond that, it was not likely that the writer of this will, the guardian appointed for his sons, or even the sons themselves, after they came of age, would take further initiative in improving the property.

Conclusion

The jurists whose decisions I have discussed in connection with the legacy of a *fundus cum instrumento* could not have accounted for all the uses to which a landowner might put property, nor could they account

---

128. The translation is that of Koenen, in *P.Oxy.* XXVII 2474: πρός τε ἡλικίας αὐτῶν γενομένων ἀποκατασταθῆναι ἑκάστῳ / αὐτῶν τὰ ὑπ' ἐμοῦ διαταγέντα αὐτοῖς σὺν τοῖς ἐσομένοις ἀπὸ τοῦ νῦν ἀπὸ τῶν ὑπαρχόντων φόροις τε καὶ ἐκφορίοις ὑγρῶν τε καὶ ξηρῶν.

for the varying attitudes toward risk and investment that testators and legatees might display. In formulating legal rules about legacies, then, the jurists did not consider all the many possible uses to which a Roman estate owner might put his or her land. But behind the simplified reality of the jurists lay a basic understanding of the economic interests of the parties involved in a legacy, and the principle underlying these economic interests involved a very restricted range of ways in which an upper-class Roman might achieve an income. The landowners from Roman Egypt whose wills I have examined seemed to have shared this basic understanding of the value of agriculture as an investment. The conception of agriculture as an investment that the jurists display in the law of legacy is the same as the conception they display in the law of tutorship. For upper-class Romans, the "profit" from an investment in agriculture could be measured chiefly in the security that it provided.

In formulating rules for the legacies of estates, the jurists were primarily concerned with defining an estate that could reasonably be expected to achieve a satisfactory level of physical productivity. The principle behind this approach to adjudicating legacies was that the landowner could most readily measure financial security in terms of access to the resources needed to produce a marketable harvest. The landowners from Roman Egypt whose wills I have examined seemed to have shared this basic understanding of economic security Although the geographical circumstances of Roman Egypt changed the conditions under which a marketable harvest might be produced, landowners there had the same basic conception of investment and profit as their counterparts elsewhere in the empire. Accounting for market conditions had little place in the way that the jurists regulated the bequest of agricultural property, but it seems clear that market conditions were a secondary consideration for Romans with little choice other than to invest their wealth in agriculture; their primary concern was to have the wherewithal to produce the crops on which their income and their security depended. To judge by the situation of the type of legatee typically envisioned by the jurists, this level of physical productivity was to be achieved with a minimum of investment, especially the type of investment that might pose risks.

The concern for security that I have ascribed to upper-class Romans is of fundamental importance for understanding the Roman economy, because it shaped the relationships among all the groups involved in the Roman agrarian economy. It particularly shaped the relationships between landowners and their tenants, both the relatively wealthy tenants and the numerous small-scale tenants on whose cultivation of the land most landowners depended for their incomes. The various arrangements that landowners made with their own slave or hired labor or struck with tenants had one overriding purpose. They were designed to keep risk at a minimum while assuring upper-class Romans access to the resources needed to produce the crops on which their income and security depended.

3
The Juristic Farm Tenant and the Management of Estates

The problem now to be addressed concerns how the overriding concern for economic security underlying upper-class Roman economic planning affected how Roman landowners contended with the legal, social, and economic institutions surrounding farm tenancy. Many upper-class landowners derived their income from leasing out to tenants their diverse and often geographically scattered holdings. Insofar as their social privileges depended on that income, the changing legal and economic relationship between landowner and tenant was one of the principal factors in defining the economy and the social structure of the Roman Empire.

The purpose of the following chapters, then, is to analyze the dynamic relationships that existed between landowners and tenants, by tracing how the Roman legal authorities conceived of the economic interests of these groups as they regulated the institution of farm tenancy. My argument is that the jurists applied the same basic assumptions about the Roman economy that we have inferred in their treatment of tutorship and the bequest of estates. Accordingly, my methodology in this investigation is similar to the one used in the preceding chapters, in that I seek to determine what "legal climate" surrounded farm tenancy in the Roman Empire.[1]

The most extensive source of evidence for economic relationships

---

1. For this type of approach in an analysis of the Roman building industry, see S.D. Martin, *The Roman Jurists and the Organization of Private Building in the Late Republic and Early Empire* (Brussels, 1989), 10–17, 138–40. L. Capogrossi Colognesi, *Ai margini della proprietà fondiaria* (Rome, 1995), 223–25 and n. 80 and idem, "Il regime degli affitti agrari," *Scienze dell'antichità, Storia, Archeologia, Antropologia* 6–7 (1992–93): 209–10 and n. 80, takes a skeptical approach toward using the writings of the jurists as evidence for economic history: he argues that little can be concluded about the historical conditions surrounding farm tenancy on the basis of the frequency of the types of cases treated by the jurists.

137

arising from farm tenancy is provided by the legal responses and other writings of the Roman jurists, compiled in the Digest, especially but by no means exclusively in section 19.2 on *locatio-conductio*, or "lease and hire." The problem with interpreting the legal sources on farm tenancy is similar to the one that we encountered with the legal sources on tutorship and legacy: it is not immediately clear to what extent such legal evidence can be used to describe historical conditions in the Roman Empire, whether in early imperial Italy or in the provinces. In formulating the legal rules for farm tenancy, the Roman jurists consistently assumed the existence of a normative farm lease, based on abstract conceptions of the landowner and the tenant, which to some extent simplified and idealized reality.[2] Accordingly, the jurists cannot be considered to have described actual historical conditions; rather, their idealized picture of the farm lease served as a legal concept designed to decide legal issues in the real world.

We should keep in mind that farm tenancy, like tutorship and the bequest of estates, was an area of the law that impinged directly on the economic interests of the classes to which the Roman jurists belonged.[3] Therefore, it seems to be a reasonable hypothesis that when the Roman jurists formulated these legal rules, they were affected by their understanding of contemporary agrarian conditions and the general economic background against which both landowners and tenants pursued their respective interests. In the following investigation, I seek to show that the juristic sources inform us about how upper-class landowners went about the task of managing their estates to achieve economic security. Consequently, the assumptions that the jurists made about the Roman economy when regulating farm tenancy allow us to infer the existence of basic economic relationships affecting landowners and tenants, even if we cannot document how prevalent particular forms of land tenure actually were.

2. For discussion of the jurists' use of "ideal" types, see the works of B.W. Frier in the bibliography, with special reference to farm tenancy (discussed in my introduction), and S.D. Martin, *The Roman Jurists and the Organization of Private Building*.

3. But cf. the view of A. Steinwenter, *Fundus cum instrumento* (Vienna and Leipzig, 1942)—discussed in my chap. 2—that the Roman jurists took an exceedingly conservative approach toward the economy, applying a static picture of economic conditions and relationships to a reality that was far more complicated. Cf. the more sanguine view of P.W. de Neeve, *Colonus* (Amsterdam, 1984), 25–27, discussed in my introduction.

Whether it is appropriate to use the legal sources as evidence for economic history can only be determined after we address two additional problems of interpretation. The first concerns the economic status of the farm tenant treated by the jurists. We can use the evidence from the Digest as a basis for drawing general conclusions about the Roman agrarian economy only if it can be shown that the Roman jurists addressed legal issues affecting a broad range of tenants. In contrast to tutorship and legacy, the law of tenancy affected relationships between parties of different social classes, including both upper-class landowners and tenants of modest social and economic status. Do the writings of the jurists inform us at all about these modest tenants?

The sources in the Digest do not generally inform us directly about the economic status of tenants envisioned in the cases under consideration. The reason for this is already familiar to us. The responses in the Digest, when they are based on actual legal cases, are usually stripped of any detail not germane to the legal point at issue; the result is that the responses are formulated as general principles useful for judges and parties in future legal disputes.[4] When the jurists do provide details about cases, they usually suggest that the tenants envisioned were leasing on a relatively large scale, that is, that they were leasing entire *fundi* with villas attached. In addition, we should expect that, in their treatment of farm tenancy, the jurists had to devote a great deal of attention to the wealthier tenants, the group most likely to avail itself of the legal system. Frier has argued this position persuasively in his study of Roman law's treatment of urban tenancy.[5] Did the jurists concern themselves exclusively with this class of tenants, as Scheidel has

---

4. For the way in which the general formulation of imperial rescripts enhanced their applicability, see D. Nörr, "Zur Reskriptenpraxis in der hohen Prinzipatszeit," *ZRG* 98 (1981): 1–46, at 41, as well as W. Turpin, "Imperial Subscriptions and the Administration of Justice," *JRS* 81 (1991): 101–18.

5. See B.W. Frier, *Landlords and Tenants in Imperial Rome* (Princeton, 1980), 39–47, and, for the argument that the Roman law of tenancy concerned primarily the "upper stratum" of tenants, his reviews in *ZRG* 100 (1983): 667–76, at 674 and n. 34, and *ZRG* 102 (1985): 564–69, at 566. Cf. R. Zimmermann, *The Law of Obligations* (Cape Town, 1990), 348–50, who accepts the conclusions of Frier. See also P.W. de Neeve, "A Roman Landowner and His Estates: Pliny the Younger," *Athenaeum* 78 (1990): 363–402, at 383–88, who argues that Pliny's tenants were "substantial" farmers (387). Capogrossi Colognesi, *Ai margini della proprietà fondiaria*, 225–37 and "Il regime," 210–16, emphasizes that Roman tenants represented a wide range of economic status.

recently suggested? In his analysis of the sources for farm tenancy in Roman Italy, Scheidel argues that the literary and legal sources provide evidence almost exclusively for the relations between the very wealthiest tenants and their landlords. In the case of the legal sources, only those wealthy tenants would have been in a position to settle differences with their landlords in a court of law. As a consequence, we remain largely uninformed over the conditions under which the vast majority of tenants in Roman Italy cultivated their land. It follows that the juristic sources, insofar as they did not address issues affecting the large numbers of humble tenants on whose careful cultivation of the land most landowners depended, can at best give us a limited account of farm tenancy in the Roman Empire.[6]

I shall argue in this and in the following chapter that the treatment of tenancy by the jurists reveals two aspects of estate ownership crucial to the interests of upper-class Romans. First, it seems clear that the Roman law of tenancy developed out of legal relationships between landowners and a relatively narrow range of tenants, those wealthy enough to negotiate their contracts freely with landowners and use the legal institutions of the state to defend their interests. When we examine how the jurists envisioned relationships between landowners and these "normative" tenants, we can trace, in the jurists' estimation at any rate, what economic goals landowners pursued in leasing their estates out to tenants. Second, however, it also seems likely that in treating tenancy, the jurists went beyond the legal issues concerning a few wealthy tenants. In theory, of course, the same law applied to all farm tenants, irrespective of social class. More important, the financial interests of upper-class Romans depended on their relationships with a broad range of tenants, including those of relatively modest status. It would be surprising if the jurists did not consider legal problems that the landowner might encounter when dealing with such tenants. A closer analysis

---

6. See W. Scheidel, *Grundpacht und Lohnarbeit in der Landwirtschaft des römischen Italien* (Frankfurt, 1994), 27–117. For other discussions of the difficulties that average people encountered in using the Roman law courts and, by implication, the unlikelihood that their concerns would be considered by the jurists, see J.M. Kelly, *Roman Litigation* (Oxford, 1966); P. Garnsey, *Social Status and Legal Privilege in the Roman Empire* (Oxford, 1970), especially 181–218; R.P. Saller, *Personal Patronage under the Early Empire* (Cambridge, 1982), especially 153–54, for the influence of patronage on the law courts; and R. MacMullen, *Corruption and Decline of Rome* (New Haven, 1988), 84–96, 132–37. See also more generally M. Peachin, *Iudex vice Caesaris* (Stuttgart, 1996), 10–91.

confirms that the jurists considered issues impinging significantly on the financial interests of landowners leasing land to a broad range of tenants. As I will argue in chapter 4, the jurists' persistent concern with legal issues connected with the tenant's security of tenure and the right of a tenant to claim a remission of rent can best be explained as the product of a continued effort to define the legal rights and obligations of small-scale tenants.

The legal evidence is difficult for the economic historian to interpret because the jurists consistently formulated legal relationships in abstract terms, describing a wide variety of arrangements in terms of conventional categories that masked the dynamism of landowner-tenant relationships. In the jurists' logic, legal principles formulated in terms of a tenant leasing a *fundus* with a villa also applied to a tenant leasing a small parcel with a humble cottage as its place of residence. We can freely admit that the terminology and legal categories used by the jurists owe their origin to cases involving relatively wealthy tenants, but this situation does not mean that the jurists did not have to deal with legal disputes involving the tenants of more humble means who made up the bulk of the farming population on the estates of Roman landowners. Despite their conservatism, then, the jurists were concerned with defining the rights and obligations of both landowners and tenants in response to the realities of the Roman economy. In this effort, the jurists displayed a coherent understanding of the economic interests of these two groups. The rules that the jurists developed as a result of this understanding provide a basis for analyzing the contribution of farm tenancy to the Roman economy.

This conclusion about the broad applicability of the evidence preserved in the Digest receives confirmation from comparative evidence for farm tenancy in Roman Egypt. As I have argued elsewhere, the papyrological evidence suggests that farmers of surprisingly humble status and modest means in Roman Egypt sought and gained the protection of legal institutions in that province when dealing with their landowners.[7] The evidence from Roman Egypt indicates that the relationships between even wealthy landowners and tenants of limited means were conducted in strict conformity to the terms of a normative lease contract. Consequently, extralegal influence the landowner might

---

7. See my "Legal Institutions and the Bargaining Power of the Tenant in Roman Egypt," *APF* 41, no. 2 (1995): 232–62.

bring to bear against his tenant was often exercised in the context of the lease contract, which was subject to enforcement by the state's legal institutions. This conclusion helps us to interpret many decisions preserved in the Digest and Justinian Code that are comprehensible only in the context of tenancy relationships involving tenants of modest status.

Using the legal sources as evidence for evaluating the institution of farm tenancy in the Roman Empire is subject to a second major complication: it is not initially clear to which geographical locations the writings of the jurists and the rescripts of the emperors about farm tenancy applied. The obvious conclusion is that the juristic sources for farm tenancy can only be used to analyze conditions where farm leases were drawn up in strict accordance with the norms of Roman private law, that is, in Italy. But matters were probably not so simple. It cannot be disputed that the writings of the classical Roman jurists were based to a large extent on legal problems in Italy, especially in the city of Rome. But on occasion, at least several jurists indicate their interest in provincial conditions in their writings.[8] For example, in his commentary on the provincial edict, Gaius discussed the circumstances under which a tenant could request a remission of rent in the event of *vis maior* (D. 19.2.25.6, 10 *ad ed. prov.*).[9] In this passage, Gaius carefully distinguished between the responsibilities of the tenant paying his rent as a fixed assessment in cash and the sharecropper. This passage represents the only direct mention of sharecropping preserved in the Digest, and it is tempting to see Gaius' concern with this form of land tenure as deriving from his provincial experience.[10] It also seems very likely that many

---

8. For the provincial influence on classical Roman law, see D. Liebs, "Römische Provinzialjurisprudenz," *ANRW* II 15 (1976): 288–362.

9. Gaius *D.* 19.2.25.6: "Vis maior, quam Graeci θεοῦ βίαν appellant non debet conductori damnosa esse, si plus, quam tolerabile est, laesi fuerint fructus: alioquin modicum damnum aequo animo ferre debet colonus, cui immodicum lucrum non aufertur. apparet autem de eo nos colono dicere, qui ad pecuniam numeratam conduxit: alioquin partiarius colonus quasi societatis iure et damnum et lucrum cum domino fundi partitur." For a discussion of *remissio mercedis*, see chap. 4.

10. Sharecropping is mentioned in the Justinian Code, however: see *CJ* 11.48.8.1 (Valentinian, Valens, Gratian, n.d.) and W. Goffart, *Caput and Colonate* (Toronto, 1974), 80. Sharecropping was the basic form of land tenure on the imperial estates in the Medjerda (ancient Bagradas) Valley in northern Tunisia: see my *The Economics of Agriculture on Roman Imperial Estates in North Africa* (Göttingen, 1988), chap. 5.

of the imperial constitutions preserved in the Justinian Code responded to petitions submitted by people in the provinces, yet the legal issues were always decided in terms of Roman law.[11]

On the basis of the available evidence, we cannot be absolutely certain that the writings of the jurists on farm tenancy considered conditions in any area of the Roman Empire outside of Italy. However, as I will argue in chapter 4, the Roman legal authorities consistently sought to describe in conventional Roman terms the legal relationships arising from a wide variety of lease arrangements, many of which were based on local custom and could well have been situated in the provinces. It therefore seems likely that conclusions based on the jurists' treatment of farm tenancy do not have to be restricted in their applicability to early imperial Italy. But the more important point is that the assumptions on the part of the Roman jurists about the economic interests of landowners and tenants provide us with the same type of evidence for the fundamental economic relationships in the Roman economy as do the jurists' assumptions about investment in and profit from agriculture. From this perspective, it does not particularly matter if the Roman legal authorities adequately responded to the needs of provincial populations when they analyzed local forms of tenancy in terms of Roman norms. It is far more important that the jurists recognized the existence of particular economic relationships between landowners and tenants. Consequently, when we analyze the jurists' treatment of farm tenancy, we are not simply investigating a form of land tenure that corresponded precisely to the norms formulated in the Digest. Instead, the economic assumptions made by the jurists provide us with a starting point to develop a model describing the fundamental relationships characterizing the Roman economy in general. The basis for this model is the way in which the jurists endeavored to describe these relationships in terms of Roman private law. My assumption, derived from an analysis of papyrological evidence for the management of estates in Roman Egypt, is that the general conditions imposed by the ancient Mediterranean economy caused relationships between landowners and tenants to share certain basic characteristics throughout the Roman Empire. The model that I am developing, then, has a heuristic function, in that it facilitates the analysis of individual cases; it cannot describe how prevalent particular forms of tenancy were.

---

11. For the process of submitting petitions to the emperor, see n. 4 in this chapter.

It is therefore my contention that the legal sources provide a basis for analyzing two critical aspects of farm tenancy. First, the jurists' assumptions about the economic interests of landowners and tenants help us to appreciate how absentee Roman landowners confronted the daunting task of devising a system of managing their estates that would minimize their risk and at the same time assure as steady and stable an income as possible. The second aspect of tenancy that the legal sources allow us to investigate was also crucial to the economic interests of Roman landowners. Achieving the stable income desired by many Roman landowners would not have been possible had this group not been able to provide for the continued investment in farm improvements needed to keep their estates productive year after year. The Roman legal authorities made decisions affecting the allocation of investment, and their treatment of farm tenancy was characterized by a struggle to balance an insistence on describing legal relationships in conventional terms against the need to recognize diverse systems of land tenure that promoted investment by the tenant. Analysis of legal decisions concerning this issue allows us to trace how the Roman legal authorities mediated a constant dynamism between landowner and tenant, with each party seeking to shape the tenancy relationship to his or her advantage.

Managerial Assumptions about Farm Tenancy

The dependence for his or her income on scattered estates presented an upper-class Roman landowner with an immediate and monumental problem of management. Accordingly, two goals had to be met by a system of management that served the needs of upper-class Romans. First, the management of often extensive and far-flung estates had to be autonomous enough to function without the constant intervention on the part of the landowner. But more important, it had to have this characteristic while at the same time being able to provide continually for an investment of resources in such a way as to maintain the long-term productivity of the estate.[12] The major managerial consideration that

---

12. For a similar formulation, see Scheidel, *Grundpacht und Lohnarbeit*, 83. For the managerial problems faced by Roman landowners, see B.W. Frier, "Law, Technology, and Social Change: The Equipping of Italian Farm Tenancies," ZRG 96 (1979): 204–28, at 215, and Capogrossi Colognesi, *Ai margini della proprietà fondiaria*, 290–92 and "Il regime," 246–47, with special reference to the

Roman landowners faced, then, was not to channel their resources in such a way as to take immediate advantage of continually changing market conditions for agricultural products, since the capacity of the Roman agrarian economy to adapt to such changing conditions was circumscribed by the overall limited possibilities for investing safely large amounts of wealth. Rather, Roman landowners were more concerned with devising an autonomous system of managing their estates that could provide for the continual investment in productivity needed to maintain the ability of estates to produce an adequate marketable harvest year after year.

The difficulties involved in managing the types of properties characteristically owned by upper-class Romans were clearly recognized by the Roman legal authorities. For example, in connection with the Roman law on tutorship, Alexander Severus issued a rescript authorizing the appointment of curators to assist the tutor in the administration of a scattered series of properties, a "late diffusum . . . patrimonium" (*CJ* 5.36.3 = 5.62.11.1, A.D. 231). Even a landowner like Pliny could play only an indirect role in the management of his scattered properties. This restriction is especially noteworthy because Pliny was himself deemed enough of an expert on agricultural matters to be consulted regularly by his circle of friends and relatives when problems beset their estates. Pliny's visits to his estates were irregular; he seems particularly to have become involved in the management of his estates when crises arose requiring his immediate attention, such as when the *negotiatores* overbid for the vintages on his Tuscan estates (*Ep.* 8.2) or when continued problems of indebtedness on the part of his tenants led Pliny to abandon the traditional system of leasing for cash rents and to insti-

---

immense managerial difficulties faced by landowners leasing their estates out to tenants. M.I. Finley, "Private Farm Tenancy in Italy before Diocletian," in M.I. Finley, ed., *Studies in Roman Property* (Cambridge, 1976), 103–21, at 117–18 is skeptical about whether tenancy solved managerial problems. For the range of choices available to Roman landowners in managing their holdings, see P. Garnsey, "Non-Slave Labour in the Roman World," in P. Garnsey, ed., *Non-Slave Labour in the Greco-Roman World* (Cambridge, 1980), 41. On the management of estates through *vilici*, see J. Carlsen, *Vilici and Roman Estate Managers until A.D. 284* (Rome, 1995), as well as Capogrossi Colognesi, "Grandi proprietari, contadini e coloni nell'Italia Romana (I–III d.C.)," in A. Giardina, ed., *Società romana e impero tardo-antico*, vol. I (Rome and Bari, 1986), 325–65, 703–23, at 332–33.

tute sharecropping on these same estates (*Ep.* 9.37).[13] Despite Pliny's willingness on occasion to become involved personally in the management of his estates, it is nevertheless clear that he did so only because he was forced to and that, under ideal conditions, his interests were best served when the management of his estates functioned more autonomously. The advantages accruing from establishing an autonomous system of management are more clear-cut in the case of absentee landowners with little interest or capacity to manage their estates directly. This type of absentee owner is represented by several members of Pliny's circle, such as Cornelius Fabatus and Suetonius Tranquillus, who regularly consulted Pliny for advice in economic matters.[14] For the purpose of the following discussion, their involvement in the management of their estates will be treated as "characteristic," in the sense that, although individual landowners might treat their agricultural investments quite differently by becoming actively engaged in the management of their land, such activity should be considered the exception rather than the rule for upper-class Romans. For such landowners, farm tenancy provided one means to simplify this task of management.[15]

13. This impression about Pliny's activities on his estates is to some extent affected by the subject matter of his letters; Pliny of course writes about what is noteworthy, rather than about everyday matters affecting his finances. But in *Ep.* 9.36, where Pliny describes his daily routine on his Tuscan estates, he mentions that his major involvement in managing the economic side of his estates consisted of settling disputes involving his tenants: "Datur et colonis, ut videtur ipsis, non satis temporis, quorum mihi agrestes querelae litteras nostras et haec urbana opera commendant" (9.36.6). Cf. *Ep.* 7.30.3, where Pliny also describes his role in settling disputes with his tenants: "Accedunt querelae rusticorum, qui auribus meis post longum tempus suo iure abutuntur." These passages offer us every reason to believe that Pliny did not closely supervise the day-to-day management of his estates or, at the very least, that neither he nor his correspondents expected most landowners to do so.

14. See, e.g., *Ep.* 1.24, in which Pliny helps Suetonius in purchasing an estate, and *Ep.* 6.30, in which Pliny seeks a manager for a villa in Campania belonging to his grandfather-in-law Cornelius Fabatus.

15. These landowners need not have been absentee. For example, Pliny's nurse (*Ep.* 6.3) may have resided on or near the estate given to her for her support, but no matter where she lived, she did not supervise the estate's cultivation. In the view of Scheidel, *Grundpacht und Lohnarbeit*, especially 27–117, 131–49, 225, and 230, farm tenancy provided landowners in early imperial Italy a useful alternative to managing their estates directly through *vilici*.

In a manner consistent with the upper-class orientation of the legal sources, the Roman jurists viewed the normative farm tenant as an independent party freely entering into a contractual relationship with his landlord.[16] The farm lease was a consensual contract, and it assumed the good faith, or bona fides, of both parties. This contractual relationship imposed mutual obligations on both parties. At the center of the relationship, of course, was the farm, or *fundus*, whose value to both landlord and tenant was measured in its capacity to produce a crop, or *fructus*.[17] The landlord, or *dominus*, undertook to provide a farm in such a condition that the tenant could enjoy it ("ut ei frui liceat"), that is by cultivating a crop.[18] For his part, the tenant, the *conductor* or *colonus*, undertook to pay his rent, or *merces*, and to return the farm and its fixed capital in good order at the end of the lease. Consistent with the picture of the tenant as an independent party freely negotiating with his landlord is the short duration of the Italian farm lease, generally assumed in the legal sources to last five years (see, e.g., Paul. D. 34.3.16, 9 *ad Plaut.*, Ulp. D. 19.2.13.11, 32 *ad ed.*).[19] Maintaining the *fundus* and its capital meant that the tenant was

---

16. For the normative Italian farm lease, see Frier, ZRG 96 (1979): 204–28, and J. Köhn, "Die Kolonen in den Rechstbestimmungen," in K.-P. Johne et al., *Die Kolonen in Italien und den westlichen Provinzen des Römischen Reiches* (Berlin, 1983), 167–257, especially 183–244. For the legal principles of what civilians have termed *locatio-conductio rei*, see Kaser, *RP* I, 565–68. T. Mayer-Maly, *Locatio Conductio* (Vienna, 1956), offers a detailed analysis of the development of Roman lease law, with a close critical analysis of relevant texts. For discussion of the development of the Roman lease law in the civilian tradition, see Zimmermann, *The Law of Obligations*, 338–83. W.E. Heitland, *Agricola* (Cambridge, 1921; reprint, Westport, Conn., 1970), 361–78, offers an interpretive survey of the evidence offered by the Digest and the Justinian Code for farm tenancy.

17. See, e.g., Nerat. apud Ulp. D. 19.2.19.2 (32 *ad ed.*), Gaius D. 19.2.25.3, 19.2.25.6 (10 *ad ed. prov.*).

18. For the principle of *ut ei frui liceat*, see, e.g., Ulp. D. 19.2.9 pr., 19.2.15.1–2 (32 *ad ed.*), 24.3.7.8 (31 *ad Sab.*), Paul. D. 19.2.24.4 (34 *ad ed.*), Afric. D. 19.2.33 (8 *quaest.*), Paul. D. 34.3.16 (9 *ad Plaut.*).

19. The following passages also assume leasing for a quinquennium, or five years: Ulp. D. 12.1.4.1 (34? *ad Sab.*), Marcellus apud Ulp. D. 19.2.9.1 (32 *ad ed.*), D. 19.2.13.11 (32 *ad ed.*), Paul. D. 19.2.24.2, 19.2.24.4 (34 *ad ed.*), D. 24.3.25.4 (36 *ad ed.*), 34.3.16 (9 *ad Plaut.*), 45.1.89 (9 *ad Plaut.*). Other passages refer to leasing *in plures annos*: Cassius apud Julian. D. 19.2.32 (4 *ex Minic.*), Paul. D. 19.2.24.5 (34 *ad ed.*), Ulp. D. 43.32.1.4 (73 *ad ed.*).

obligated to perform normal agricultural duties, *opera rustica*, in a timely fashion.[20]

The independent status of the tenant is also presupposed in the principle that he would pay his rent in cash. Formally the *locatio-conductio* contract required that the rent be paid in money, and cash is almost universally assumed as the form of the rent payment. Other forms of land tenure, such as sharecropping and leasing for fixed rents in kind, are certainly mentioned in the legal sources, but only rarely.[21] As we have seen, Gaius distinguished between the tenant paying a cash rent and the sharecropper when considering the tenant's right to an abatement of rent (*D*.19.2.25.6). Similarly Paul, in discussing the tenant's right to bring an action for the theft of his crops, restricted this capacity to the tenant paying a fixed rent in cash (*D*. 47.2.26.1, 9 *ad Sab*.). Pliny, however, described his plan to institute sharecropping on his Tuscan estates in terms suggesting that his correspondent would only be generally familiar with this form of land tenure, rather than having firsthand knowledge of it (*Ep*. 9.37.2–4). Later rescripts refer to leasing for fixed rents in kind, which was of course the standard type of lease for grain land in Egypt.[22] Clearly the jurists were aware of a range of forms

---

20. See Gaius *D.* 19.2.25.3 (10 *ad ed. prov.*), Ulp. *D.* 19.2.11.2, cf. 19.2.15.3 (32 *ad ed.*). Cf. Iav. *D.* 19.2.51 pr. (11 *epist.*), [Paul.] *Sent.* 2.18.2. The obligation of the tenant to cultivate the farm derived from bona fides rather than from a normative *lex locationis*, so this obligation was intrinsic to the contractual relationship: see Mayer-Maly, *Locatio Conductio*, 177–81, with discussion of these passages.

21. See also Ulp. *D.* 19.2.19.3 (32 *ad ed.*), where instead of cash the tenant is to provide the landowner with a fixed quantity of crops valued at a predetermined price. Frier, *ZRG* 96 (1979): 219 n. 78, points out that in his discussion of leasing, Columella (1.7.1–2) seems to be acquainted with no other system besides leasing for cash rents. Of course, tenants might be required to perform services or make other payments over and above the rent. Thus Ulpian describes how account must be made of such additional payments in connection with dotal property at the time of divorce (*D.* 24.3.7.8, 31 *ad Sab.*). Cf. also the labor services described by Columella (1.7.2). These services represented a charge added on to the conventional cash rent: see Finley, "Private Farm Tenancy in Italy," 119–21. Egyptian leases often included payments in addition to the basic charge in kind or in cash for each unit of land; for the norms of Egyptian leases, see J. Herrmann, *Studien zur Bodenpacht im Recht der graeco-aegyptischen Papyri* (Munich, 1958), 114–22.

22. A rescript of Alexander Severus (*CJ* 4.65.8, A.D. 231) refers to leasing for *certis annuis quantitatibus*, while a rescript of Diocletian and Maximian refers to leasing an olive crop for a fixed amount of oil (*CJ* 4.65.21, A.D. 293).

of land tenure, but for the purpose of developing legal rules about farm tenancy, they restricted themselves to considering a "normative" lease for a fixed rent in cash.

The jurists also presupposed that the tenant disposed over substantial resources, which he brought into the tenancy. In the allocation of capital in the lease contract, the landlord was responsible to maintain the fixed capital of the lease, such as, for example, oil or winepresses (see Nerat. apud Ulp. D. 19.2.19.2).[23] The tenant was expected to provide the movable capital needed for the cultivation of the farm. The tenant pledged this movable capital, the *invecta aut illata,* as security for the rent and the condition of the tenancy (see, e.g., Labeo D. 20.6.14, 5 *poster. a Iav. epit.;* cf. Ulp. D. 43.33.2, 70/73 *ad ed.*). The tenant did not lose the use of his property pledged as security, but the lessor had a claim on it if the tenant defaulted on his contractual obligations.[24] The emperor Alexander Severus later affirmed the old rule that whatever property the tenant brought into the *fundus* in accordance with the wishes of the landowner would be pledged as security (*CJ* 4.65.5, A.D. 223). The property pledged in this way might include, for example, a slave woman, or *ancilla* (Julian. D. 43.33.1 pr., 49 *dig.*); by analogy other slaves and livestock brought by the tenant into the farm would also be pledged. The tenant envisioned in this case was a slaveholder, but that should come as no surprise. The legal question in this response concerning the status of the offspring only arose because of the gender of the slave, so, in the logic of this response at least, it was completely conventional for tenants to pledge slaves as security.[25] That assumption stands behind a decision by Paul concerning a slave assigned to a farm tenant on *aestimatum* (D. 19.2.54.2, 5 *resp.*); more often the tenant could be expected to provide his own slaves.[26] Although the Digest includes

---

23. For the allocation of capital in the Roman farm lease, see Frier, *ZRG* 96 (1979): 204–28. See also my discussion in chap. 4.

24. This type of pledge is termed both *pignus* and *hypotheca:* see Kaser, *RP* I, 457–73, especially 464. The *actio Serviana* of the second century B.C. provided the lessor with a legal recourse in rem against the property of a defaulting tenant: see Kaser, 459, 472–73.

25. Julian. D. 43.33.1 pr.: "Si colonus ancillam in fundo pignoris nomine duxerit et eam vendiderit, quod apud emptorem ex ea natum erit, eius adprehendendi gratia utile interdictum [utilem actionem Iul. *Lenel*] reddi oportet."

26. A number of cases concerning the tenant's liability when the villa is held under lease envision the tenant as owning slaves (they are the culprits in these cases): Alf. D. 19.2.30.4 (3 *dig. a Paul. epit.*), Sab., Nerat., Urseius Ferox, and Proc.

no explicit statement on this matter, it seems clear that the normative tenant provided livestock, which was also pledged as security. References to livestock are rare in the legal sources on tenancy, but Ulpian, in defining the grounds for allowing the tenant an *actio ex conducto*, included the failure of the landlord to provide a stable for the tenant's flocks, or *greges*, as a violation of the landlord's obligation to provide the tenant with a usable farm (Ulp. D. 19.2.15.1, 32 *ad ed.*).[27] Finally, the crops remained the property of the landlord as long as they were attached to the soil. When they were harvested, they became the property of the tenant, but the landlord customarily still retained a lien over them until the rent was paid.[28]

Other texts envision the tenant as supplying substantial resources. In some instances the resources that the tenant brought to his farm were sufficiently valuable to become an object of legal dispute. For example, Paul, in ruling on a case involving the legacy with its *instrumentum* of a *fundus* leased out to a tenant, included in the legacy the *instrumentum* belonging to the tenant if the testator had none on the bequeathed property (D. 33.7.24, 3 *ad Nerat.*). In this legacy *per damnationem*, either the heir would have to acquire the *instrumentum* belonging to the *colonus* and pass this on to the legatee, or he or she would have to compensate the legatee for the value of the *instrumentum*. The principles in this decision were applied to an admittedly abstract set of circum-

---

apud Ulp. D. 9.2.28.9, 11 (18 *ad ed., Coll.* 12.7.7., 9.), Ulp. D. 19.2.11.4 (32 *ad ed.*). For the tenant's owning slaves, see Frier, ZRG 96 (1979): 217–18; see also Capogrossi Colognesi, *Ai margini della proprietà fondiaria*, 230–31 n. 86 and "Il regime," 212–13 n. 86. In Ulp. D. 19.2.11.4, it is a question of the burning of a *villa urbana*, which was part of a rural estate: see De Neeve, *Colonus*, 7 and n. 27, with discussion of other views. Cf. Mayer-Maly, *Locatio Conductio*, 197. For analysis of how the classical jurists treated the tenant's liability in these cases, see Mayer-Maly, 198–202.

27. On Mayer-Maly's objections to the authenticity of the circumstances in this text giving the tenant an action, see n. 31 in this chapter.

28. Pomp. D. 20.2.7 pr. (13 *ex var. lect.*): "In praediis rusticis fructus qui ibi nascuntur tacite intelleguntur pignori esse domino fundi locati, etiamsi nominatim id non convenerit." Cf. Afric. D. 47.2.62.8 (8 *quaest.*): "Locavi tibi fundum, et (ut adsolet) convenit, uti fructus ob mercedem pignori mihi essent." Cf. also Paul. D. 19.2.24.1 (34 *ad ed.*), as well as Papin. D. 19.2.53 (11 *resp.*) for the pledging of *fructus* in the lease of *publica praedia*. On tacit pledging of property, see W. Schuller, "Zum *Pignus Tacitum*," Labeo 15 (1969): 267–84, and Köhn, "Die Kolonen in den Rechtsbestimmungen," 202–7, as well as Kaser, *RP* I, 464–65 and n. 18.

stances, but the jurist's ruling does envision the possibility that the tenant might provide all of the *instrumentum* used in cultivating a *fundus*. As we have seen (in chap. 2), this *instrumentum* typically included all the movable farm equipment, as well as slaves and livestock.[29]

In some cases, the jurists clearly envisioned tenants leasing a self-contained estate, or *fundus,* complete with both its productive side, or *ager,* and its villa. One example is a case in which the involvement of the tenant is purely incidental to the point of law in question. Thus Ulpian ruled on the ownership of a herd of swine snatched away from their keeper by wolves but subsequently recovered by "the tenant of a neighboring estate" [vicinae villae colonus] (D. 41.1.44, 19 *ad ed.*). The villa typically represented the farmhouse attached to an estate, so any case involving this issue probably envisioned a large-scale lease of an entire estate.[30] Most generally, Ulpian numbered the failure of the landlord to repair a villa as a failure to provide the tenant with a *fundus* that he could enjoy, "ut puta si re quam conduxit frui ei non liceat," and accordingly he regarded this failure as grounds for an *actio ex conducto* on the part of the tenant (D. 19.2.15.1, 32 *ad ed.*).[31] The ruling by Gaius about the obligation of tenants to maintain villas also suggests the scale of farm that a Roman tenant might lease; Gaius considered maintaining the villa as one of the tenant's primary duties, along with performing the *opera rustica* in a timely fashion (D. 19.2.25.3, 10 *ad ed. prov.*).[32]

The term *villa* in these cases must refer to the kind of farmhouse to be

---

29. See De Neeve, *Colonus,* 170 n. 248, for further references to the resources of *coloni.*

30. See De Neeve, *Colonus,* 82–86, 159 n. 192, for tenants leasing whole *fundi,* and, for the meaning of the term *fundus,* idem, "Fundus as Economic Unit," *RHD* 52 (1984): 3–19. See also Scheidel, *Grundpacht und Lohnarbeit,* 72–73 n. 201, who argues that the legal sources always envision a single *colonus* on a *fundus.*

31. Mayer-Maly, *Locatio Conductio,* 152–53, considers this enumeration of grounds in this passage for which the lessor might be liable to the tenant to be the result of postclassical reworking of Ulpian's text, but he considers the legal principles implicit in the enumeration to be classical Roman law.

32. Gaius D. 19.2.25.3: "Conductor omnia secundum legem conductionis facere debet. et ante omnia colonus curare debet, ut opera rustica suo quoque tempore faciat, ne intempestiva cultura deteriorem fundum faceret. praeterea villarum curam agere debet, ut eas incorruptas habeat." Mayer-Maly, *Locatio Conductio,* 180–81, considers this passage to have been reworked in postclassical times, with the phrases *ne . . . faciat* and *ut . . . habeat* likely interpolations. See De Neeve, *Colonus,* 51–52 and 209 nn. 39–40, on this passage and for further references of *coloni* leasing villas.

found on estates like the ones that Pliny had in mind when he discussed with his correspondents various aspects of life on a country estate. Such villas need hardly have been on the same lavish scale as, for example, the ones that Pliny maintained on his estates at Tifernum Tiberinum (*Ep.* 5.6) or on his property at Laurentum (*Ep.* 2.17), but the reference to specialized slaves in the employ of the tenant does indicate that the tenant in effect took the place of a landowner as the occupant of a villa. It was probably such tenants as these who might be represented by guarantors, or *fideiussores*. They were responsible for the performance of contractual obligations by tenants; such guarantors were responsible for the tenant's paying his rent (Labeo *D.* 20.6.14, 5 *poster. a Iav. epit.,* Paul. *D.* 19.2.54 pr., 5 *resp.*) and for the tenant's maintaining the *instrumentum* of the farm (Papin. *D.* 46.1.52.2, 11 *resp.*).[33] In some circumstances, tenants might provide a *cautio*, a formal promise in the form of a stipulation, as a means of assuring their ability to fulfill their lease obligations. For example, Scaevola considered whether the legacy of *reliqua colonorum*, or the arrears of tenants, should include the arrears of tenants who had left their tenancy after the termination of the lease but had provided a *cautio* (*D.* 33.7.20.3, 3 *resp.*).[34]

We should expect tenants able to provide guarantors or to cover the costs of a *cautio* to have had a relatively strong legal position vis-à-vis their landlords. We might also include in this category of tenant the purchaser of a *fundus*, discussed by Iavolenus, who leased it until the full price was paid (*D.*19.2.31, 11 *epist.*) and the seller of a *fundus* who, as part of the terms of the sale, kept it under lease for an agreed on rent,

---

33. Papin. *D.* 46.1.52.2: "Fideiussores a colonis datos etiam ob pecuniam dotis praediorum teneri convenit, cum ea quoque species locationis vinculum ad se trahat. . . ." Other passages concerning guarantors for tenants include Ulp. *D.* 19.2.13.11 (32 *ad ed.*), discussing the role of the guarantor in the tacit renewal of a lease; Paul. *D.* 47.2.86 (2 *manualium*), denying the *fideiussor* of a *colonus* an *actio furti;* and Paul. *D.* 46.1.58 pr. (21 *quaest.*), where a landlord accepts a *fideiussor.*

34. On guarantors, see Kaser, *RP* I, 660–66; see 539 on the *cautio,* which might include the provision of security. On guarantors of leases in Roman Egypt, see *P.Oxy.* IV 707, and the large-scale lease of the wealthy Claudia Isidora's property in the Small Oasis (*P.Oxy.* XIV 1630). Lessees of public property also had to provide *cautiones* (Papin. *D.* 50.8.5 pr., 1 *resp.; D.* 19.2.53, 11 *resp.*). According to Herrmann, *Bodenpacht,* 150–53, guarantors were rare in Egyptian leases, but there are numerous lease contracts in which several tenants, leasing in common, served as mutual sureties.

certa mercede (Paul. *D.* 19.1.21.4, 33 *ad ed.*).³⁵ The ability of a tenant to act on an equal basis legally with his landlord is also envisioned in a case discussed by Iavolenus, concerning the legacy to a tenant of a usufruct over the *fundus* that he was cultivating (*D.* 33.2.30.1, 2 *ex poster. Lab.*). Iavolenus ruled that the tenant could sue the heir to be released from his lease obligations. As we will see in chapter 4, the tenant who was the beneficiary of his original landlord's will was not likely to have belonged to the same social class as the original landlord and his heir. Nevertheless, this case does suggest that the tenant had some sort of long-term relationship with the writer of the will in question, one not defined by the lease contract. The tenant in this case, moreover, had sufficient resources to sue the heir for the fulfillment of the legacy.

In his comments about legal questions arising from an actual will, the jurist Paul provides a more concrete hint about the resources that a large-scale tenant might bring to his farm or estate (Paul. *D.* 32.27.2, 2 *decret.*).³⁶ A certain Iulius (or Iulianus) Severus had left a legacy of fifty thousand sesterces to his *alumnus*, or foster son, and he intended this money to be paid out of funds that his tenant Iulius Maurus, himself a legatee, owed as rent for his *fundus*. Maurus probably did not represent the same social class as his landlord, but the expectation that he could pay a substantial legacy out of his back rent suggests the large scale of the farm that he was leasing.

The conventional tenant capable of leasing a whole estate had to be relatively wealthy, and a number of rulings in the Digest accordingly presuppose that tenants could deal with their landlords on a legally and socially equal basis. For example, in a discussion of the legal consequences of a disagreement between landlord and tenant, Iavolenus assumed that both parties were free to negotiate lease terms (Iav. *D.* 19.2.51 pr., 11 *epist.*).³⁷ Later, Julian (*D.* 2.14.56, 6 *ad Minic.*) and Pomponius (*D.* 19.2.52, 31 *ad Quint. Muc.*) made the same assumption about the ability of both parties to negotiate the lease contract. The consequences of the legal and social equality of landlord and tenant are expressed explicitly in a case, discussed by Paul, in which a landlord and a tenant mutually stipulated to pay penalties if either party

---

35. See, for the same case, Hermog. *D.* 18.1.75 (2 *iuris epit.*).
36. On this case, see C. Sanfilippo, *Pauli Decretorum Libri Tres* (Milan, 1938), 77–81.
37. Iav. *D.* 19.2.51 pr.: "respondit: in huiusmodi obligationibus id maxime spectare debemus, quod inter utramque partem convenit. . . ."

reneged on the lease obligations: the tenant stipulated that he would remain in the tenancy, and the landlord undertook not to expel the tenant from the tenancy (D. 19.2.54.1, 5 *resp.*).[38] The social equality of tenant and landowner is clear in a decision by Julian concerning the claims on the part of a wife to income from dotal property after a divorce. In this case, a husband leased out to the mother of his wife a *fundus* given to him by his wife as part of her dowry (D. 23.4.22, 2 *ad Urs. Fer.*). Another response by Julian (15 *dig.*) envisioned a similar social equality of landowner and tenant (apud Ulp. D. 19.2.15.9, 32 *ad ed.*; cf. Julian. D. 19.2.18, 15 *dig.*). In this case, the tutor of an underage heir of a tenant decided that it was in the best interests of his pupil to reject the inheritance. When the pupil later regained his inheritance, Julian ruled that he could not sue to regain his lease. The lessee in this case was wealthy enough that the affairs of his underage heir had to be managed by a tutor.

Each tenant represented in these decisions was leasing a complete estate. The type of tenant figuring in Gaius' prescriptions about maintaining villas might correspond to the mother-in-law leasing the dotal estate in the case discussed by Julian (D. 23.4.22) or to the tenant whom the same jurist hypothesized as purchasing a *fundus* from the supposed heir of his landlord (D. 41.3.33.1, 44 *dig.*). A case discussed by Paul also envisions a tenant leasing on a substantial scale (D. 26.7.46 pr., 9 *resp.*). This tenant had incurred substantial arrears while leasing property belonging to a pupil. After growing to maturity, the pupil appointed the tenant as his procurator, or business manager.[39]

We might also consider as belonging to this category the owner of a *fundus* whom the second-century jurist Pomponius described as leasing it back from a usufructuary (apud Ulp. D. 7.4.29 pr., 17 *ad Sab.*). In this case, the economic relationship that I have hypothesized as underlying leases between landowners and large-scale tenants is readily inferred. The usufructuary's only interest in the *fundus* in question was to gain a safe income, to achieve which he leased the property that was placed at his disposal en bloc to a large-scale tenant, who happened to be the owner himself. The kind of tenant envisioned in Gaius' strictures about maintaining villas might also be represented in the tenant whom Iav-

---

38. Cf. discussion later in this chapter on the provisions by the lessor in certain Egyptian leases not to compel the tenant to remain in the leasehold.

39. The tenant is variously described as leasing a *fundus Cornelianus* and *praedia* belonging to the pupil.

olenus, and before him Labeo and Trebatius, envisioned as purchasing one-half of a *fundus* while leasing the other half from the seller for ten years (D. 18.1.79, 5 *ex poster. Lab.*). We also see such a tenant in the seller of a *fundus* who, in a decision discussed by Paul, retained the property under lease (D. 19.1.21.4, 33 *ad ed.*). In these two cases, the assumption that tenant and landowner might represent the same social and economic class is taken to its logical extreme, with the same person alternating between the roles of landowner and lessee.

On paper, the leasing out of estates under the conditions I have described was well suited to serve the interests of upper-class landowners seeking to minimize their risk and involvement in the day-to-day operations of their estates. The payment of a cash rent meant that the tenant bore the bulk of the risk for the size of the harvest and for the market price of the crops. The expectation that the tenant could pay a rent in cash, moreover, presupposes his ability to sell some of the crops that he produced, so that the landowner's own involvement in this troublesome task might be reduced.[40] The landlord for his part had ample means to enforce the tenant's fulfilling his lease obligations. As we have seen, the landlord could make losses caused by the tenant good by confiscating the movable property pledged as security. Likewise, the landlord might be able to exercise a great deal of leverage over the tenant by virtue of his lien on the tenant's crops. A tenant in default of his contractual obligations faced the possibility that the landlord might confiscate the crops or, less dramatically, only allow the tenant to sell them after the tenant had arranged to settle outstanding obligations. In addition, in theory at least, the limited duration of the lease gave the landlord flexibility in dealing with tenants. The landowner of course had the right to dismiss, even during the lease period, a tenant who had failed to meet his contractual obligations. In any case, at the termination of the lease contract, the landowner could dismiss an unsatisfactory tenant or, to settle outstanding obligations, negotiate any future lease on terms less favorable to the tenant.

---

40. For the distribution of the risk in leasing for cash rents, leasing for fixed rents in kind, and sharecropping, see, e.g., S.N.S. Cheung, *The Theory of Share Tenancy* (Chicago, 1969), 72–79. For the involvement of the tenant in the commercial sector by marketing his surplus, see L. Capogrossi Colognesi, "Grandi proprietari, contadini e coloni," 335, and idem, *Ai margini della proprietà fondiaria*, 202–3, 222–37, especially 229–30 and "Il regime," 196–97, 208–16, especially 212–13; Capogrossi Colognesi's views are also discussed in my introduction.

The threat to take these steps or to confiscate the property pledged as security enabled the landowner to enforce a degree of compliance on the part of the tenant. The short-term lease also lowered the landowner's obligations toward his tenant in the event that he alienated the property.[41] The form of tenancy envisioned in the normative rules in the Digest corresponds at least generally to the system of leasing attested in the few historical examples where we can trace the terms under which actual Roman landowners leased to tenants. Thus, before introducing sharecropping, Pliny leased his estates out for fixed terms of five years in exchange for cash rents, while the movable property brought onto his estates by his tenants was pledged as security for the rent. The farms occupied by Pliny's tenants were of substantial enough size to require slaves to work them. Similarly Columella envisioned leasing his estates out under short-term leases for cash rents (1.7.1–2).

The way in which Roman landowners saw tenancy as providing a solution to managerial concerns can be traced in Columella's brief series of recommendations concerning the proper treatment of tenants (1.7.1–7). To be sure, Columella sought to encourage his contemporaries to take a greater interest in their estates, so for him the ideal estate was one whose cultivation the landowner, either resident there or at least able to visit frequently, might supervise directly.[42] Accordingly, after emphasizing that the landowner could profit the most from his tenants by fostering their long-term ability to cultivate their farms productively (1.7.1–4), Columella asserted that land cultivated directly, with the owner himself or his *vilicus* supervising slave labor, could

---

41. This is an observation made orally by Prof. Richard Helmholz, at the conference of the Roman Law Society of America, Champaign, Ill., April 1988. The obligations of the owner toward the tenant at the sale or bequest of property are also discussed in chap. 4.

42. All the ancient agronomists emphasize the importance of active supervision on the part of the landowner to productive agriculture: see, e.g., Mago in Plin. *Nat.* 18.35 (= Col. 1.1.18); Cato *Agr* 1.1, 2.1 f., 4; Plin. *Nat.* 18.31 (= Cato 4), 18.43; Col. 1 pref. 12, 1.1.18–20, 1.8.20 (see H.B. Ash, *Columella de Re Rustica*, vol. I, Loeb Edition [London and Cambridge, Mass., 1977; first ed., 1941] 36 n. a). For a discussion of Columella's general purposes in his treatise, see R. Martin, *Recherches sur les agronomes latins et leurs conceptions économiques et sociales* (Paris, 1971) 289–373. For a detailed analysis of Columella's recommendations about leasing, see W. Scheidel, "Pächter und Grundpacht bei Columella (*Colonus*-Studien II)," *Athenaeum* 81 (1993): 391–439. See also idem, *Grundpacht und Lohnarbeit*, 83–117.

always provide a higher return than land cultivated indirectly through tenants (1.7.5).[43] The only factor that interfered with this principle was the inattention on the part of the landowner, which allowed potentially rapacious slave bailiffs to bring an estate to ruin (ibid.).[44] Columella, however, recognized that the direct involvement of the landowner in the cultivation of his estates was an ideal not always realized in practice. As a result, he recommended leasing out to tenants distant estates that the landowner could not easily visit (1.7.6–7). For Columella, the advantages of leasing were most apparent in farms cultivated with grain, since, in his view, the tenant could do little harm to the farm's capital, whereas the dishonest slave *vilici* could, through lazy or dishonest management, fritter away the landowner's income. But Columella also applied his recommendations concerning leasing to other types of land, including vineyards.[45] For the purpose of exploiting distant estates, then, tenancy allowed the absentee landowner to avoid many costs of management, since the tenants themselves had an incentive to make sure that the land was cultivated efficiently. In Columella's view, tenancy represented the second-best alternative after direct management by the landowner, and he only grudgingly recommended leasing, as preferable to entrusting one's fortune to the greedy and indolent hands of unsupervised slave bailiffs.[46] For landowners less dedicated to the ideal of owner cultivation, the managerial advantages provided by tenancy must have been crucial.

---

43. Col. 1.7.5: "ceterum cum mediocris adest et salubritas et terrae bonitas, numquam non ex agro plus sua cuique cura reddidit quam coloni, numquam non etiam vilici...."

44. Col. 1.7.5: "... nisi si maxima vel neglegentia servi vel rapacitas intervenit. Quae utraque peccata plerumque vitio domini vel committi vel foveri nihil dubium est, cum liceat aut cavere ne talis praeficiatur negotio, aut iam praepositus ut summoveatur curare."

45. Col. 1.7.6: "In longinquis tamen fundis, in quos non est facilis excursus patris familiae, cum omne genus agri tolerabilius sit sub liberis colonis quam sub vilicis servis habere, tum praecipue frumentarium, quem et minime, sicut vineas et arbustum, colonus evertere potest et maxime vexant servi...." Cf. Col. 1.7.7: "Quare talis generis praedium, si, ut dixi, domini praesentia cariturum est, censeo locandum." Cf. the very different view of P.W. de Neeve, *Colonus*, 93–95, who argues that Columella primarily recommended the leasing of land producing grain and the direct exploitation of vineyards.

46. Cf. Col. 1.1.20 on the dangers and evils in turning over distant estates to the depredations of slaves, and cf. Pliny the Elder's famous condemnation of *latifundia* (NH 18.35).

I have shown in my examination of the concept of profit in agriculture that the jurists often assumed that the beneficiaries of wills would exploit their estates by leasing them out to tenants. This assumption on the part of the jurists was apparently matched by beneficiaries of wills in the real world, to the extent that their activity is reflected in a small corpus of papyrological documents from Roman Egypt. For the beneficiaries envisioned by the jurists, the annual rent paid by the tenant could for all intents and purposes be equated with the profit that an agricultural property might provide.[47]

This supposition draws support from a small corpus of papyri attesting how guardians, or *epitropoi,* in Roman Egypt managed the land belonging to their wards. One must recognize, however, that most of these documents concern guardians of Greco-Egyptian background, who were not necessarily exercising their responsibilities in accordance with the norms of Roman private law. Even so, the regulation of "non-Roman" guardians in Egypt was increasingly assimilated to the norms of Roman law during the early principate, so that we are justified in using the papyri to supplement the juristic evidence.[48]

To judge by the papyrological evidence, the Roman administration in Egypt clearly viewed tenancy as the best way to assure orphans of a stable income. Accordingly, guardians were normally expected to lease out to tenants the land belonging to their wards, and, having done this, they were responsible to exact the rent and enforce the tenant's other contractual obligations. To illustrate how leasing was envisioned as serving the interests of underage orphans, let us consider the cases of property belonging to two sets of orphans at Hermopolis in the first century A.D. We know of the situations of the orphaned children of a certain Sarapion son of Kastor and of Didyme daughter of Hermias, from two documents containing publicly posted lease offers (*P.Amh.* II 85 = *M.Chr.* 274, *P.Amh.* II 86, both A.D. 78).[49] The *exegetes,* the civic offi-

---

47. See chap. 2.
48. For discussion of guardianship in Roman Egypt, see J.-U. Krause, *Witwen und Waisen im römischen Reich,* vol. III, *Rechtliche und soziale Stellung von Waisen* (Stuttgart, 1995), 104–7, as well as R. Taubenschlag, *The Law of Greco-Roman Egypt in the Light of the Papyri, 332 B.C.– 640 A.D.,* 2d ed. (Warsaw, 1955), 157–58, and E. Seidl, *Rechtsgeschichte Ägyptens als römischer Provinz* (St. Augustin, 1973), 223–24.
49. I discuss these documents in *Management and Investment on Estates in Roman Egypt during the Early Empire* (Bonn, 1992), 129. For the managerial advantages from leasing gained by landowners in Roman Egypt, see also J. Rowlandson, *Landowners and Tenants in Roman Egypt* (Oxford, 1996), 259–79.

cial commonly charged with overseeing the performance of guardians, would post such lease offers for ten days, inviting other prospective tenants to submit higher bids. If no higher bid was forthcoming, the party making the original offer to lease was awarded the contract.[50] This system of leasing was designed to reduce the possibility of fraud, but the leases themselves indicate the efforts of the administration to protect the interests of individuals not capable of managing their own finances. In the cases of the children of Sarapion and Didyme, although the land was to be cultivated with grain, for which a rent in kind was customarily exacted, the lessees were required to pay fixed rents in cash of 600 and 260 drachmas per year respectively (*P.Amh.* II 85, lines 9–13; 86, lines 9–11, 14–16).[51] In addition, the lessees, in contrast to the much more common practice in Egyptian farm leases of the landowner paying the taxes, took on themselves the task of paying the public taxes due on the land, including the *artabia*, or land tax, and the *naubion*, a charge for the maintenance of public irrigation facilities. The point of these leases clearly was not to make the most lucrative use possible of the land but rather to guarantee a safe and steady income for the orphans, and leasing provided the authorities with a ready means of achieving this purpose.[52]

50. For the process, see L. Mitteis in L. Mitteis and U. Wilcken, *Grundzüge und Chrestomathie der Papyruskunde*, vol. II (Leipzig, 1912; reprint, Hildesheim, 1963), 307, and B.P. Grenfell and A.S. Hunt, *P.Amh.* II 85, introd., p. 105. Seidl, *Rechtsgeschichte Ägyptens*, 224, in discussing SB XVI 12557 (= SB VI 9049, A.D. 222–35), an oath by a tutor to fulfill his duties, argues that the leasing of the entire property of a ward was characteristic of Athenian law. On this topic, see R. Osborne, "Social and Economic Implications of the Leasing of Land and Property in Classical and Hellenistic Greece," *Chiron* 18 (1988): 279–323, at 304–319. Cf. *P.Heid.* IV 337 (ca. A.D. 78), an offer to lease land belonging to two orphans; it is addressed to the same Hermias serving as *exegetes* of the Hermopolite nome in *P.Amh.* II 85–86. In this case, the prospective lessee lost the lease, since he was outbid; thus the term ὑ(περ)εβλήθη is added in a second hand at the end. On this document, see B. Kraut, "Seven Heidelberg Papyri concerning the Office of Exegetes," *ZPE* 55 (1984): 167–90, at 178–80.

51. In Roman Egypt, the vast majority of leases for grain land involved fixed payments of grain for each unit of land. See D. Hennig, "Untersuchungen zur Bodenpacht im ptolemäisch-römischen Ägypten" (Ph.D. diss., Ludwig-Maximilians-Universität zu München, 1967), 26–28.

52. Each lease offer did include a provision for renegotiating the rent in the event of ἀβροχία, that is, insufficient flooding by the Nile. Such provisions are common but by no means the rule in farm leases in Roman Egypt; see Herrmann, *Bodenpacht*, 143–45, 161–62, 174.

Several papyri allow us to follow how individual guardians fulfilled their responsibilities of overseeing the management of their wards' land. One instructive example is a detailed report submitted to the *exegetes* in which two guardians account for how they exploited the land belonging to the orphaned children of a certain Horion son of Philon (*P.Heid.* IV 336, A.D. 141–42, Hermopolite nome).[53] Horion had died, apparently leaving his two sons heirs to three-fourths of his estate, with another heir receiving the other fourth. The extant document must represent an example of the type of report that guardians routinely had to file. The first part describes how the land belonging to the minor sons was exploited over a period of two years. The land consisted of thirteen *arourae* of land in two locations, to be cultivated with a rotation of several crops, including wheat, barley, vegetable seed, and fodder. As was customary in Roman Egypt, rents in kind were charged for all of these crops, except for fodder, for which a rent in cash was exacted. The land was divided into two parcels, of five and eight *arourae*, respectively. The first parcel of five *arourae* was leased out on a two-year cycle of wheat and fodder. In the first year of the lease for the second parcel, five *arourae* were to be cultivated with vegetable seed and three with barley; in the following year five *arourae* were to be cultivated with wheat and barley, while the other three *arourae* were to be cultivated with vegetable seed.[54] The land was to produce a total gross revenue for two years of 105 *artabae* of wheat, 30 *artabae* of barley, 24 *artabae* of vegetable seed, and 180 drachmas. From this total were deducted 40 *artabae* of wheat and 14 drachmas for the *artabia* and *naubion* taxes, and the sons' three-fourths share of the remainder came to 48 1/3 *artabae* of wheat (the total should be 48 3/4), 22 1/2 *artabae* of barley (which should be 24 1/2), 18 *artabae* of vegetable seed, and 122 drachmas. The final part of the document accounts for property that was now in possession of the sons' mother but was eventually to pass to the sons' ownership.

For our purposes, it is significant that all of the land bequeathed by the deceased father was let out to tenants. Even if guardians were not formally required to lease out the land belonging to their wards, they could most conveniently fulfill their fiduciary obligations to their

---

53. On this document, see Kraut, *ZPE* 55 (1984): 167–78.
54. For crop rotation in Egyptian farm leases, see Rowlandson, *Landowners and Tenants in Roman Egypt*, 236–47.

wards by doing so. Such considerations surely stand behind the decision by many a guardian to lease out the land belonging to a ward. For example, a third-century lease offer attests how a guardian of a daughter from a relatively wealthy family leased out his ward's lands (*P.Diog.* 29, A.D. 225). The daughter in question was named Aurelia Kopria, and she was a member of an extended family from Antinoopolis that included several veterans. This family enjoyed the privilege of Antinoite citizenship, but it also had property in the Arsinoite nome (the Fayum).⁵⁵ The third-century document is an offer to lease for three years eight and one-half *arourae* of grain land near Philadelphia belonging to Aurelia Kopria. The offer was submitted to Aurelia Kopria's guardian; she herself is styled underage, or *aphelix*, and was apparently about seventeen years old.⁵⁶

A similar approach to managing land must have influenced the way in which two wealthy brothers resident at Oxyrhynchus appointed their own business manager, or *phrontistes,* to take over their duties as guardians for a nephew and niece (*P.Oxy.* IV 727, A.D. 154). The two brothers, C. Marcius Apion, alias Diogenes, and C. Marcius Apolinarius, alias Iulianus, were Roman citizens, and a sign of their high status was that they were making provisions for the management of their wards' property during their absence from Egypt. The *phrontistes,* whose name was Ophelas, managed the brothers property in the Oxyrhynchite nome (lines 12–14). He was to have a number of duties in connection with this guardianship, in particular exacting rents from tenants and arranging sales of crops.⁵⁷ Clearly, the two wards were to

---

55. Aurelia Kopria was the daughter of Marcus Lucretius Diogenes II and Ammonarion II. For the affairs of this family, see the archive edited by P. Schubert, *Les archives de Marcus Lucretius Diogenes et textes apparentés* [= *P. Diog.*] (Bonn, 1990), ad loc. and pp. 7–18.

56. For her age, see Schubert, *Les archives de Marcus Lucretius Diogenes,* 13. For leases of land belonging to the minor Ptollarion son of Ptollarion, who was from a wealthy family with extensive holdings in land in the Fayum during the second century, see my *Management and Investment,* 89–90.

57. *P.Oxy.* IV 727, lines 12 ff.: . . . συνεστακέναι τὸν προγεγραμμένον Ὠφελᾶν / ὄντα καὶ τῶν ὑπαρχόντων αὐτοῖς ἐν τῷ Ὀξυρυγχεί/τῃ νομῷ φροντιστὴν καὶ κατὰ τήνδε τὴν συνχώρησιν / φροντιοῦντα καὶ ἐπιμελησόμενον ὧν καὶ αὐτοὶ ἐπι/τροπεύουσιν ἀφηλίκων ἑαυτῶν . . .; lines 18 ff.: . . . ἔτι δὲ καὶ ἀπαιτήσοντα φόρους / καὶ ἐγμ[ι]σθώσοντα ἃ ἐὰν [δ]έον ἦν καὶ καταστησόμενον / πρὸς οὓς ἐὰν δέῃ καὶ γένη διαπωλήσαντα ἃ ἐὰν δέον / ᾗ τῇ αὐτοῦ πίστει. . . .

receive their incomes from rents and crops sold on the market; these crops in all likelihood represented the rents in kind exacted from the wards' tenants.

In Roman Egypt, tenancy also apparently provided a solution to the problem of managing lands to other classes of landowners unable to supervise their finances personally. One such group consisted of soldiers on active military service, and a number of documents indicate that such soldiers regularly leased out land that they owned, entrusting relatives and friends to oversee the collection of the rents and the enforcement of the tenants' obligations. For example, a certain Apollodoros wrote a letter to a soldier named Terentianus in which he described how he had lowered the rent owed by Terentianus' tenant, who was also the brother of the soldier (*P.Mich.* VIII 464, Karanis, second century A.D.).[58] Especially illustrative of the situation in which soldiers on active service might find themselves is a lease arranged for land belonging to C. Iulius Apollinarius, a member of a well-known family with several generations in the army (*P.Mich.* IX 562, A.D. 119).[59] A certain Sabeinos son of Sokrates arranged to lease from Apollinarius for three years a series of parcels scattered among several villages in the Fayum. This land included three *arourae* of grain land and an olive grove consisting of two *arourae* at the village of Karanis, three-fourths of an *aroura* of land at Bacchias, an additional two *arourae* of grain land at Hiera, and an olive grove consisting of one aroura at Alkias. An unspecified rent is described in the contract as paid in advance, and, since it is designated by the term *phoroi*, it was apparently a rent in cash. The lessee was to be responsible for performing all the labor on the land in question, to pay the taxes, and to return the olive groves at the end of the lease in full working condition. A number of exceptional features in this lease suggest that the lessor was taking somewhat extraordinary

---

58. This letter is included in the collection of J.L. White, *Light from Ancient Letters* (Philadelphia, 1986), no. 101. For further discussion of the affairs of Terentianus, see R. Alston, *Soldier and Society in Roman Egypt* (London and New York, 1995), 135–37.

59. The affairs of Apollinarius and his family are preserved in a small archive: see Alston, *Soldier and Society in Roman Egypt,* 134–35. Apollinarius served in the Roman army that occupied Arabia Petraea. For discussion of this archive's significance to Roman military history, see K. Strobel, "Zu Fragen der frühen Geschichte der römischen Provinz Arabia und zu einigen Problemen der Legionslokation im Osten des Imperium Romanum zu Beginn des 2. Jh.n.Chr.," *ZPE* 71 (1988): 251–80, at 257–60.

steps to assure himself of the most stable income possible with the fewest possible managerial concerns while he was outside of Egypt on active military service. It was not common for an individual farm lease in Roman Egypt to encompass a number of parcels of land with different crops cultivated, and it was also unusual that a rent in cash be exacted for grain land. Clearly, the lessee was taking over all managerial responsibilities for the land belonging to Apollinarius, and accordingly the lessee undertook to pay the taxes and also to pay a rent in cash in advance.[60] Apollonarius was a landowner with no capacity to supervise the cultivation of his land personally, and during his absence he sought to keep his income as stable as possible by imposing on his tenant all managerial responsibilities. Presumably Apollonarius paid a premium for his tenant's service in the form of a lower rent, but he saw tenancy as the surest way of safeguarding his income.

The importance to landowners of retaining the services of tenants sometimes resulted in landowners attempting to force unwilling tenants to remain on their farms. Cassius and Julian both addressed the problem of determining the tenant's rights when someone receiving a *fundus* through legacy tried to compel the tenant to remain in the lease (Cassius apud Julian. D. 19.2.32, 4 *ex Minic.*). The *legatarius* envisioned in this case could most conveniently gain an income from the estate in question by retaining the services of the sitting tenant. The common view of Cassius and Julian was that the tenant could not be compelled to cultivate the farm, because the heir had no interest in the farm's cultivation. The imperial government was apparently faced time and again with the problem of defining the rights of tenants whom landlords sought to force to remain in their tenancies after the expiration of the lease contract. Thus the emperor Philip cited numerous previous rescripts on this subject when he vindicated the tenant's freedom to give up a lease at the end of the contract period (*CJ* 4.65.11, A.D. 244).[61]

---

60. This lease was not antichretic or a *datio in solutum* as were many prodomatic leases (leases with the rent paid in advance), since it is specified that the lessee is responsible for both the labor and the taxes.

61. *CJ* 4.65.11: "Invitos conductores seu heredes eorum post tempora locationis impleta non esse retinendos saepe rescriptum est." Earlier, Hadrian had issued a constitution decrying the practice of forcing lessees of fiscal property, or *vectigalia*, to renew their leases (Callist. D. 49.14.3.6, 3 *de iure fisci*). The administrators of *ousiai* in first-century Egypt apparently also resorted to compulsory leasing; the most famous example is under the reign of Nero, and in his

This supposition about the importance of tenancy to upper-class landowners draws support from a number of texts in the Digest that seem to assume that the landowner's ability to profit from an estate depended on his ability to maintain the continued presence of a tenant there. This hypothesis is difficult to test, but we might expect to discern some indication of this when agricultural properties were bought and sold. The condition of the tenants was a crucial factor affecting Pliny's decision whether or not to purchase an estate (*Ep.* 3.19). A good example of this type of consideration in the legal sources is found in a response of Iavolenus. The case concerns a landowner who sold off half of a *fundus* under the condition that the purchaser agree to lease for ten years the half retained by the landowner (*D.* 18.1.79, 5 *ex poster. Lab.*). Iavolenus, going against the previous opinions of Trebatius and Labeo, granted the seller an *actio ex vendito* if the purchaser failed to live up to his obligation to lease the property in question. Iavolenus was concerned with defining what the proper remedy was for the seller/lessor of the property in question, whether his claim was based on the lease or on the sale. In ruling that the claim could be made on the basis of the sale contract, Iavolenus showed that the guarantee that the landowner could lease out a farm whose ownership he retained constituted an important consideration in the act of sale. Clearly the seller and other landowners like him are envisioned as having a vital interest in maintaining the continued and long-term presence of a tenant.[62]

According to this reasoning, the assurance of the continued presence of a tenant would have been an important factor in determining the value and thus the sale price of an estate. Such an assumption at least

---

edict of A.D. 68 the prefect Tiberius Iulius Alexander promised to put an end to the practice. See G. Chalon, *L'édit de Tiberius Julius Alexander* (Olten and Lausanne, 1964), lines 10–15 (= W. Dittenberger, *OGIS* 669, *BGU* VII 1563, lines 26–41), and my *Management and Investment*, 29 n. 40. For compulsion on the part of lessors in private lease arrangements in Roman Egypt to keep the tenant in place, see H.-J. Rupprecht, "Die Beendigung von Vertragsverhältnissen, Überlegungen zur Rechtswirklichkeit anhand der Pacht," *JJP* 20 (1990): 119–28. We might compare the apparently successful efforts of a certain Isidoros son of Psophthis to resist the efforts of an administrator of an *ousia* belonging to Livia Augusta to force him to take up a lease against his will in A.D. 9–10; see A.E. Hanson, "The Archive of Isidoros of Psophthis and P. Ostorius Scapula, Praefectus Aegypti," *BASP* 21 (1984): 77–87, and, earlier, "Two Copies of a Petition to the Prefect," *ZPE* 47 (1982): 233–43.

62. Cf. the inverse situation in Paul. *D.* 19.2.21.4 (33 *ad ed.*), where the seller of a *fundus* retains the right to lease it.

stands behind a legal response made by Papinian (*Frag. Vat.* 13, 3 *resp.*) and quoted by Hermogenianus, an *epitomator* of the late third or early fourth century (*D.* 19.1.49 pr., 2 *iuris epit.*).[63] This case concerns the liability toward the purchaser of a property on the part of a seller who, to raise the price, feigned the presence of a tenant. The seller was liable to the purchaser of the estate on an *actio ex empto* even if he guaranteed the payment of five years' rent by the tenant.[64] The assumption behind this decision is that the tenant was expected to have a continued tenure on the estate, despite the short-term duration of his lease, and that he was not easily replaced. The presence of a tenant was therefore an important factor affecting the ability of the landowner to achieve a long-term income commensurate with the purchase price of his property; the tenant's value went beyond the rent that he would pay during the remainder of the lease. At any rate, this case would also seem to presuppose that for many landowners leasing to a tenant represented the preferred option and that direct cultivation through a slave *vilicus* was only resorted to when leasing proved impossible.

This same assumption figures in cases in other areas of the law. For example, it was a factor affecting a case decided by the second-century jurist Scaevola (*D.* 20.1.32, 5 *resp.*). In this case, a debtor had pledged certain agricultural properties as security, and Scaevola ruled that slaves assigned to the cultivation of one of these properties were to be included as part of the security. The way in which the legal question is formulated is noteworthy: part of the debtor's property was without tenants, and he assigned this part to the management of an *actor* along with slaves to perform the agricultural labor.[65] Finally, an assumption about the importance of the presence of a tenant can be inferred from a decision by Proculus concerning the obligations of a seller of a *fundus* who contracted as part of the sale to provide to the purchaser whatever rent he might have exacted from the tenant (*D.* 18.1.68 pr., 6 *epist.*). In this case, the seller was obliged to exercise diligence in exacting the rent

---

63. On the career of Hermoginianus, see my introduction.
64. Hermog. *D.* 19.1.49 pr.: ". . . nec defenditur, si, quo facilius excogitata fraus occultetur, colonum et quinquenni pensiones in fidem suam recipiat." Cf. Papin. *Frag. Vat.* 13: ". . . nec idcirco recte defenditur, si, quo facilius excogitatam fraudem retineret, colonum et quinque annorum mercedes in fidem suam recipiat."
65. Scaev. *D.* 20.1.32: ". . . eorum praediorum pars sine colonis fuit eaque actori suo colenda debitor ita [interea *Mommsen*] tradidit adsignatis et servis culturae necessariis. . . ."

from the tenant. Clearly purchasers of estates might acquire their properties under any number of circumstances, and the mere presence of a tenant on the estate in this transaction does not allow us to draw many conclusions about the overall importance of tenants to upper-class Romans when they invested in estates. But the important point is that this type of sale is at least consistent with a purchaser's desire to acquire an estate with a tenant in place.[66] Formally, as Ulpian states, the lessor who sold a farm retained the right to lease payments from the sitting tenant (Ulp. *D.* 19.1.13.11, 32 *ad ed.*).[67]

Servi Quasi Coloni

An assumption that landowners used tenancy as a means of simplifying the task of managing estates seems implicit in many of the decisions concerning *servi quasi coloni*.[68] A *servus quasi colonus* was a slave belonging to a landowner, whom the landowner set in charge of a farm or even of an entire estate as a kind of tenant. The slave managed the estate and paid the landowner a rent out of his *peculium*. *Servi quasi coloni* occupied the attention of the jurists primarily in connection with the legacy of *fundus cum instrumento*, that is, the legacy of a farm with the slaves, equipment, and livestock needed to cultivate it.[69] The prin-

---

66. The same principle is treated by Celsus (8 *dig.*) apud Ulp. *D.* 19.1.13.16 (32 *ad ed.*).

67. Ulp. *D.* 19.1.13.11: "Si in locatis ager fuit, pensiones utique ei cedent qui locaverat: idem et in praediis urbanis. . . ."

68. For discussion of the institution of *servi quasi coloni*, with a review of the important sources, see G. Giliberti, *Servus Quasi Colonus*, 2d ed. (Naples, 1988), who sees *servi quasi coloni* as occupying a position midway between slaves and tenants; cf. the review of Frier, *ZRG* 100 (1983): 667–76, at 667–71, who emphasizes the managerial advantage gained by leasing to such tenants. For a similar view of *servi quasi coloni*, see also Scheidel, *Grundpacht und Lohnarbeit*, 131–49, as well as idem, "Sklaven und Freigelassene als Pächter und ihre ökonomische Funktion in der römischen Landwirtschaft (Colonus-Studien III)," in H. Sancisi-Weerdenburg et al., eds., *De Agricultura* (Amsterdam, 1993), 182–96. Scheidel is skeptical about the historical deductions that may be drawn from the surviving evidence for this institution. D. Vera, "Schiavitù rurale e colonato nell'Italia imperiale," *Scienze dell'antichità, Storia, Archeologia, Antropologia* 6–7 (1992–93): 291–339, argues persuasively that the slavery attested in late imperial Italy involved slaves cultivating individual farms like free tenants.

69. For discussion of the details surrounding the legacy of *fundus cum instrumento*, see chap. 2.

cipal legal question that the jurists had to address was whether a *servus quasi colonus* was to be counted among the slaves included in a legacy of *fundus cum instrumento*. Such a legacy was generally deemed to include not only the slaves who served as farm laborers but other slaves as well, including those serving the villa, and, most importantly for our purposes, the slave *vilicus* actually managing the *fundus*. But the jurists persistently excluded from this legacy slaves who managed a *fundus* as if they were tenants, paying their owner a rent and making a profit for their *peculium* from the revenues provided by the estate. For example, in a ruling concerning what constituted the equipment of an estate in a legacy of a *fundus cum instrumento,* Ulpian, following Labeo and Pegasus, excluded from the legacy a slave who functioned as a tenant even if he customarily supervised the estate's slave staff (D. 33.7.12.3, 20 *ad Sab.*).[70] The criterion according to which the jurists judged the case was that only a conventional *vilicus* could be considered included in a legacy of a *fundus cum instrumento*. The slave tenant in this case might be envisioned as fulfilling two functions on the estate. He performed the ordinary duties of a *vilicus* in supervising the other slaves, while at the same time leasing an individual farm on the estate. But we should also imagine that the slave tenant in question might supervise the slave staff of the estate as part of his lease relationship with the estate owner. In other words, the estate was to be worked by slave labor but managed by a slave who held the estate in a quasi lease. Clearly other estates might be worked in a similar fashion but be leased out to conventional free tenants rather than to *servi quasi coloni*. The assignment of an estate to a slave bailiff, a *servus quasi colonus*, or to a free tenant was a managerial choice made by the estate owner, and it had little to do with the way in which the land comprising the estate was actually farmed—that is, whether through slave labor, hired free labor, free tenants, or some combination of all three.

We might hypothesize that managing a property through a *servus quasi colonus* provided advantages over the more conventional system of managing directly through a slave *vilicus*, in that the former had a financial incentive to manage his property well. The *servus quasi colonus* could turn a profit on his farm in the same way that a free tenant might, and he could thereby increase his *peculium*. In managing an estate in

---

70. Ulp. *D*. 33.7.12.3: "servus, qui quasi colonus in agro erat . . . etiamsi solitus fuerat et familiae imperare."

this way, the landowner would still have to make the same investments in farm capital and improvements as he would if administering his estate through a *vilicus*. For instance, in the first century B.C., Alfenus discussed the case of a landowner who leased a *fundus* out to his slave and provided his newly installed tenant with the oxen, or *boves*, needed for cultivating the property (*D*. 15.3.16, 2 *dig*.). Here the landowner bore an expense that the tenant might have to cover under ordinary leasing. In connection with this case, Alfenus ruled that the landowner was responsible for all the debts incurred by the slave in purchasing new animals. By analogy, the landlord would bear the responsibility for other debts incurred by the tenant.[71] Leasing to a *servus quasi colonus* did not allow the landowner to pass on to his tenant the burden of risk involved in cultivating the property in question in the same way that leasing to a free tenant did, since legally, at least, the landlord would ultimately bear some responsibility for any contractual obligations entered into by the slave tenant. We should imagine that managing through a *servus quasi colonus* was seen as advantageous because it might allow an absentee landowner to avoid some of the pitfalls involved in entrusting estates to conventional *vilici*, whom Columella at any rate condemned for being inherently untrustworthy (1.7.6–7). This advantage can be illustrated in a rescript issued by the emperor Gordian to a former slave named Chrestus, who himself had served his owner as a *servus quasi colonus* (*CJ* 4.14.5, A.D. 243). In this service, Chrestus had, according to his own description of the situation, cultivated several *fundi* belonging to his owner and for some reason accumulated substantial arrears. The owner confiscated Chrestus' *peculium*, but at some point she also gave Chrestus his freedom. Later, someone, in all likelihood the owner's heir, sought to sue Chrestus for these arrears, and the emperor ruled that any property that Chrestus had acquired after his manumission could not be disturbed to satisfy debts that he had accumulated while still a slave.[72] The owner had confis-

---

71. Under certain circumstances, the slaveholder could restrict his liability to the amount of the *peculium* when sued in the *actio de peculio*, on which see Kaser, *RP* I, 606–7. In the present case, the owner was liable to a person selling oxen to the slave on the *actio in rem verso* (Kaser, 607), since the provision of oxen benefited the owner.

72. *CJ* 4.14.5: "Si, ut adlegas, antequam a domina manumittereris, fundos eius coluisti posteaque adempto peculio libertate donatus es, ob reliqua, si qua pridem contracta sunt, res bonorum, quas postea propriis laboribus quaesisti, inquietari minime possunt."

cated her slave's *peculium* as a means of making good the tenant's arrears, and other landowners surely used the possibility of taking this drastic step as a way of promoting appropriate management of the land by the *servus quasi colonus* or at least obtaining compensation for improper management.

Leasing to a *servus quasi colonus*, then, represented one of the several options that Roman landowners had in managing their estates. It was not simply a form of agricultural labor.[73] Several responses by the jurists concerning the legacies of estates confirm that, in their conception at least, *servi quasi coloni* made a significant contribution in the management of upper-class properties. All these responses presuppose the presence on the estates in question of *servi quasi coloni* serving as large-scale managers rather than as cultivators occupying individual farms. As I already stated, the universal doctrine of the jurists was that such slaves were to be excluded from the legacy if the slave paid the landowner a rent, whereas slaves managing directly under the landowner's supervision were to be included in the legacy. Thus Scaevola, followed later by Paul, excluded from a legacy of *instrumentum* a slave manager not directly responsible to the landowner but instead paying rent, comparing him to a conventional tenant: "respondit, si non fide dominica, sed mercede, ut extranei coloni solent, fundum coluisset, non deberi" (Scaev. D. 33.7.20.1, 3 *resp.*; cf. Paul. D. 33.7.18.4, 2 *ad Vitell.*). The general principle is enunciated by Ulpian,

---

73. Scheidel, *Grundpacht und Lohnarbeit*, 131–49, views *servi quasi coloni* as represented in the legal sources solely as a managerial alternative to conventional slave *vilici*, not as a class of tenant laborers cultivating the land. A similar managerial function of *vilici* and tenants is suggested in several decisions concerning the legacy of *reliqua colonorum*, or tenants' arrears. See Scaev. D. 33.7.20.3 (3 *resp.*): "praedia . . . cum . . . reliquis colonorum et vilicorum"; Paul. D. 32.97 (2 *decret.*), concerning the legacy of *praedia* to an *actor*, where the legacy included "reliqua . . . tam sua quam colonorum"; and Papin. D. 32.91 pr. (7 *resp.*), concerning the legacy of "praedia . . . cum reliquis actorum et colonorum." It is not possible to prove this point conclusively, but as I will argue later in this chapter, it seems likely that the jurists envisioned estates as being divided into individual farms that might in some cases be managed by slave bailiffs and in others by tenants. *Actores* and *vilici* might in many cases be synonymous terms, but in the early empire the *actor* was usually a figure with greater responsibilities: see J.-J. Aubert, *Business Managers in Ancient Rome* (Leiden, 1994), 117–200, especially 186–96, as well as Carlsen, *Vilici and Roman Estate Managers*, 70–80, 121–42. For the difficulties in interpreting legacies that mention both *vilici* and *coloni*, see Capogrossi Colognesi, *Ai margini della proprietà fondiaria*, 276–78 n. 136 and "Il regime," 238–39 n. 136.

who excluded *servi quasi coloni* from the legacy of a *fundus cum instrumento*. He followed the earlier doctrine of Labeo and Pegasus, who excluded such a slave even if his task was to manage the slave labor force cultivating the land (which would of course be included in a legacy): "etiamsi solitus fuerat et familiae imperare" (Ulp. D. 33.7.12.4, 20 *ad Sab.*). In this latter ruling, the *servus quasi colonus* is envisioned simply as a manager of an estate, rather than as a small-scale tenant actually cultivating the land. This same assumption informed Paul's treatment of the issue, where, in citing the earlier response of Scaevola already mentioned, the third-century jurist used the term *vilicus* to describe the slave in question (D. 33.7.18.4).[74]

The responsibilities of a *servus quasi colonus* might in many cases have been considerable. Many tenants of this class must have performed a function comparable to that of an *actor* of slave status named Antiochus who belonged to a landowner named Hosidius; the provisions of the latter's will are known because of a court case on which Paul commented (D. 32.97, 2 *decret.*).[75] Hosidius granted Antiochus his freedom and, in addition, bequeathed to his former slave certain properties, his *peculium,* and also the debts that Antiochus and the tenants on the estate had owed Hosidius: ". . . praedia certa et peculium et reliqua relegaverat tam sua quam colonorum. . . ." Antiochus, then, apparently served as the manager of a large estate on which the bulk of the land was leased out among tenant cultivators. His responsibilities would have included collecting the rent from the tenants and enforcing their contractual obligations. At the same time, however, the existence of debts suggests that Antiochus managed this estate under some sort of quasi-contractual arrangement with his owner. Quite possibly, Antiochus paid his owner a rent for his right to manage the estate, and he was of course in a position to turn a profit from this undertaking. Scaevola seems to have envisioned a slave with similar responsibilities when he ruled that for a certain slave manager to receive the liberty requested for him by his former owner through a trust, the slave first had to comply with the wishes of the owner by rendering accounts. The slave was to repay money that he had stolen; his responsibilities included, among other things, collecting the rent owed by tenants and watching over the storehouses on the estate, the *horrea* and *apothecae*

---

74. Paul. D. 33.7.18.4: "si non pensionis certa quantitate, sed fide dominica coleretur"; cf. De Neeve, *Colonus,* 101 n. 168.

75. This text is also discussed in chap. 2.

(Scaev. *D.* 40.7.40 pr., 24 *dig.*).⁷⁶ A *servus quasi colonus* could also accumulate debt from cultivating a farm assigned to him, not just by collecting rent. At least Scaevola conceived of this situation, in the ruling already mentioned concerning the criteria by which a slave manager might be included in a legacy: "... Stichus servus, qui praedium unum ex his coluit et reliquatus amplam summam ..." (*D.* 33.7.20.1, 3 *resp.*).

To judge by the cases in the Digest involving *servi quasi coloni*, Roman landowners viewed tenancy as the preferred solution to the problem of managing estates. This preference is demonstrated in the jurists' consideration of cases in which individual farms on a larger estate were leased out to slave tenants, even when conventional tenants were to be found on the same estate. Arguably, landowners would be most likely to resort to this expedient when it was not possible to find a sufficient number of free tenants to cultivate all the land within an estate; if this supposition is true, leasing to *servi quasi coloni* represented a second-best alternative to leasing to conventional tenants, but it was preferable to managing a farm directly through a slave bailiff. Several descriptions by the jurists of admittedly hypothetical estates bequeathed in legacy suggest that the jurists conceived of landowners guided by such considerations. The estates standing behind the jurists' hypothetical descriptions would have been cultivated in a variety of ways, with some farms divided among free and slave tenants, and with others exploited under the management of slave bailiffs.⁷⁷ Free tenants, of course, could not be bequeathed, and slave tenants, unless the testator made a specific provision concerning them in the will, did not count as part of a legacy of a *fundus cum instrumento*.

We learn of the presence of various categories of tenants on estates from the jurists' discussion of legal questions about the status of the *reliqua colonorum*, the back rent and other debts owned by tenants. For example, Paul's discussion whether a legacy of a *fundus* might include the arrears of the tenants as well as the slaves, *reliqua colonorum et man-*

---

76. See De Neeve, *Colonus,* 159 n. 192, on this text as attesting tenancy on "compact" estates.

77. On slaves working side by side with tenants, see Vera, *Athenaeum* 83, (1995): 352–56; De Neeve, *Colonus,* 72, 163 and n. 213, 165–66; Capogrossi Colognesi, "Grandi proprietari, contadini e coloni," 344–48; idem, *Ai margini della proprietà fondiaria,* 274–88 and "Il regime," 237–44; and P. Rosafio, "Slaves and *Coloni* in the Villa System," in J. Carlsen et al., eds., *Landuse in the Roman Empire* (Rome, 1994), 145–58.

cipia, suggests what might have been a common type of arrangement (*D.* 32.78.3, 2 *ad Vitell.*).[78] In Paul's conception, an estate could be expected to have tenant farmers working individual parcels as well as slave labor cultivating other lands. The slaves may have included household slaves serving the villa, but they should also have included agricultural slaves. Some of the agricultural slaves may even have worked on farms occupied by tenants, if the tenants did not supply their slaves themselves.[79]

A similar organization of an estate seems presupposed in Scaevola's discussion of a legacy of *fundi* equipped with their farm implements, furniture, livestock, bailiffs, the tenants' arrears, and a storehouse: "'Seiae fundos quos reliqui, ita ut sunt instructi rustico instrumento suppellectile pecore et vilicis cum reliquis colonorum et apotheca habere volo'" (*D.* 33.7.20 pr., 3 *resp.*). Since more than one *fundus* was bequeathed, it is not clear whether the *vilici* were to be envisioned as managing estates divided among tenants or as supervising the cultivation of individual *fundi*. In all likelihood, the estates envisioned in this case consisted of autonomous individual farms, some of which will have been managed by *vilici,* with others being leased out to tenants. Another ruling on the disposition of slaves in a legacy also indicates the presence of tenants on an estate worked at least in part by slaves. In this case, a landowner bequeathed seaside properties to a foster parent; the legacy included slaves, *instrumentum,* crops, and the arrears of the tenants (Scaev. *D.* 33.7.27 pr.-1, 6/16 *dig.*).[80] Similarly, the estate envisioned in a third decision by Scaevola (*D.* 33.7.27.1, 6/16 *dig.*) was cultivated by a combination of tenants and slaves.[81]

---

78. Paul. *D.* 32.78.3: "'Peto, ut fundum meum Campanianum Genesiae alumnae meae adscribatis ducentorum aureorum ita uti est.' quaero, an fundo et reliqua colonorum et mancipia, si qua mortis tempore in eo fuerint, debeantur. respondit reliqua quidem coloroum non legata: cetera vero videri illis verbis 'ita uti est' data."

79. For much the same case, see Scaev. *D.* 32.101.1 (16 *dig.*): "peto fundum meum ita, uti est, alumnae meae dari." Cf. the slaves who worked on the farms occupied by tenants on the estate that Pliny was considering purchasing at Tifernum Tiberinum (*Ep.* 3.19.6–7).

80. Scaev. *D.* 33.7.27 pr.: "Praedia maritima cum servis qui ibi erunt et omni instrumento et fructibus qui ibi erunt et reliquis colonorum nutritori suo legavit."

81. Scaev. *D.* 33.7.21.1: "Adfini suo ita legavit: 'fundum Cornelianum Titio ita ut est instructus cum omnibus rebus et mancipiis et reliquis colonorum dari volo.'"

Still another case discussed by Scaevola provides additional evidence that the jurists conceived of estates managed by tenants of both free and servile status (*D.* 33.7.20.3, 3 *resp.*). This legacy concerned *praedia instructa*, a characterization of property more inclusive than the *fundus cum instrumento*, since in the legal concept of *instructum*, the legacy included not only the equipment, slaves, and livestock needed for cultivation but also the equipment and slaves required for serving the owner during his visits (see Ulp. *D.* 33.7.12.27, 20 *ad Sab.*).[82] The legacy in this particular case included, in addition to equipment and livestock, the arrears of both tenants and bailiffs, the *peculia* belonging to the slaves, and the *actor*.[83] This estate is then envisioned as having been under the general administration of an *actor* who we know, because of his inclusion in the legacy, was a slave, while the individual farms were divided among tenants of both free and servile status. Since the *vilici* owed arrears, they must be viewed as *servi quasi coloni*. Slave tenants in some cases clearly might be assigned the responsibility for a very large unit of land when smaller units were leased out to free tenants, but there is no reason to suppose that the farms managed by slaves were envisioned as being much different from those leased by the free tenants, in terms of their size and the crops raised there. The slaves included in this particular legacy must have been assigned to the charge of the *servi quasi coloni*, but, to judge at least by the example of Pliny's estates, slaves may also have been used to cultivate the farms managed by conventional tenants (*Ep.* 3.19.6–7). Papinian envisioned an estate managed in this way when he discussed the legacy of *praedia* along with the debts owed by the tenants and by other managers of servile status, or *actores* (. . . "cum reliquis actorum et colonorum," *D.* 32.91. pr., 7 *resp.*), The extent of the responsibilities of a slave *vilicus* or *actor* leasing his part of an estate is revealed in the case discussed by Scaevola concerning the status of a slave assigned the cultivation of an individual farm of a larger estate (*D.* 33.7.20.1). This slave tenant, who was assigned the use of slaves and oxen, had far-reaching responsibilities, since it fell to him to acquire additional livestock needed to cultivate his land.

---

82. On the legacy of the *fundus instructus*, see chap. 2.

83. Scaev. *D.* 33.7.20.3: "praedia ut instructa sunt cum dotibus et reliquis colonorum et vilicorum et mancipiis et pecore omni legavit et peculiis et cum actore. . . ."

Types of Tenants

There was clearly a wide range in the resources that tenants leasing from upper-class landowners might have at their disposal. On one end of the scale, we might find a tenant like the Maurus discussed in the passage from Paul's *Decreta* (Paul. D. 32.27.2, 2 *decret.*). We might compare the *urbanus colonus,* whose disadvantages as a tenant Columella emphasizes, following the combined authority of the Sasernae, agronomists of the early first century B.C., and of his own elderly and wealthy contemporary L. Volusius Saturninus (1.7.3–4).[84] The tenants envisioned by Columella and his authorities may well have included a type markedly different from the tenants to be found on the estates of a landowner like Pliny. Pliny's tenants owned slaves and cultivated substantial farms, but there is no indication in Pliny's letters that they did not live on or at least near their farms or, more important, that they did not depend on their individual farms for their livelihood. The "city tenant" in Columella's discussion, by contrast, could have rented several farms at the same time and so need not have been dependent for his livelihood on the production of any individual farm, or, more significantly, on his contractual relationship with any single landlord. The "city tenant" may indeed have been a substantial landowner himself. His ability to lease substantial farms or estates would have allowed him to extend his agricultural income, making the most intensive use possible of the productive resources available to him, in particular slaves and livestock, without having to invest his capital in the purchase of additional land.

Columella viewed leasing to a "city tenant" as disadvantageous because such a tenant, taking a less direct interest in his holding, preferred to manage the leased property through a slave bailiff and so subjected the estate owner's land to all the disadvantages of absentee ownership with none of the advantages accruing from leasing.[85] We should imagine that Columella was reacting to what he considered to be a practice that many contemporary landowners pursued despite the dis-

---

84. On the Sasernae and the *urbanus colonus,* see J. Kolendo, *Le traité d'agronomie des Saserna* (Wroclaw, 1973), 40–41. On the identity of Columella's Volusius, see Scheidel, *Grundpacht und Lohnarbeit,* 56–57 n. 136.

85. For Columella's recommendations concerning leasing, see my discussion earlier in this chapter. On the *urbanus colonus,* see Scheidel, *Grundpacht und Lohnarbeit,* 109–15.

advantages that he saw it bringing to agriculture. To judge by Columella's emphasis on the point, it seems clear that some of his contemporaries endeavored to solve the problem of managing their estates by leasing to large-scale contractors. These landowners were interested primarily in attaining a steady income with as little bother as possible, and Columella saw the practice as deleterious to agriculture because the contractor could not be expected to provide the same care to his farm as a tenant dependent on it for his livelihood.

In this connection, we might consider a wealthy tenant envisioned by Labeo as employing slaves who supervised the renting of land (apud Marcianus D. 32.65 pr., 7 inst.). In this case, the wealthy individual had excluded from his will slaves involved in business activities, *exceptis negotiatoribus*. Labeo ruled that the slaves that were to be considered as excluded from this type of legacy included those in charge of such operations as purchasing, leasing, and renting.[86] For Labeo it would have been unexceptional if an individual were at the same time both a tenant and a landlord of agricultural property. The dearth of evidence makes it impossible to demonstrate the existence of tenants leasing under these particular conditions, but tenancy did fulfill this function for modest and even substantial landowners in Egypt during the early empire. There, cultivators sought to extend their resources over as large a cultivated area as possible by leasing parcels in addition to the land that they owned; the result was that the boundaries between landowner and tenant were often blurred.[87]

But wealthy tenants must have been comparatively rare, and the jurists treatment of them only allows us to infer the existence of a type

---

86. D. 32.65 pr.: "Labeo scripsit eos legato exceptos videri, qui praepositi essent negotii exercendi causa, veluti qui ad emendum locandum conducendum praepositi essent. . . ."

87. See especially the family of Sarapion, who owned a substantial amount of land in the Hermopolite nome, perhaps several hundred *arourae*, but also increased their income from agriculture by engaging in a number of leasing activities, including leasing and subletting land belonging to large estate owners, leasing land to cultivate themselves, and leasing out some of their own land to tenants. I discuss the economic activities of the family of Sarapion in *Management and Investment*, 67–72; for the documents in the Sarapion archive, see J. Schwartz, *Les archives de Sarapion et ses fils* (Cairo, 1961). Rowlandson, *Landowners and Tenants in Roman Egypt*, 212–13, 257–58, 264–66, argues that, although the lines between landowners and tenants might be blurred, growing stratification of wealth led to an increasing gulf between landowners and tenants in Roman Egypt.

of leasing "characteristic" of the Roman Empire.[88] This form of land tenure presumably played some historical role in the Roman agrarian economy, but it is clear that the vast majority of tenants were of relatively humble status, leasing on a much smaller scale.[89] A more likely common pattern can be seen in the organization of Pliny's estates at Tifernum Tiberinum. These estates consisted of a number of farms of varying sizes, which were not all contiguous.[90] A business manager, or procurator, had general charge of Pliny's property, and individual properties were managed by *actores*, a term that is here probably synonymous with *vilici* (*Ep.* 3.19.2). Many, and in all likelihood most, of the individual farms comprising these properties, or *praedia*, were cultivated by tenants. A comparable organization of an estate seems to be presupposed in Scaevola's decision that property brought onto a farm already pledged as security for a loan would also be counted as part of the pledge (*D.* 20.1.32, 5 *resp.*). The owner of the property in this case is envisioned as having assigned a part of it not cultivated by tenants to the management of an *actor*; he also is to have supplied this *actor* with the slaves necessary for cultivating the land in question.[91] In this case,

---

88. For large-scale tenants to have played an important role in Italian farm tenancy would imply that agricultural land was difficult to purchase, even for relatively wealthy farmers, so that farmers whom we might otherwise have expected to be landowners were forced to rent land.

89. As Capogrossi Colognesi, *Ai margini della proprietà fondiaria*, 200–2 and "Il regime," 194–96 argues, large-scale tenants in classical Italy are more likely to have been "middlemen," comparable to the *conductores* on the North African imperial estates or the *misthotai* leasing *ousiai* in Julio-Claudian Egypt, rather than large-scale agricultural entrepreneurs who leased in land for the purpose of producing and marketing crops. Capogrossi Colognesi (220–45 and "Il regime," 206–21) also emphasizes that Italian farm tenants disposed over a wide range of resources, with many completely dependent on the landowner for the provision of needed equipment, livestock, and even slaves.

90. For this aspect of estates in early imperial Italy in general and Pliny's estates in particular, see n. 4 in my introduction. As is emphasized by Capogrossi Colognesi, *Ai margini della proprietà fondiaria*, 245–74 and "Il regime," 221–37, with reference to comparative evidence from modern Italy, the scattered farms comprising a larger estate might be cultivated with a variety of crops and would represent an economic unit primarily in the sense that they were subjected to a unified managerial hierarchy.

91. *D.* 20.1.32: "eorum praediorum pars sine colonis fuit eaque actori suo colenda debitor ita [interea Mo.] tradidit adsignatis et servis culturae necessariis." On this text, see also Capogrossi Colognesi, *Ai margini della proprietà fondiaria*, 269–70 and "Il regime," 234–35.

Scaevola apparently envisioned an estate consisting of a number of individual farms, cultivated separately. On such an estate, the farms probably only constituted an economic unit in the sense that they were owned by one person. A tenant like the Maurus I have previously discussed, then, might have cultivated one of the larger units within an estate like this; the service of such a tenant would consist of taking over the management of a series of lands that otherwise would have been assigned to a slave bailiff. A large-scale tenant like Maurus might employ his own slaves, similar to the slaves supervised by a *vilicus*. He might also lease out some of his land to subtenants.[92] The individual farms comprising such an estate, moreover, need not all have been subjected to the same form of management; some might be leased out to tenants, while others might be assigned to the management of a slave bailiff in charge of slave labor.

Although this discussion has centered on what appear to have been large-scale tenants, we should not conclude on this basis that the jurists only considered legal problems involving wealthy tenants. Instead, in keeping with their overall approach in treating legal problems involving private property, the jurists remained very abstract in their use of terminology and legal categories. Such terms as *fundi* and *villae* provided a convenient way to describe a wide range of tenancy arrangements, from the leasing of a huge estate by a Roman aristocrat to the cultivation of a modest farm by a small-scale tenant. Indeed, this terminology that the jurists used in describing forms of tenancy is well suited to describing tenancy on Pliny's estates. Pliny's tenants, who could hardly be characterized as large-scale entrepreneurs, could be described as leasing individual *fundi,* and, if farmhouses were included in the bargain, they would be characterized as *villae*. Pliny's tenants also provided the movable capital needed to cultivate their farms, including slaves and probably also livestock (*Ep.* 3.19.6–7).[93] The jurists

---

92. The subleasing of *fundi* is treated in the legal sources; see Heitland, *Agricola*, 364 and n. 3, with references. Cf. T. Frank, *Rome and Italy of the Empire,* in *An Economic Survey of Ancient Rome*, vol. V (Baltimore, 1940), 179–80, on Plin. *Ep.* 3.19; Frank views the estate at Tifernum Tiberinum as let out to large-scale *conductores*.

93. In *Ep.* 10.8.5, Pliny discusses the leasing out of his estates: "Agrorum enim, quos in eadem regione possideo, locatio, cum alioqui CCCC excedat, adeo non potest differri, ut proximam putationem novus colonus facere debeat. praeterea continuae sterilitates cogunt me de remissionibus cogitare; quarum rationem nisi praesens inire non possum." On the basis of this passage, De

did not extend or adapt the legal categories with which they worked to accommodate various forms of farm tenancy. Rather, they persisted in describing a wide range of arrangements in terms of a standard contract. Their purpose in doing so was to keep the contractual relationship as the central principle defining the relationship between landowner and tenant.[94] Valuable for the economic historian is the fact that the jurists are also consistent in their assumptions about the economic value of leasing to tenants and that their assumptions in this area are consistent with the economic principles that they apply in other areas of the law impinging on the financial interests of upper-class Romans. The terms under which the jurists envisioned the normative tenant leasing his holding suggest a basic economic orientation on the part of the upper-class landowner. This landowner, financially conservative to the extreme, sought to avoid to the extent possible direct involvement in the management of his or her estates and at the same time benefited when the tenant could contribute to the costs of invest-

---

Neeve, *Athenaeum* 78 (1990): 379–83, 392–93, especially 392–93, argues that Pliny was concerned in this case with leasing his *agri* to a single tenant and that the figure of four hundred thousand sesterces represented the rent paid by this tenant, not for a single year to be sure, but rather for the entire five-year *lustrum*; cf., earlier, De Neeve, *Colonus*, 17, 82, 166 and n. 228. Accordingly, in De Neeve's view, Pliny's tenants, at least before the institution of sharecropping, were of relatively high economic status, with the wealthiest recruited from the local decurion class. For a similar view, see Scheidel, *Grundpacht und Lohnarbeit*, 65. But the singular *colonus* must be a collective describing the tenants on the estate as a whole, or else the following reference to remissions in the plural would not make sense. We might also question why Pliny would need to provide Trajan with details about how his estates were leased; his point was to emphasize the financial urgency of his presence at Tifernum Tiberinum. The possibility of granting remissions adds to this urgency, since his rents provided a substantial income. For *novus colonus* as a collective, see Johne, in Johne et al., *Kolonen*, 135–36; R. Duncan-Jones, *JRS* 76 (1986): 296; and the other references in De Neeve, *Athenaeum* 78 (1990): 380 n. 91. For a different view, see N. Brockmeyer, "Arbeitsorganisation und ökonomisches Denken in der Gutswirtschaft des römischen Reiches" (Ph.D. diss., Ruhr-Universität Bochum, 1968), 387, who views the *novus colonus* as a large-scale lessee, like the *conductores* on the North African imperial estates. For skepticism about the accuracy of four hundred thousand sesterces as the amount of Pliny's rent, see W. Scheidel, "Finances, Figures and Fiction," *CQ* 46, no. 1 (1996): 222–38, at 228.

94. For discussion of this concern of the jurists, see Zimmermann, *The Law of Obligations*, 348–50, 379, who emphasizes their concerns with legal arrangements among the upper classes.

ing in the estate's continued productivity. As we will see in chapter 4, the allocation of these costs became fertile ground for disputes between landowner and tenant.

Conclusion

The evidence examined in this chapter does not allow us to say how common particular forms of land tenure were, in Italy or in other parts of the empire. Despite this limitation, the jurists' treatment of tenancy does allow us to infer several important characteristics of the Roman estate economy. It seems clear that tenancy represented to many absentee landowners a major part of the solution to problems inherent in managing extensive and distant estates. Landowners characteristically kept their estates divided into smaller and more manageable units;[95] the importance of keeping the task of running large estates at manageable proportions surely led many landowners to assign units within such estates to the care of slave tenants when a suitable number of free tenants could not be found. From this perspective, the leasing to *servi quasi coloni* served primarily to help landowners overcome the difficult problem of managing diverse and scattered properties. The *servus quasi colonus* replaced the conventional bailiff in supervising the cultivation of a large portion of the estate, and the actual organization of the labor within the estate was independent of the managerial structure at top.

The basic orientation toward economic matters that the jurists envision for landowners in their treatment of farm tenancy, then, is the same as that they envision in their treatment of tutorship and the bequest of agricultural properties. The type of leasing envisioned by the jurists in their treatment of farm tenancy seems well suited to the needs of financially cautious landowners who looked for security above all else from their holdings. But another factor also affected the security that landowners derived from their estates, namely, their ability to keep their estates productive for the long term. It is in terms of productivity that tenancy made its most important contribution to the

---

95. The restricted size of the units typically characterizing an estate in early imperial Italy was partly a function of the means by which estates were acquired, with *latifundia* typically consisting of smaller *fundi* gradually absorbed under the ownership of a single proprietor. Often the old boundaries of a formerly independent *fundus* absorbed into a larger estate might be kept intact. For this process of building estates, see my introduction.

financial interests of upper-class landowners: the regular investments in farm improvements needed to keep an estate productive provided a major part of the solution to the problem of keeping at a minimum the landowner's share of the risks and expenses involved in cultivating large estates. The efforts on the part of the jurists and imperial government to treat this issue are the subject of chapter 4.

4
The Allocation of Resources in Farm Tenancy

In this chapter, I investigate how the jurists addressed an issue of crucial importance to the economic interests of cautious upper-class Romans, namely, the allocation of the costs and risks of farm tenancy. As we have seen in chapter 3, farm tenancy, as conceived by the jurists, was well suited to serve the economic interests of upper-class Romans who valued their investments in agriculture primarily for the economic security that they provided. In the jurists' treatment of farm tenancy, leasing provided a way to simplify the task of managing diverse and scattered holdings. The tenants took on themselves the responsibility to manage individual properties, and through their rent they furnished a disposable income.

It is clear, however, that many landowners depended for their income on leasing out their lands to numerous tenants of modest status and that the relationship between such landowners and tenants was far more dynamic and complicated than would appear if we limited our investigation to the abstract tenant of the normative Roman lease. To what extent do the legal sources inform us about this dynamic economic relationship? In previous studies of the Roman economy, I have argued that landowners enhanced their security when their tenants remained on their land for the long term and continued to invest their own resources in the improvements needed to keep it productive. The question as to which party would cover the costs of such improvements thus became a crucial issue in defining the economic relationships between Roman landowners and tenants. For their part, landowners sought to adapt the legal, social, and economic institutions surrounding farm tenancy so that they could impose the costs of investment on the tenants. Tenants able to cover the costs of needed equipment, draft animals, and labor exercised considerable bargaining power with their landlords. This struggle over allocating the costs and risks of farming can be traced in several settings, including in the steps taken by the imperial government in exploiting its estates in North Africa, in the relationships of Pliny the Younger with his tenants, and also in farm tenancy in Roman Egypt.[1]

To summarize my argument at the outset, the legal evidence preserved in the Digest and in the Justinian Code suggests that the Roman legal authorities contended with this struggle over covering the costs and risks of tenancy. The very nature of the legal evidence, however, makes it difficult to trace such a struggle. Legal decisions concerning farm tenancy tend to be formulated in terms of general principles, and they rarely reveal much about the parties involved in a dispute. Moreover, the legal sources do not allow us to trace over a period of time how any one landowner dealt with a group of tenants. Nevertheless, the jurists did devote attention to several issues crucial to the economic relationships between landowners and small-scale tenants, including the tenant's security of tenure, the tenant's right to compensation for improvements to the farm held under lease, and the distribution of risk. When we consider their treatment of these issues against a broad understanding of their assumptions about the economic interests of landowners, we can appreciate how the Roman jurists adapted legal principles to the realities of the Roman economy.

The jurists remained very conservative in their regulation of farm tenancy, in that they were reluctant to reformulate existing legal categories to accommodate the varying conditions under which tenants occupied their land. Thus the jurists constantly struggled with the difficulty of fitting a dynamic economic reality into conventional legal cat-

---

1. See my *The Economics of Agriculture on Roman Imperial Estates in North Africa* (Göttingen, 1988); "Allocation of Risk and Investment on the Estates of Pliny the Younger," *Chiron* 18 (1988): 15–42; and *Management and Investment on Estates in Roman Egypt during the Early Empire* (Bonn, 1992), chap. 4. This aspect of Greek and Roman tenancy is emphasized by L. Foxhall, "The Dependent Tenant: Land Leasing and Labour in Italy and Greece," *JRS* 80 (1990): 97–114, and D. Vera, "Schiavitù rurale e colonato nell'Italia imperiale," *Scienze dell'antichità, Storia, Archeologia, Antropologia* 6–7 (1992–93): 291–339, at 323–24, 335–38; cf., for Egypt, J. Rowlandson, *Landowners and Tenants in Roman Egypt* (Oxford, 1996), 272–79. For a different view of the importance of farm tenancy in the Roman economy, cf. D. Rathbone, "More (or Less?) Economic Rationalism in Roman Agriculture," *JRA* 7 (1994): 432–36, at 435, who argues that it served landowners primarily as a convenient means to organize labor; cf. R.S. Bagnall, "Managing Estates in Roman Egypt: A Review Article," *BASP* 30 (1993): 127–35.

egories. Part of the reason for this struggle is that the jurists were not in the business of prescribing contractual terms as part of an effort to implement a social policy, such as governmental protection of the tenant. In a formal sense, both landowner and tenant were free to negotiate contractual terms as they saw fit; the task of the jurists was to define and enforce contracts in terms of conventional Roman law. But it is clear that the jurists did recognize the existence of the dynamic relationships between landowner and tenant that I have described. The jurists' conservative approach can be seen in their treatment of the tenant cultivating a farm for a long period of time and continually investing his own resources in the land. Instead of formally recognizing this type of tenancy arrangement, the jurists persisted in describing the rights and duties of landlord and tenant in terms of a normative short-term lease against a cash rent, which remained the basis for the Roman law of farm tenancy.

Long-Term Lease Relationships

It is difficult to trace in the Digest and in other legal sources the contribution of resources that the tenant made. There certainly were wealthy tenants capable of taking over the management of an entire estate, and their services were undoubtedly valued by upper-class landowners. But the majority of tenants cultivating the estates of upper-class landowners were of much more modest means, and landowners had to adapt the institution of farm tenancy to channel the resources of such tenants into continued investment in an estate's productivity. Under this circumstance, we should expect many landowner-tenant relationships to have lasted far beyond the short-term lease that is envisioned as the norm by the Roman jurists. To judge by the parallels from the Egyptian evidence, where short-term farm leases were also the legal norm, tenants who contributed to their landlords' financial interest by carefully managing and tending their farms surely had far more secure tenure on their land than is suggested by the terms of the short-term lease envisioned by the Roman jurists.[2]

---

2. For enduring relationships between landowners and tenants in Roman Egypt, see my *Management and Investment,* 140–58, and "Legal Institutions and the Bargaining Power of the Tenant in Roman Egypt," *APF* 41, no. 2 (1995): 232–62, at 238–50; cf. Rowlandson, *Landowners and Tenants in Roman Egypt,* 252–59, on the duration of leases.

By examining the jurists' treatment of the tenant's security of tenure, we can trace the conflict between the jurists' conservatism and the need to recognize economic reality. To be sure, no tenant would want to undergo the expense and effort of improving his farm if he had no assurance that he would be able to benefit from his investment. And yet the jurists' treatment of farm tenancy was somewhat paradoxical. The jurists persistently defined the normative farm lease as a short-term one, lasting five years, and in this scheme the tenant seeking to remain in his tenancy for longer periods of time apparently met with severe legal disadvantages. At first sight, it might be argued that Roman farm tenants, lacking the right of possession, always had to contend with the insecurity of their tenure. Their tenure was especially insecure if the land that they were leasing was alienated, whether by sale or bequest. The new owner was not required to honor a lease contract existing between the tenants and the previous owner, although he was obliged to provide compensation for tenants evicted from their holdings. But this compensation would be of little solace if the farms that they were leasing provided the evicted tenants with their only means of securing their livelihood.[3]

Lacking the right of possession, the tenant could, subject to compensation, be evicted from his tenancy without notice, and the tenure of the tenant was, in theory at least, threatened every time an estate changed hands. The principle that sale (or any other change of ownership) ends a lease relationship, *emptio tollit locatum*, was basic to classical Roman lease law. As Mayer-Maly has argued, the classical jurists did not maintain this principle out of any concern to assert the economic power of landowners over tenants. Rather, in the jurists' view, it would compromise the rights of a property owner if the purchaser of a *fundus* leased out to a tenant were required to take over the seller's obligations under the lease. Otherwise, a property owner would be subject to obligations contracted by a third party.[4] But even if the legal principle restricting

---

3. On the lack of *possessio* on the part of tenants, see J. Köhn, "Die Kolonen in den Rechstbestimmungen,"in K.-P. Johne et al., *Die Kolonen in Italien und den westlichen Provinzen des Römischen Reiches* (Berlin, 1983), 167–257, at 191–95. For the treatment of the tenure of the tenant (in particular the apartment tenant) in Roman law and in the later civilian tradition, see R. Zimmermann, *The Law of Obligations* (Cape Town, 1990), 377–83.

4. On whether the principle *emptio tollit locatum* was classical Roman law, see T. Mayer-Maly, *Locatio Conductio* (Vienna, 1956), 42–60, especially 45–46. Postclassical Roman law strengthened the rights of tenants to remain on their farms that were alienated: see Mayer-Maly, 55.

the rights of a tenant at the alienation of their farm did not result from any economic concern on the part of the jurists, the tenant's situation could still be precarious. Indeed, Finley identifies this insecurity of tenure as a major stumbling block to promoting investment by the tenant and general progress in Roman agriculture.[5]

A number of responses in the Digest offer a grim picture of the plight of tenants as a consequence of the alienation of the land that they were leasing. For example, if a *fundus* leased out to a tenant was bequeathed to another party unwilling to honor the testator's original contractual obligations, the tenant could sue the heir (Cassius apud Julian. D. 19.2.32, 4 *ex Minic.*), but the evicted tenant could only expect to gain some compensation for his inconvenience, not restoration to the land in question. Eviction was not the only danger that the tenant faced. For example, the new owner of a farm who chose to allow the tenant to remain in his holding could impose less favorable lease terms. Also, if a farm was alienated, the tenant did not even exercise control over the disposition of the crops that he had harvested. In connection with an estate leased out to a tenant by a now deceased testator, Ulpian ruled that such crops would belong to the heir or the legatee. The tenant could still of course sue the heir for his inconvenience in not being allowed to sell the crops (Ulp. D. 30.120.2, 2 *resp.* = *Frag. Vat.* 44).[6]

The new owner of an estate, then, had a great deal of freedom in dealing with sitting tenants, but the jurists do seem to have recognized efforts on the part of both tenants and landowners to afford tenants greater security than was provided by the classical Roman lease. For example, Paul discussed a lease with a provision that any lessor evicting the tenant from his holding would have to pay a substantial penalty (D. 19.2.54.1, 5 *resp.*). In this case, the tenant was aware of the unfortunate consequences of his lacking the right of *possessio*. Earlier, Gaius had considered it an obligation of a landowner selling a farm to make sure that the purchaser would maintain for the tenant the same terms

---

5. See M.I. Finley, "Private Farm Tenancy in Italy before Diocletian," in M.I. Finley, ed., *Studies in Roman Property* (Cambridge, 1976), 103–21, at 109.

6. Ulp. D. 30.120.2: "[r(espondit) Aurelio Felici *Vat.*] fructus ex fundo pure legato [per vindicationem pure relicto *Vat.*] post aditam hereditatem a legatario perceptos ad ipsum pertinere, colonum autem cum herede ex conducto habere actionem." Cf. Valerian, Gallienus, *CJ* 4.65.15 (A.D. 259): "Si fundo a locatore expulsa es, agere ex conducto potes poenamque, quam praestari rupta conventionis fide placuit, exigere ac retinere potes."

of tenure (*D.* 19.2.25.1, 10 *ad ed. prov.*).⁷ This recommendation does not mean that lessors selling properties could create any binding legal obligation on the part of the purchaser toward a tenant. Rather, Gaius was probably concerned simply with avoiding unnecessary lawsuits, so he urged that the lessor selling a property not infringe the tenant's contractual rights.⁸ These considerations were not purely academic, as is indicated by a constitution of Alexander Severus, in which that emperor repeated the principles expressed by Gaius (*CJ* 4.65.9, A.D. 234).⁹

The imperial authorities, then, had to deal with legal problems involving the rights of the tenant when the farm that he was leasing changed ownership. Several responses in the Digest suggest how some of these legal problems might arise in the real world. Mela and Ulpian considered the rights of the tenant in what probably was not an altogether infrequent situation. A widow hoped to regain through the institution of *relegatio* a *fundus* that she had made part of her dowry and that her now late husband had leased out to a tenant for a fixed period (Ulp. *D.* 33.4.1.15, 19 *ad Sab.*).¹⁰ In addition, Sabinus and later Paul treated a similar case involving the return to the wife at divorce of dotal property that the husband had leased out to a tenant (Paul. *D.* 24.3.25.4, 36 *ad ed.*).¹¹ In the former case, Mela ruled that the widow could not regain ownership of the *fundus* unless she guaranteed that the tenant would be

---

7. Gaius *D.* 19.2.25.1: "Qui fundum fruendum vel habitationem alicui locavit, si aliqua ex causa fundum vel aedes vendat, curare debet, ut apud emptorem quoque eodem pactione et colono frui et inquilino habitare liceat: [alioquin prohibitus is aget cum eo ex conducto]." For discussion of this text, see Mayer-Maly, *Locatio Conductio,* 42–46, who considers the final clause to be a postclassical gloss (44).

8. Mayer-Maly, *Locatio Conductio,* 44.

9. *CJ* 4.65.9: "Emptori quidem necesse non est stare [sinere *suppl.* Mommsen *cum* BΣ] colonum, cui prior dominus locavit, nisi ea lege emit. verum si probetur aliquo pacto consensisse, ut in eadem conductione maneat, quamvis sine scripto, bonae fidei iudicio ei quod placuit parere cogitur." This constitution is addressed to Aurelius Fuscus, *miles.* It is not clear whether this individual was the purchaser of the property or the tenant. For discussion of this constitution, see Mayer-Maly, *Locatio Conductio,* 44–46.

10. *Relegatio* is the legacy to a surviving wife by her husband of property that was brought into the marriage as part of the wife's dowry; see Kaser, *RP* I, 337–38, 751; cf. S. Treggiari, *Roman Marriage* (Oxford, 1991), 387–88.

11. For discussion of these two texts, see Mayer-Maly, *Locatio Conductio,* 49–50.

able to use it for the remainder of his lease. The widow, however, would be entitled to the tenant's rent payments.[12] In the latter case, the now divorced wife had to promise to observe the tenant's rights, so that the husband would no longer have any liability toward the tenant.[13] But the jurists clearly had to pay attention to the much more common problem that both Gaius and the imperial government addressed in the responses I have previously discussed, namely, the rights of a sitting tenant when the property that he was leasing was alienated. Determining the rights of the tenant and the obligations of the landlord was not always a completely straightforward matter, however, since the two parties might make agreements in which one party or the other might bargain away his rights. For example, Paul discussed a case in which the landlord agreed to pay a penalty if the tenant was expelled from the *fundus* held under lease before the expiration of the contract, while the tenant for his part also agreed to pay the same penalty if he left the lease early (D. 19.2.54.1, 5 *resp.*).[14] Servius, Tubero, and Ulpian all addressed the question of the tenant's rights when a seller of a farm contracted with the purchaser to allow the tenant to remain in his tenancy (Ulp. D. 19.1.13.30, 32 *ad ed.*). If the purchaser failed to live up to his end of the bargain, the seller was entitled to sue him *ex vendito* for his breech of the contract of sale. This protection was important for the seller, because the tenant could bring an action against the seller for failing to provide a cultivable farm for the duration of the lease period.

A case whose discussion by Paul is preserved in detail in the Digest illustrates a more concrete version of the broad principle formulated by Gaius (D. 49.14.50, 3 *decret.*). This case concerns the sale of an estate by the imperial Fiscus to a private individual, under the condition that the

---

12. Mela apud Ulp. D. 33.4.1.15: "Ibidem Mela coniungit, si fundus in dote fuit locatus a marito ad certum tempus, uxorem non alias fundum ex relegatione consequi, quam si caverit se passuram colonum frui, dummodo ipsa pensiones percipiat."

13. Sab. apud Paul. D. 24.3.25.4: "Si vir in quinquennio locaverit fundum et post primum forte annum divortium intervenerit, Sabinus ait non alias fundum mulieri reddi oportere, quam si caverit, si quid praeter unius anni locationem maritus damnatus sit, id [a *ins. edd.*] se praestatum iri: sed et mulieri cavendum, quidquid praeter primum annum ex locatione vir consecutus fuerit, se ei restiturum. . . ." For this explanation of the text, see Mayer-Maly, *Locatio Conductio*, 49.

14. The legal question that Paul answers in this passage concerns the rights of the lessor to expel the tenant for failing to pay rent.

Fiscus could sell the estate to another party if it received a higher bid within a specified time (*in diem addictio*).[15] In this case, the original purchaser matched a subsequent higher bid. The legal question discussed by Paul concerned the ownership of the crops harvested during the time intervening between the purchaser's original bid and his subsequent, successful bid. This case was heard by the emperor's *consilium*, and it evoked conflicting responses among the jurists present. The conventional wisdom was that the crops belonged to the seller, which was the opinion of the procurator representing the Fiscus, Valerius Patruinus. But the case was made complicated by two factors. First, the estate consisted at least in part of vineyards, and both of the eventual purchaser's bids had been submitted during the vintage season. Second, the estate was leased out to a tenant. As a result of the first complication, the conventional wisdom about the ownership of the crops was rejected, so the crops would have been awarded to the purchaser. But Papinian and Messius considered it unfair that the tenant should lose the vintage, so they ruled that the tenant should be allowed to harvest and dispose of the vintage, while the purchaser of the estate should receive the rent for the current year. In this way, the Fiscus would not be liable to a suit filed by the tenant for failing to provide him with a property that he could cultivate.[16] There is no reason to suppose that

---

15. On *in diem addictio*, see A. Berger, *Encyclopedic Dictionary of Roman Law* (Philadelphia, 1953; reprint, 1989), s.v., and Kaser, *RP* I, 561. The original purchaser had the right to match any higher bid. For the process involved in judging this case, see T. Honoré, *Emperors and Lawyers*, 2d ed. (Oxford, 1994), 24; for detailed commentary, see Mayer-Maly, *Locatio Conductio*, 50–53, who considers the passage to be largely free of interpolation, comparing Pomp. apud Ulp. *D.* 18.2.6.1 (28 *ad Sab.*), which also addresses the ownership of the crop when a farm is sold. See also C. Sanfilippo, *Pauli Decretorum Libri Tres* (Milan, 1938), 112–19.

16. Paul. *D.* 49.14.50: "Papinianus et Messius [novam sententiam induxerunt], quia sub colono erant praedia, iniquum esse <existimabant> fructus ei auferri universos: sed colonum quidem percipere eos debere, emptorem vero pensionem eius anni accepturum, ne fiscus colono teneretur, quod ei frui non licuisset: atque si hoc ipsum in emendo convenisset. pronuntiavit tamen secundum illorum opinionem, [quod [si *Mo.*] quidem domino colerentur, universos fructus habere: si vero sub colono], pensionem <emptorem> accipere. Tryphonino suggerente, quid putaret de aridis fructibus, qui ante percepti in praediis fuissent, respondit, si nondum dies pensionis venisset, cum addicta sunt, eos quoque emptorem accepturum." Sanfilippo's suggested deletions and emendations to the text are indicated by brackets. The purchaser did, then,

the new owner of this estate intended to expel the sitting tenant or tenants.[17] The legal issues in this case rather concerned the obligations of the Fiscus, both toward the purchaser of the estate and toward the tenant. As Paul indicates, the solution that Papinian and Messius proposed was innovative,[18] but it also seems likely that similar situations must have arisen time and again when estates leased out to tenants were sold or otherwise changed ownership. We should imagine that the seller and purchaser would often have made the kind of agreement that Gaius recommended to avoid legal difficulties comparable to the ones in the present case.

The recommendation of Gaius suggests how economic considerations might induce landowners to provide their tenants with greater security than strict lease law required. We can trace the bargaining power that tenants were able to exercise when we consider the situation on the estates at Tifernum Tiberinum whose purchase Pliny the Younger considered (*Ep.* 3.19). There is no question that Pliny's tenants, although their leases were limited to the conventional five years, enjoyed security of tenure to a remarkable degree. This derived at least in part from the bargaining power they exercised as a result of their continuing investment in maintaining their farms.

The rents paid by tenants comprised the lion's share of the income that Pliny derived from his estates, and he depended on his tenants' continued productivity. When increasing indebtedness made them unable to pay their rent, Pliny could not avail himself of either of the customary remedies open to a landlord. He could not dismiss his tenants at the end of their leases, nor could he compensate for lost rent by confiscating the property that they had pledged as security. Neither measure would have served his interests. Tenants with resources adequate to cultivate a farm in the way that Pliny desired were apparently in short supply, so Pliny had to take steps to maintain his tenants'

---

acquire ownership of the grain crops harvested but still on the estate. These crops would generally have been harvested in the spring, while the vintage was harvested in the fall.

17. It is not clear whether the property involved in this case was leased out to a single tenant or whether the phrase *sub colono* simply indicated that the estate was leased out to an unspecified number of tenants. If so, this would be a similar usage of the singular form of *colonus* as in Plin. *Ep.* 10.8.5, where the term *colonus* clearly refers to the large number of tenants leasing his estates at Tifernum Tiberinum. This question is discussed in chap. 3.

18. But see Sanfilippo's alterations to the text, in n. 16 in this chapter.

access to needed resources. Under circumstances such as these, even the threat of confiscating the tenants' pledges would be of little help to a landowner like Pliny. He could not use such a threat to enforce the tenants' obligation to cultivate their farms in accordance with prescribed norms, since that step would only have handcuffed his tenants, who were aware themselves of this situation (see "Investment by Tenants" later in this chapter). Pliny tried to solve the problem of tenant indebtedness while preserving their access to resources. First, he granted remissions of rent (*Ep.* 9.37.2, 10.8.5). Then, when this measure did not prove successful, he abandoned the traditional and convenient system of leasing for cash rents and in its place instituted sharecropping (9.37).[19]

It is difficult to trace the existence of this type of security of tenure in the legal sources. Classical Roman law defined the farm lease as a contractual arrangement normally lasting five years, and no special legal problems arose as a result of the landlord and tenant negotiating to extend the lease. Even so, it is clear that the jurists did recognize the existence of long-term tenancy arrangements. In the second century, Pomponius was concerned with defining the rights of the tenant in a lease that was to last as long as the lessor so desired. Such a lease arrangement, in Pomponius' formulation, ended with the death of the lessor (*D.* 19.2.4, 16 *ad Sab.*).[20] Probably far more characteristic were farm leases that, despite being formally of short duration, lasted indefinitely. We can infer the jurists' recognition of this type of situation from the attention that they gave to cases in which landowners made their tenants beneficiaries of their wills. The jurists chiefly dealt with the legal problems involved when tenants received what they paid as rent for their farms through legacy (e.g., Paul. *D.* 34.3.16, 9 *ad Plaut.*) or when they received usufructs over their farms (Iav. *D.* 33.2.30.1, 2 *ex*

---

19. For different interpretations of Pliny's decision to institute sharecropping, see P.W. de Neeve, "A Roman Landowner and His Estates: Pliny the Younger," *Athenaeum* 78 (1990): 363–402, and P. Rosafio, "Rural Labour Organization in Pliny the Younger," *ARID* 21 (1993): 67–79.

20. Pomp. *D.* 19.2.4: "Locatio precariive rogatio ita facta, quoad is, qui eam locasset dedissetve, vellet, morte eius qui locavit [deditve *ins. Hal.*] tollitur." On the passage, see S. Waszynski, *Die Bodenpacht* (Leipzig and Berlin, 1905), 93, who compares farm leases from Byzantine Egypt whose term depended on the will of the lessor, ἐφ᾽ ὅσον χρόνον βούλει; cf. also Rowlandson, *Landowners and Tenants in Roman Egypt*, 252 n. 142.

poster. Lab., Julian. *D.* 7.1.34.1, 35 *dig.;* cf. Paul. *D.* 19.2.24.5, 34 *ad ed.*). Since the jurists dealt repeatedly with the legal consequences of a bequest to a tenant, we have reason to suspect that they were not simply engaging in an academic exercise in applying the law (here the law of legacy) but rather addressing a legal situation that to some extent characterized Roman upper-class social life.

Clearly, under normal circumstances, a landowner would only make a bequest to a tenant with whom he enjoyed some kind of enduring social bond. A typical arrangement underlies a case discussed by Iavolenus and previously by Labeo. A landowner had bequeathed through legacy to his tenant the usufruct over the *fundus* that the latter was leasing (*D.* 33.2.30.1, 2 *ex poster. Lab.*). In Iavolenus' response, the tenant had an action to compel the heir to release him from the obligation to fulfill the lease requirements. A somewhat more complicated legal situation is addressed in a response of the third-century jurist Paul, who was considering a case previously commented on by the first-century jurists Nerva and Atilicinius (*D.* 34.3.16, 9 *ad Plaut.*). In this case, a landowner provided in his will that a tenant currently leasing land from him should be freed from his obligation to pay rent.[21] Paul addressed this same situation in another place, where he described a trust in which the testator required the heir to free the tenant from his contractual obligations (*D.* 19.2.24.5, 34 *ad ed.*).[22] In this case, the testator had leased the farm to the tenant for several years and then died before the lease term ended. This type of bequest put the tenant in a very safe position, for, as Paul ruled in both cases, the heir could not circumvent the testator's clear intentions by expelling the tenant from the

---

21. Paul. *D.* 34.3.16: "Ei qui fundum in quinquennium locaveram legavi, quidquid eum mihi dare facere oportet oportebitve, ut sineret heres sibi habere Nerva Atilicinius, si heres prohiberet eum frui, ex conducto, si iure locationis quid retineret, et testamento fore obligatum aiunt, quia nihil interesset, peteretur an retineret: totam enim locationem videri." Cf. also Paul. *D.* 34.3.18, continuing the passage already cited in this note: ". . . et praeterea placuit agere posse colonum cum herede ex testamento, ut liberetur ex conductione: quod rectissime dicitur." On Nerva and Atilicinus see Berger, *Encyclopedic Dictionary of Roman Law*, s.v., and W. Kunkel, *Herkunft und soziale Stellung der römischen Juristen*, 2d ed. (Graz, Vienna, and Cologne 1967), 129–30.

22. Paul. *D.* 19.2.24.5: "Qui in plures annos fundum locaverat, testamento suo damnavit heredem, ut conductorem liberaret. si non patiatur heres eum reliquo tempore frui, est ex conducto actio: quod si patiatur nec mercedes remittat, ex testamento tenetur."

farm. In this case, the tenant, although lacking the right of possession, nevertheless could sue to be allowed to use the farm; the tenant would at least be entitled to compensation to any losses imposed on him by having his tenure on the farm curtailed.[23] If the heir allowed the tenant to remain on the *fundus* but did not remit the rent, the tenant had an *actio ex testamento* to compel the heir to fulfill the trust.

In this last case, we see that underlying the juristic discussion was an assumption that an enduring social relationship with significant economic implications could be adapted to the norms of Roman lease law: the landowner and tenant were connected to one another by an unspecified social bond, although the lease itself was formulated in completely conventional terms. It may be worthwhile to consider the likely background of the type of tenant envisioned in these responses. As Champlin has recently demonstrated on the basis of a careful analysis of wills recorded in the Digest, the circle of connections on whom upper-class Romans would bestow such a trust included especially wives, relatives other than the heirs, and dependent members of the household, such as freedmen and freedwomen.[24] In theory at least, an upper-class landowner could promote the careful long-term cultivation of his or her land more readily by leasing to a freedman than by leasing to a tenant of free status, since the social dependency of the freedman would make it easier for the landowner to enforce norms of behavior.[25] There would be no need for either Paul or the compilers of the Digest to mention that the tenant in question was a freedman, since the tenant's status had no bearing on the legal point at issue. One cannot press the point too far on the basis of evidence like this, but it seems probable that the legal cases discussed here arose in a social milieu in which landowners characteristically could exercise some social pressure on tenants to remain on their farms and cultivate them carefully. The ten-

---

23. The tenant could reclaim his economic loss, "id quod sui interest."

24. See E. Champlin, *Final Judgments* (Berkeley and Los Angeles, 1991), 103–68, on the circle of acquaintances remembered in wills, and 131–54, for freedmen and freedwomen.

25. See S. Treggiari, *Roman Freedmen in the Late Republic* (Oxford, 1969), 68–81, on the legal bond between freedmen and their patrons, and 208–28, on the social bond. For leasing to freedmen, see W. Scheidel, "Sklaven und Freigelassene als Pächter und ihre ökonomische Funktion in der römischen Landwirtschaft (Colonus-Studien III)," in H. Sancisi-Weerdenburg et al., eds., *De Agricultura* (Amsterdam, 1993), 182–96, and idem, *Grundpacht und Lohnarbeit in der Landwirtschaft des römischen Italien* (Frankfurt, 1994), 131–49.

ants, in turn, might look forward to some sort of reward for their compliance.

The Digest does provide at least one historical example of a tenant who enjoyed a relationship with his landlord close enough to allow him to become a beneficiary of the latter's will. I am referring to Iulius Maurus, the tenant of Iulius (or Iulianus) Severus; I discussed him in chapter 3 as an example of a tenant with considerable wealth (Paul. *D.* 32.27.2, 2 *decret.*). This individual was a legatee, receiving some unspecified property from his former landlord. But his former landlord had also left a legacy of fifty thousand sesterces to his foster son, and he wanted his former tenant to provide this money from the arrears that he owed as rent for the *fundus* that he had leased. Maurus may have been atypical to the extent that, to judge by his arrears, he was a large-scale tenant. But he clearly must have occupied his land long enough to develop his personal connection with the landowner, who thought enough of him to leave him a bequest, and who depended on him to pay out a trust.

Although it is certain that landowners could gain concessions from their tenants by applying social pressure, the jurists do seem to assume the existence of tenants in long-term lease relationships who had the capacity to negotiate with their landlords over the costs of maintaining the farm. A good example of such an assumption can be seen in a case, discussed by the second-century jurist Julian, concerning the legacy to a tenant of the usufruct over the *fundus* held under lease (*D.* 7.1.34.1, 35 *dig.*). The tenant could vindicate his usufruct, but in so doing he would terminate the lease that he held over the farm. As a consequence, he could sue the heir for compensation for unexhausted improvements, that is, improvements that he had made on the farm and that raised the value of the farm for lease.[26] It is significant for our purposes that this case presupposes the existence of a tenant who would regularly invest his own resources in maintaining the farm. Evidently, this tenant must have enjoyed some sort of continuing social relationship with the landowner, since only under such a circumstance is a legacy to a tenant possible. We must freely admit that such a social bond could afford the

---

26. Julian. *D.* 7.1.34.1: "Si colono tuo usum fructum fundi legaveris, usum fructum vindicabit et cum herede tuo aget ex conducto et consequetur, ut neque mercedes praestet et impensas, quas in culturam fecerat, recipiat." For the economic significance of compensation for unexhausted improvements, see J.M. Currie, *The Theory of Land Tenure* (Cambridge, 1981), 69–80.

landowner a greater degree of control over the tenant, but this conclusion does not call into question the importance that the tenant's contribution of resources had for the economic interests of the landowner.[27] The social bond between landowner and tenant accounts for the legacy, but the improvements that the tenant made in the farm were the product of the convergence of the economic interests of the two parties in the lease.

Investment by Tenants

The tenants in these cases must be envisioned as enjoying a greater security of tenure than we might expect on the basis of the classical Roman farm lease. Such tenants could contribute to the financial interests of landowners when they remained on their estates for the long term, cultivating their farms productively and continually investing in improvements maintaining or enhancing productivity. In fact, it is fair to say that both landowners and tenants sought to promote this situation, despite the terms under which farm leases were theoretically negotiated. Under these conditions, tenants with resources at their disposal would exercise great bargaining power, minimizing the disadvantages of their legal situation. The crucial question for landowners and tenants was not whether tenants could remain on their farms but rather which party would pay for the improvements needed to keep the tenant's farm productive. Both landowners and tenants benefited from the institution of farm tenancy, and they depended on one another for their respective economic security.

Let us now consider in greater detail the investment that a tenant might make in his farm. In the normative Roman farm lease, the landlord was expected to provide the fixed and permanent capital; we learn

---

27. For discussion of the dependence inherent in a private Roman lease relationship, see E. Lo Cascio, "Considerazioni sulla struttura e sulla dinamica dell'affitto agrario in età imperiale," in H. Sancisi-Weerdenburg et al., eds., *De Agricultura* (Amsterdam, 1993), 296–316, at 310. This theme is also emphasized by L. Capogrossi Colognesi, "Grandi proprietari, contadini e coloni nell'Italia Romana (I–III d.C.)," in A. Giardina, ed., *Società romana e impero tardo-antico*, vol. I (Rome and Bari, 1986), 325–65, 703–23, especially 339–44, and idem, *Ai margini della proprietà fondiaria* (Rome, 1995): 230–34, 236–37, 282–88, 296–300 as well as "Il regime degli affitti agrari," *Scienze dell'antichità, Storia, Archeologia, Antropologia* 6–7 (1992–93): 212–14, 215–16, 241–44, 249–52. See also Foxhall, *JRS* 80 (1990): 97–114.

this from a letter of the first-century jurist Neratius quoted by Ulpian (*D*. 19.2.19.2, 32 *ad ed.*). In a discussion of a vineyard leased out to a tenant, Neratius describes the *instrumentum* that the landlord was to provide; this equipment included storage jars set in the ground and winepresses—in short, the equipment that was permanently part of the *fundus*. The tenant for his part was responsible for materials that needed periodic replacement, such as ropes, baskets, and the like.[28] The terms under which the tenants on Pliny's estates cultivated their land seem to have corresponded generally to those of the normative Roman farm lease. These tenants leased their land for five-year periods, and they paid a fixed annual rent in cash. Pliny says nothing directly about the allocation of capital on his estates, but his discussion of the plight of the tenants on the estate whose purchase he was considering suggests that they contributed considerable resources (*Ep*. 3.19.6–7). Because the tenants had fallen chronically behind on their rent, the previous owner, to compensate himself for lost rent, repeatedly confiscated their pledges. We can assume that these pledges included slaves, since Pliny saw the need to reequip these tenants with slaves as a major stumbling block toward making the estate profitable.[29] And the tenants probably provided livestock as well.

In the case of Pliny's estates, we can detect a struggle between landowner and tenant over the allocation of resources, since Pliny's

---

28. For a detailed discussion of this passage and an analysis of the allocation of capital in the normative Roman farm lease, see B.W. Frier, "Law, Technology, and Social Change: The Equipping of Italian Farm Tenancies," ZRG 96 (1979): 204–28. Capogrossi Colognesi, *Ai margini della proprietà fondiaria*, especially 220–45 and "Il regime," 206–21, discusses the range of resources that Italian farm tenants brought to their farms. In his view, many contributed virtually no resources at all other than their labor and were therefore dependent on their landlords. For the resources contributed by tenants in Roman Egypt, see Rowlandson, *Landowners and Tenants in Roman Egypt*, 213–28.

29. Plin. *Ep*. 3.19.6–7: "Sed haec felicitas terrae imbecillis cultoribus fatigatur. nam possessor prior saepius vendidit pignora, et dum reliqua colonorum minuit ad tempus, vires in posterum exhausit, quarum defectione rursus reliqua creverunt. sunt ergo instruendi, eo pluris quod frugi, mancipiis; nam nec ipse usquam vinctos habeo nec ibi quisquam." For a different interpretation of the situation on this estate, see Capogrossi Colognesi, "Grandi proprietari, contadini e coloni," 354, 719 nn. 78–79, who argues that Pliny considered replacing the failed tenants on the estates with slaves, in particular with *servi quasi coloni*. Against this view, see De Neeve, *Athenaeum* 78 (1990): 385–86, with further bibliography.

tenants, reduced to chronic indebtedness, cultivated their land in such a destructive manner that he was forced to give up the time-honored system of cash rents and institute sharecropping (*Ep.* 9.37).[30] This decision on Pliny's part represented an effort to preserve the existing system of contributing resources. If Pliny did not take decisive steps to lower permanently the tenants' risks for poor crops, he would always find himself in the situation of the previous owner of the estate mentioned in *Epistulae* 3.19, having to choose between granting remissions of rent and pressing his legal claim for the rent owed at the cost of assuming the expense of maintaining the tenants' farms.

At issue for Pliny was preserving what was most important for him in the landlord-tenant relationship, namely, the continued interest of the tenants in investing in their farms to maintain the productivity of his estates. This continued investment by his tenants was so important for Pliny that he was willing to sacrifice the convenience of receiving a yearly cash rent from his tenants and instead introduced sharecropping. He took this step despite the fact that leasing to sharecroppers would increase considerably the costs of managing his estates.[31] If Pliny's economic planning was at all typical for the Roman aristocracy—if not in terms of individual decisions, at least in terms of his general economic goals—then we can hypothesize that Roman landowners were generally best served by the institution of farm tenancy when they could find tenants who remained on their estates for long periods and continually invested their own resources in keeping their farms productive. The yearly cash income was less important for the landowner than investment for continued productivity.

---

30. Pliny describes the destructive way in which his tenants cultivated their land in the letter announcing his plan to institute sharecropping, *Ep.* 9.37.2: "... rapiunt etiam consumuntque quod natum est, ut qui iam putent se non sibi parcere." For peasants using a similar strategy to force concessions from their landlords in feudal Poland, see W. Kula, *An Economic Theory of the Feudal System*, trans. L. Garner (London, 1976), 48, 64–65.

31. Pliny recognized these costs, for he described the additional managerial personnel that he would have to employ as a result of instituting sharecropping: "Medendi una ratio, si non nummo sed partibus locem ac deinde ex meis aliquos operis exactores, custodes fructibus ponam" (*Ep.* 9.37.3). For the managerial costs in sharecropping and in other forms of tenancy, see S.N.S. Cheung, *The Theory of Share Tenancy* (Chicago, 1969), 66–72, and Currie, *The Economic Theory of Land Tenure*, 113–14.

We can see a similar attitude toward farm tenancy expressed by the Roman agronomist Columella. Columella was concerned with the lack of interest shown toward agriculture by his upper-class contemporaries, and he accordingly sought to convince them to take a direct interest in the management and intensive cultivation of their estates.[32] But Columella recognized that personal management of estates was impossible for many landowners, so he also offered recommendations about the best ways for Roman landowners to deal with their tenants. Columella's recommendations boil down to two important principles. First, the landowner should above all be concerned with the quality of the work done by the tenant. Accordingly, the landowner was to avoid the shortsightedness of insisting on timely payments of rents or other dues and was to provide instead, to the extent possible, conditions under which the tenant could work his land productively; the landowner's long-term revenues, after all, depended on the continued diligent cultivation of his estates (1.7.1–2).[33] Second, the efforts of the landowner to promote this type of agriculture would be most successful when the landowner found tenants who could be relied on to remain on their farms year after year (1.7.3–7). That Columella postulated this principle does not mean that he ever contemplated binding tenants to estates with perpetual leases; it is doubtful whether that would have been legally possible for a private landowner in the early

---

32. Columella's views about leasing are also discussed in chap. 3, n. 42.

33. Col. 1.7.1–2: "Comiter agat cum colonis facilemque se praebeat, et avarius opus exigat quam pensiones, quoniam et minus id offendit et tamen in universum magis prodest. Nam ubi sedulo colitur ager plerumque compendium, numquam, nisi si caeli maior vis aut praedonis incessit, detrimentum adfert, eoque remissionem colonus petere non audet. Sed nec dominus in unaquaque re, cui colonum obligaverit, tenax esse iuris sui debet, sicut in diebus pecuniarum vel lignis et ceteris parvis accessionibus exigendis, quarum cura maiorem molestiam quam impensam rusticis adfert; nec sane est vindicandum nobis quicquid licet, nam summum ius antiqui summam putabant crucem." For the interpretation of the term *opus,* see Finley, "Private Farm Tenancy in Italy," 119–21. For detailed discussion of Columella's recommendations about handling tenants, see W. Scheidel, "Pächter und Grundpacht bei Columella (*Colonus*-Studien II)," *Athenaeum* 81 (1993): 391–439, and *Grundpacht und Lohnarbeit,* 83–117. On Columella's approach to handling requests for remission of rent on the part of his tenants, see P.W. de Neeve, "Remissio Mercedis," *ZRG* 100 (1983): 296–339, at 311, 324–26, 331, with further literature.

empire.³⁴ Rather, Columella recommended that landowners avoid leasing to a tenant who would take no personal interest in his farm. He suggested that landowners instead take care to cultivate relations with tenants whose livelihoods depended on the careful, long-term cultivation of their land. Leasing under such conditions provided the best way, short of taking a direct personal interest in the cultivation of the estate, of maintaining the capacity of an estate to produce the harvest on which a landowner's income ultimately depended.

This consideration was especially important for the production of valuable crops, such as wine, which required a substantial outlay of labor and capital. Indeed, one attitude prevailing among upper-class landowners was that viticulture carried too many risks for a landowner to be able to rely on it for a stable income.³⁵ Columella was fighting against this attitude when he made his well-known calculations about the profits readily achievable from viticulture (3.3); he sought to convince his contemporaries to take a greater interest in the intensive cultivation of their estates.³⁶ Viticulture, nevertheless, contributed significantly to the income of upper-class Romans: Pliny represents just a

---

34. For a different view, see N. Brockmeyer, "Arbeitsorganisation und ökonomisches Denken in der Gutswirtschaft des römischen Reiches" (Ph.D. diss., Ruhr-Universität Bochum, 1968), 172–73. For discussion of the origins of perpetual leasing, which in the classical period of Roman law primarily concerned municipal and state land, see Kaser, *RP* I, 455–56, and Mayer-Maly, *Locatio Conductio*, 25–26, 70–71. For a contrasting interpretation of Columella's views about the value of tenancy, see P.W. de Neeve, *Colonus* (Amsterdam, 1984), 93–95.

35. On upper-class attitudes toward the risks involved in viticulture, see A. Aymard, "Les capitalistes romaines et la viticulture italienne," *Annales ESC* 2 (1947): 257–65; N. Purcell, "Wine and Wealth in Ancient Italy," *JRS* 75 (1985): 1–19; and A. Tchernia, *Le vin de l'Italie romaine* (Rome, 1986), 215–21.

36. For a critique of Columella's calculations about the profits from viticulture, see R. Duncan-Jones, *The Economy of the Roman Empire*, 2d ed. (Cambridge, 1982), 33–59, 376–77. In general on Columella's methods, see R.H. Macve, "Some Glosses on Ste. Croix's 'Greek and Roman Accounting,'" in P. Cartledge and F.D. Harvey, eds., *Crux* (Exeter, 1985), 233–64. For more favorable appraisals of Columella's accounting methods, see A. Carandini, "Columella's Vineyard and the Rationality of the Roman Economy," *Opus* 2 (1983): 172–204, and J. Love, "The Character of the Roman Agricultural Estate in the Light of Max Weber's Economic Sociology," *Chiron* 16 (1986): 99–146, at 124–29. On Columella's purpose and intended audience, see Tchernia, *Le vin de l'Italie romaine*, 209–21.

single example of the many upper-class Romans who drew much of their income from the sales of wine produced on their estates (*Ep.* 4.6.1, 8.2). Moreover, Pliny's tenants were themselves engaged in viticulture, and it seems beyond doubt that this was the case on the estates of numerous other upper-class landowners (e.g., *Ep.* 10.8.5). How did Pliny and other comparable landowners divide the substantial costs for viticulture with their tenants?

The Digest provides us with some indirect evidence for how the costs of maintaining vineyards might be allocated. This evidence makes clear that tenants regularly invested their own resources in planting and cultivating vines, even when they had not previously contracted with their landlords to do so. This type of investment became a legal issue when the tenant, on leaving his tenancy, sought compensation from the landlord for the improvements that he had made and that continued to enhance the value of the property. In Roman law, the tenant was entitled to compensation for such "unexhausted improvements."[37] This claim remained valid, even when the tenant undertook the improvement on his own initiative, not at the instigation of the landlord (Paul. D. 19.2.55.1, 2 *sent.*).[38] Of course, the easiest way for the landowner and tenant to cooperate in investing in viticulture was to come to an explicit understanding, and Paul's emphasis on the right of the landlord to sue for the completion of tasks that the tenant had contracted to perform suggests that this practice was common (D. 19.2.24.3, 34 *ad ed.*).[39] Indeed, in this response (assuming that the text is

---

37. For discussion of "unexhausted improvements," see Currie, *The Theory of Land Tenure*, 69–80. On the legal basis of the tenant's claim to compensation for unexhausted improvements in Roman law, see Mayer-Maly, *Locatio Conductio*, 170–72, as well as W.E. Heitland, *Agricola* (Cambridge, 1921; reprint, Westport, Conn., 1970), 366–67.

38. Paul. D. 19.2.55.1: "In conducto fundo si conductor sua opera aliquid necessario vel utiliter auxerit vel aedificaverit vel instituerit, cum id non convenisset, ad recipienda ea quae impendit ex conducto cum domino fundi experiri potest"; this passage is also quoted in [Paul.] *Sent.* 2.18.4. Cf. also Scaev D. 19.2.61 pr. (7 *dig.*, discussed later in this chapter under "The Tenant's Profit").

39. Paul. D. 19.2.24.2–3: " Si [domus vel] fundus in quinquennium [pensionibus] locatus sit, potest dominus, si deseruerit [habitationem vel] fundi culturam colonus [vel inquilinus] cum [eis] <eo> statim agere. [Sed et] de his, quae praesenti die praestare [debuerunt] <debuit>, [velut [ut *ins. Hoffmann*] opus aliquod efficerent, propagationes facerent], agere similiter potest." Mayer-Maly, *Locatio Conductio*, 139, follows Solazzi and Beseler in arguing that this

genuine), Paul mentioned the planting of new vines as one of the projects that the tenant might contract to perform. It is possible that Roman leases regularly included a provision, common in Egyptian leases for vineyards, whereby the tenant agreed to maintain the productivity of a vineyard by planting a set number of new vines each year.[40] When this type of work was explicitly provided for in the lease contract, a tenant failing to perform it was liable to be sued.

However, the situation is likely to have been far more complicated if tenants typically remained in their tenancies for longer periods of time, renewing their leases through the institution of *relocatio tacita* (discussed in more detail later in this section), and not formally renegotiating their lease terms each year. Tenants cultivating their land under these circumstances, as I will argue later, had every incentive to renew the vineyards and the other plantations on the farm, especially if they had reasonable confidence that they would be able to remain on the farm and enjoy the fruits of their investment. As far as the type of estates envisioned by the jurists in the Digest is concerned, such tacit lease agreements were probably much more the typical pattern than the detailed written leases that characterized tenancy in Roman Egypt. Under this circumstance, the landowner and the tenant would have an implicit understanding that the tenant could remain in his tenancy indefinitely, as long as he cultivated the farm in a manner satisfactory to the landowner. The short duration in the Roman farm lease would not necessarily have meant that the tenant's tenure was precarious, although, as we shall see, the short-term lease provided the landowner with a means of exercising some leverage against the tenant to secure more advantageous lease terms.

In this situation, legal problems might arise if, contrary to his expectation, the tenant lost his lease, even after having invested in substantial improvements on the farm. The second-century jurist Scaevola seems to have been addressing just such a situation when he ruled that a tenant expelled from his lease for failing to pay rent was still entitled to compensation for any improvements that he had made (*D.* 19.2.61 pr., 7 *dig.*). The circumstances in this case are curious, for the tenant had planted new vines on his farm, which raised the value of the farm and

---

passage is substantially interpolated; what Mayer-Maly considers interpolated is in brackets. Even so, the passage attests the lessor's right to sue for the completion of agreed on tasks. In Mayer-Maly's view, this passage also attests the tenant's obligation to cultivate the farm held under lease.

40. The maintenance of vineyards is also discussed in chap. 2.

with that the rent. The tenant, however, did not pay his rent and lost his lease. In Scaevola's ruling, the tenant would only be obliged to pay any further rent if he received compensation for his expenses.⁴¹ When the lessor set the rent for the farm in its newly improved condition, then he had to take into account the expenses that the tenant had borne. In the real world, the situation would not be so clear-cut. Tenants secure in their tenure would have continually made improvements, with the confidence that they could remain in their leases without having to pay an increased rent. The ruling emphasizes that, theoretically at least, the tenant was entitled to legal protection if the landlord stood to profit from the tenant's improvements without providing compensation. However, the only reason for this court case was the fact that the tenant lost his lease after the rent was increased; otherwise the tenant's planting of the vines appears to have been unremarkable.

This supposition draws confirmation from a similar case, discussed by both Marcellus and Ulpian, concerning the rights of a tenant leasing from a beneficiary of a usufruct (D. 19.2.9.1, 32 *ad ed.*). In this case, the *fructuarius* leased out a *fundus* to a tenant for a five-year term but died before the lease period had come to an end. Marcellus ruled that the heir of the *fructuarius* was in no way bound to observe the existing lease terms, since the death of the beneficiary always terminated a usufruct.⁴² It is important for our purposes that the tenant could not claim compensation for the improvements that he had made under the assumption that he would cultivate the farm for the full five years, since the ending of the usufruct by the death of the *fructuarius* was an event that the tenant should have foreseen as a possibility.⁴³ The tenant was, however, entitled to compensation for his expenses if the *fructuar-*

---

41. Scaev. *D.* 19.2.61 pr.: "Colonus, cum lege locationis non esset comprehensum, ut vineas poneret, nihilo minus in fundo vineas instituit et propter earum fructum denis amplius <aureis> annuis ager locari coeperat. quaesitum est, si dominus istum colonum fundi [fundo *dett.*] eiectum pensionum debitarum nomine conveniat, an sumptus utiliter factos in vineis instituendis reputare possit [opposita doli mali exceptione *Iust.? Pernice*]. respondit vel expensas consecuturum vel nihil amplius praestaturum."

42. Marcellus apud Ulp. *D.* 19.2.9.1: "Hic subiungi potest, quod Marcellus libro sexto digestorum scripsit: si fructuarius locaverit fundum in quinquennium et decesserit, heredem eius non teneri, ut frui praestet, non magis quam insula exusta teneretur locator conductori."

43. Marcellus apud Ulp. *D.* 19.2.9.1: "idem [sc. Marcellus] quaerit, si sumptus fecit in fundum quasi quinquennio fruiturus, an recipiat? et ait non recepturum, quia hoc evenire posse prospicere debuit."

ius represented himself as the owner of the farm with clear title, and Ulpian cites a rescript by Septimius Severus and Caracalla to that effect (ibid.).

In this case the *fructuarius* was imitating the practice of estate owners who, seeking to minimize their own costs, relied on their tenants to maintain their land. The rescript by the emperors indicates again that the question addressed by Marcellus and later by Ulpian was not purely academic, and that disputes might arise over the tenant's understanding of his own rights to the farm that he was leasing. But such disputes were only likely to arise if a farm tenant characteristically used his own resources to maintain or to improve the tenancy with the expectation of profiting from his investment himself. Indeed, the discussion by Julian, previously mentioned, concerning the rights of a tenant whose landlord bequeathed to him a usufruct over the *fundus* held under lease presupposes that the tenant as a matter of course invested his own resources. The tenant could vindicate the usufruct and at the same time sue the heir *ex conducto* to be released from the obligation to pay rent and to be compensated for his expenses. The tenant was to receive compensation for his investment because it resulted, in effect, in improvements not yet exhausted when the bequest of the *fundus* ended the lease.[44] These improvements increased the value of the legacy that the tenant was to receive, but the tenant also played the role of a contractor who is entitled to compensation because his contract is prematurely terminated.

The cultivation of an estate, to sum up at this point, depended on a dynamic partnership between landowner and tenants. The tenant's contribution of labor and other resources, especially livestock and even slaves, made it possible for the landowner to pursue greater financial security. Clearly a landowner managing an estate in this way renounced a certain amount of revenue for each unit of land, in that he had to pay a premium to the tenant for his share in the investment and managerial costs. This relationship also provided benefits for the tenant, who gained access to land and expensive capital equipment, such

---

44. Julian. D. 7.1.34.1 (35 *dig*.), quoted in n. 26 in this chapter. Other passages that seem to refer to investment by the tenant are Labeo apud Paul. D. 39.3.5 (49 *ad ed*.) and Julian. apud Ulp. D. 39.3.4.2–3 (53 *ad ed*.). Both of these passages concern the liability of the landlord and the tenant under the interdict *quod vi aut clam* and the *actio aquae pluviae arcendae* for damage to a neighbor's property resulting from an improvement made by the tenant.

as winepresses. A small farmer restricted to farming his own land might well lack the resources to pay for such important equipment and as a result might be excluded from cultivating cash crops, such as wine, on a scale large enough to be profitable. The need to invest in improvements to the farm every year made the relationship between landowner and tenant dynamic; without continuing investment of labor and other resources, the capacity of a farm to produce an adequate harvest would be compromised. Each party in the landowner-tenant relationship sought to profit by imposing as much as possible on the other party the costs and risks involved in this investment.

This struggle over allocating the costs of investment provides the context for a rescript by the emperor Antoninus Pius cited by Ulpian.[45] This rescript answered a petition by a tenant, who sought a remission of rent because of the advanced age of the vines on the farm held under lease (Ulp. D. 19.2.15.5, 32 ad ed.). The emperor refused this request, characterizing it as "revolutionary," or *nova res*.[46] In Roman law, the tenant was entitled to a remission of rent (*remissio mercedis*) in the event of an unforeseeable disaster that made it impossible for him to fulfill his lease. (I discuss this in more detail later under "Distribution of Risk between Landlord and Tenant.") The jurists termed such a disaster *vis maior*. Ulpian (and Pomponius), quoting Servius, listed a number of circumstances that would qualify as *vis maior*, including a destructive infestation of birds, an attack by the enemy, an earthquake, a blight (*uredo*), an unaccustomed heat wave, or even a fire (Ulp. D. 19.2.15.1-3, 32 ad ed.). But the tenant was not entitled to a remission for the normal risks in agriculture. Servius (or his commentators Pomponius and Ulpian) provided as examples of this type of risk the organic deterioration of wine, the infestation of a grain crop by birds or weeds, and the

---

45. The author of this rescript is considered by most scholars to have been Antoninus Pius, rather than Caracalla: see B.W. Frier, "Law, Economics, and Disasters down on the Farm: 'Remissio Mercedis' Revisited," *BIDR*, 3d ser., 31–32 (1989–90): 237–70, at 259 n. 100, with bibliography, and De Neeve, *ZRG* 100 (1983): 308-9 and n. 40. H. Ankum, "*Remissio mercedis*," *RIDA* 19 (1972): 219–38, at 231 n. 23, takes the emperor to be Caracalla, as does Mayer-Maly, *Locatio Conductio*, 143.

46. Ulp. D. 19.2.15.5: "Cum quidam de fructuum exiguitate quereretur, non esse rationem eius habendam rescripto divi Antonini continetur. item alio rescripto ita continetur: 'Novam rem desideras, ut propter vetustatem vinearum remissio tibi detur.'"

theft of crops by a passing army.⁴⁷ These hazards, *vitia ex re*, are distinguished from *vis maior* in that they represent the normal risks associated with agriculture that any farmer could be expected to have taken into account before entering the lease contract. Nevertheless, beginning in the second century A.D. at least, the jurists undertook to resolve an ambiguity in the law by justifying claims for remission of rent on the basis of a disastrously poor crop, *ob sterilitatem*.⁴⁸ The tenant could certainly lose an entire crop to catastrophic weather conditions, but his claim was not so clear-cut in the event of less dramatic but still damaging changes in the weather that made it impossible for him to produce a crop large enough for him to be able to pay his rent. Landlords who granted such remissions did not compromise their claim to a full rent.⁴⁹ Under no circumstances, however, was the tenant legally entitled to a remission of rent simply because of a poor crop yield, and Ulpian cites a second rescript, also of Antoninus Pius, to that effect (*D*. 19.2.15.5).

The tenant answered by the rescript of Antoninus Pius concerning the age of the vines based his petition for a remission of rent on the claim that the advanced age of his vines made it impossible for him to produce a vintage adequate to allow him to pay his rent. Ulpian cited this rescript to emphasize that a poor harvest alone did not justify a claim for remission of rent.⁵⁰ In this particular case, the tenant had no

---

47. Servius apud Ulp. *D*. 19.2.15.2: "sed et si uredo fructum oleae corruperit aut solis fervore non adsueto id acciderit, damnum domini futurum: si vero nihil extra consuetudinem acciderit, damnum coloni esse. idemque dicendum, si exercitus praeteriens per lasciviam aliquid abstulit." Cf. F. Sitzia, "Considerazioni in tema di *periculum locatoris* e *di remissio mercedis*," in *Studi in memoria di Giuliana D'Amelio*, vol. I (Milan, 1978), 331–61, at 338–39, who convincingly argues that the theft by the passing army should be viewed as a *vitium ex re*; it was a foreseeable risk for which the tenant was responsible.

48. See my fuller treatment of *remissio mercedis* and *sterilitas* later in this chapter.

49. De Neeve, ZRG 100 (1983): 312–13 and n. 56, offers a somewhat different interpretation: the lack of a sufficient vintage can be considered to have arisen from the tenant's failure to have made adequate provision for the renewal of the vines—the tenant could be entitled to a remission of rent only if there were no failure on his part in his duty to cultivate the farm. On this rescript, see also Sitzia, "Considerazione," 357. See my additional discussion in the subsequent text.

50. For the processes by which petitions such as these were answered and then came to be quoted by the jurists, see W. Turpin, "Imperial Subscriptions and the Administration of Justice," *JRS* 81 (1991): 101–18, and for this text, see

claim because the advanced age of the vines was an eminently foreseeable problem inherent in viticulture. In the logic of the rescript, the tenant should have considered the condition of the vines when negotiating his lease contract; any sensible farmer would consider the condition of a farm before entering into a lease.

The legal basis for the tenant's claim in this case seems so flimsy, however, at least as it is reported by Ulpian, that we might well ask why the emperor had to respond to this petition and similar petitions for remissions of rent. The answer to this question becomes more apparent if we set aside for a moment the terms of the normative Roman farm lease and consider instead the more likely conditions under which many tenants occupied their land. We can well imagine that tenants remaining on their farms for extended periods of time regularly had to face the problem of restoring to full working order aged equipment and plantations within their farm. This must have been a particularly difficult problem with viticulture, but it also affected olive orchards and arboriculture in general. It is expensive to plant new vines and fruit trees, and they require substantial amounts of time before they can produce a crop. But any farm contains stock and equipment requiring periodic renewal; in a long-term lease relationship, disputes might arise between landowner and tenant as to who was to pay for the maintenance of storage or irrigation facilities or of other installations crucial to the productivity of a farm.

The "revolutionary" attempt on the part of the tenant in this case to seek a remission of rent can therefore be viewed as an effort to use political and legal channels to resolve a conflict that characterized the relationships between landowners and tenants in general. The tenant in this case used legal channels in an attempt to reduce his own share of the costs of renewing the vines on the farm that he leased. One likely scenario is that he withheld his rent, for which his landlord sued in a local court. The landlord probably won the suit, and the tenant then made his appeal to the emperor.[51] For other tenants, the costs of renew-

---

105. What is called a "rescript" here may actually have been a *subscriptio* to a written petition. In general on imperial rescripts, see D. Nörr, "Zur Reskriptenpraxis in der hohen Prinzipatszeit," ZRG 98 (1981): 1–46.

51. The significance to tenants of the *libellus* procedure and *cognitio extraordinaria* in seeking remissions of rent and in otherwise defending their rights is also discussed later in this chapter under "Distribution of Risk between Landlord and Tenant."

ing the farm will have (at least implicitly) formed part of their negotiations with their landlords. The treatment of this question by the imperial authorities remained, from a juridical point of view, very conservative. In this rescript, the emperor interpreted the rights and duties of the tenant as legally enforceable only in terms of the short-term lease, and in doing so he acted consistently with the general approach that the imperial government took toward developing legal rules for farm tenancy.

This rescript, then, took little notice of the economic reality that lay behind the tenant's petition. The imperial authorities did account for the economic realities of the farm lease with the principle of tacit renewal of the lease, or *relocatio tacita*. According to this principle, if a tenant remained on his farm and continued to cultivate it after the expiration of his lease contract, his lease was considered to be renewed for one year, under the same conditions as before. The willingness of the lessor to allow the tenant to remain is assumed; it is only by taking positive action that the landlord could terminate the lease. The tenant would be obligated to pay the same rent as before. In addition, he continued to pledge to the lessor the property that had previously served as security for the rent and the condition of the farm (Ulp. D. 19.2.13.11, 32 *ad ed.*, 19.2.14, 71 *ad ed.*).[52] This tacit renewing of the lease contract could go on indefinitely, as long as the tenant continued to cultivate his farm and the lessor did not take any action to object. Significant for understanding the approach of the imperial authorities toward farm leases is that the lease was considered to be renewed each time for one year only (Ulp. D. 19.2.13.11).[53] Again, the imperial government continually faced the problem of defining the rights of a tenant who culti-

---

52. Ulp. D. 19.2.13.11: "Qui impleto tempore conductionis remansit in conductione, non solum reconduxisse videbitur, sed etiam pignora videntur durare obligata." The tenant could not assume, however, that anyone standing as surety for him would continue to do so: "sed hoc ita verum est, si non alius pro eo in priore conductione res obligaverat: huius enim novus consensus erit necessarius" (ibid.). For discussion of *relocatio tacita*, see Zimmermann, *The Law of Obligations*, 356–57.

53. Ulp. D. 19.2.13.11: "Quod autem diximus taciturnitate utriusque partis colonum reconduxisse videri, ita accipiendum est, ut in ipso anno, quo tacuerunt, videantur eandem locationem renovasse, non etiam in sequentibus annis, etsi lustrum forte ab initio conductioni praestitutum. sed et si secundo quoque anno post finitum lustrum nihil fuerit contrarium actum, eandem videri locationem in illo anno permansisse: hoc enim ipso, quo tacuerunt, consensisse videntur. et hoc deinceps in unoquoque anno observandum est."

vated a farm without negotiating a formal lease contract. We should imagine that disputes might arise about the rent owed by the tenant, not to mention about the tenant's other duties and obligations. To judge by the formulation of Ulpian, the imperial authorities formally recognized only the short-term lease, and they endeavored to describe the tenure of tenants occupying their farms for indefinite periods in terms of this type of lease. Accordingly, the imperial authorities found a way to describe in terms of the normative farm lease a wide variety of traditional arrangements involving long-term occupation of the land.[54]

The scope of the problem that the imperial authorities faced in developing this definition is suggested by a rescript from the reign of Valerian and Gallienus (*CJ* 4.65.16, A.D. 260). In this rescript, the emperors enjoined lessors to observe the terms of the lease and not to demand any payments beyond what had been contracted for. The emperors then repeated the principle that if a tenant remained in his tenancy after the expiration of the lease, the lease would be considered to be renewed, with the tenant's property remaining pledged.[55] Rulings of

---

Mayer-Maly, *Locatio Conductio,* 216–22, considers the legal principle of *relocatio tacita* to have had a late classical origin. This principle served the fiscal interests of the Roman state, which wished to keep farmland under continuous cultivation (218; see also later in this chapter under "Distribution of Risk between Landlord and Tenant"). Mayer-Maly (220 ff.) considers Ulp. *D.* 19.2.13.11 and Ulp. *D.* 19.2.14 (the latter of which is to be connected with Ulp. *D.* 43.26.4.4, 43.26.6 pr., 71 *ad ed.*) to have been largely reworked in postclassical times. In the former case, Ulpian was originally concerned solely with the status of pledged property in a tacit renewal of the lease, while in the second passage his concern was the tacit renewal of *precarium*. See also Köhn, "Die Kolonen in den Rechtsbestimmungen," 237–38, who cites Celsus *D.* 47.2.68.5 (12 *dig.*) and Ulp. *D.* 12.1.4.1 (34? *ad Sab.*) on the question of the tenant harvesting the crop after the termination of the lease against the will of the lessor.

54. W. Goffart, *Caput and Colonate* (Toronto, 1974), 69 considers customary leases of indefinite duration based on *relocatio tacita* to have been a "widespread type of tenancy" in the Roman Empire. Heitland, *Agricola,* 364 views tacit renewal of the lease as deriving from local custom. My point is that *relocatio tacita* provided the legal framework for the jurists to describe many forms of tenancy in conventional Roman legal terms.

55. *CJ* 4.65.16: "Legem quidem conductionis servari oportet nec pensionum nomine amplius quam convenit reposci. sin autem tempus, in quo locatus fundus fuerat, sit exactum et in eadem locatione conductor permanserit, tacito consensu eandem locationem una cum vinculo pignoris renovare videtur." Constantine may also have been concerned with such tenants in a constitution forbidding landowners to raise the rents of their *coloni* (*CJ* 11.50.1, A.D. 325); see Goffart, *Caput and Colonate,* 68–69 and n. 6.

this type probably protected the tenant, in as much as the landlord was prohibited from arbitrarily and retroactively raising the rent or imposing other obligations on a sitting tenant. Certainly a tenant could not be dismissed after he had planted and began to cultivate the current year's crop. Without such protection, tenants could find themselves at the mercy of a landlord who would use the threat of dismissal as a bargaining chip to gain greater concessions. At any rate, the tacit renewal of the lease provided Roman authorities with a means of describing a wide variety of tenure arrangements in terms of the normative Roman lease. These tenure arrangements shared one characteristic, the tenant's long-term occupation and cultivation of a farm.

In Roman Egypt we can trace the same type of relationship that the Roman jurists classified as *relocatio tacita*. In that province, written leases were almost always contracted for a limited period, generally for shorter periods than the five years that characterized the normative Italian lease.[56] Individual lease documents generally do not permit many conclusions about the ongoing relationships among the parties involved, but several documents announcing a tenant's intention to quit a lease imply the existence of a long-term lease between the tenant in question and the landowner. One example is a petition submitted by a certain Isidoros son of Nikandros, who held a four-year lease over eighty-eight *arourae* of catoecic, or private, land near the village of Theadelphia in the Fayum (*P.Strasb.* 511, A.D. 169). The owner of the land, one Sabeinos, alias Thrakides, who as a former *gymnasiarch* ranked among the local elite, had recently died, and the tenant was now petitioning the *strategos* to inform the lessor's sister and heir of his intention not to continue in the lease. There is no indication that any irregular circumstances had arisen in connection with the lease, and the existence of this and similar announcements to terminate leases suggests that, under ordinary circumstances, one would expect a lease to be renewed under the same terms. More to the point, some written leases included guarantees on the part of the lessor not to compel the tenant to remain in the contract at the end of the lease period. Such

---

56. For the terms of farm leases in Roman Egypt, see D. Hennig, "Untersuchungen zur Bodenpacht im ptolemäisch-römischen Ägypten" (Ph.D. diss., Ludwig-Maximilians-Universität zu München, 1967), and J. Herrmann, *Studien zur Bodenpacht im Recht der graeco-aegyptischen Papyri* (Munich, 1958). For the tacit renewal of leases in Egypt, see Herrmann, 170, and Rowlandson, *Landowners and Tenants in Roman Egypt*, 253.

guarantees, attested primarily in connection with leases for state land, also suggest that tenants often remained in a holding after the expiration of the original written contract, without any new written contract taking its place.[57]

The Tenant's Profit

Up to this point, we have considered largely the landowner's interests when examining the jurists' treatment of tenancy. The assumptions that the jurists made about the economic background to their rules for farm tenancy allow us to infer the existence of certain characteristic approaches to managing private estates. But we also need to examine how the institution of farm tenancy worked from the tenant's perspective; in other words, we need to consider how the tenant in general expected to profit from leasing land. This task is difficult, since the legal sources offer us little direct evidence for this important issue, but the jurists did make certain assumptions about how the tenant profited from leasing a farm. An examination of these assumptions affirms that the Roman legal authorities were aware of the importance of the market for the tenant's interests, but in their treatment of legal issues connected with this concern, they maintained the same conservatism that they displayed in other aspects of the law of tenancy.

It seems clear that even small-scale tenants had regular recourse to the market, if only to sell a small surplus. Indeed, in a comprehensive study of fairs and markets in the Roman Empire, De Ligt has amassed a considerable array of comparative evidence from medieval Europe, the Ottoman Empire, and premodern China. This evidence indicates that it is highly likely that even small-scale tenants, largely producing to cover their own subsistence needs, depended on the market to sell surpluses and to purchase needed commodities. Accordingly, it was a characteristic feature of the Roman economy for landowners to seek to exercise some control over their tenant's access to markets. Farmers often used periodic markets (*nundinae*) as a setting to sell surpluses.

---

57. Two examples are *SB* X 10533 (A.D. 171, provenance unknown) = A. Tripolitis, "Some Princeton Papyri Reconsidered," *BASP* 5 (1968): 7–16, at 7 ff., and *P.Oxy.* X 1279 (A.D. 139), an application for a five-year lease of state land, formerly cleruchic, or private, land but now classified as *ge hypologos*. I discuss these announcements to quit leases in greater detail in *APF* 41, no. 2 (1995): 238–50.

Apparently, landowners commonly petitioned to the senate or the emperor for the right to hold periodic markets within their estates. The maintenance of a market on an estate, attested particularly in North Africa, provided the landowner with an important means of exercising greater economic control over tenants, especially in their commercial relationships with traders from outside the estate.[58] In addition, the interests of landowners might conflict with those of neighboring towns, which were concerned above all with maintaining an adequate and inexpensive food supply and will have been wary of any interference with their access to crops on the market. This tension, De Ligt argues, explains the resistance on the part of the council of the northern Italian town of Vicetia to a petition by the praetorian senator L. Bellicius Sollers, which is discussed by Pliny (*Ep.* 5.4, 5.13). This individual sought permission from the senate to establish a periodic market on his own estates, which were located in the vicinity of Vicetia. According to De Ligt's reconstruction, Sollers sought to establish the market on his estates to maintain more direct control over the disposition of his tenants' crops. The town, by contrast, feared that this control would compromise its own access to a ready food supply.[59]

The tenant's dependence on the market is consistent with the jurists' assumption that the tenant paid his rent in cash, as we have seen in chapter 3. The normative tenant of the jurists could freely bargain with

---

58. See L. de Ligt, *Fairs and Markets in the Roman Empire* (Amsterdam, 1993). For the involvement of even relatively poor tenants in the market, see the view of Capogrossi Colognesi, discussed in my introduction. In the view of H.W. Pleket, "Wirtschaft," in *Handbuch der Europäischen Wirtschafts- und Sozialgeschichte*, vol. I (Stuttgart, 1990), 87–88, small independent farmers produced for the market to a limited extent; small holders also occasionally leased additional land to cultivate. Taxes collected in cash may have also increased the small farmers' dependence on the market; see Pleket, loc. cit., and K. Hopkins, "Taxes and Trade in the Roman Empire (200 B.C.–A.D. 400)," *JRS* 70 (1980): 101–25. For markets on estates and the landowner's interest in maintaining them, see De Ligt, 155–98. For a contrasting interpretation of the North African estate markets, see B.D. Shaw, "Rural Markets in North Africa and the Political Economy of the Roman Empire," *AntAfr* 17 (1981): 37–83. I discuss North African estate markets in *The Economics of Agriculture*, 215–20.

59. For discussion of L. Bellicius Sollers, see De Ligt, *Fairs and Markets*, 202–24, and idem, "The Nundinae of L. Bellicius Sollers," in H. Sancisi-Weerdenburg et al., eds., *De Agricultura* (Amsterdam, 1993), 238–62. For the legal regulations surrounding markets, see J. Frayn, *Markets and Fairs in Roman Italy* (Oxford, 1993), 117–44.

the landlord over the terms of the lease, and he was capable of using legal institutions to defend his interests. The jurists did of course recognize the existence of other forms of land tenure, including sharecropping. But in general, the jurists simplified reality by considering primarily the tenant paying his rent in fixed cash assessments. In dealing with the form of the rent, we may suspect, the jurists followed the same general procedure they followed when defining the tenant's rights in terms of the length of the lease.

This assumption on the part of the jurists involved more than simply the abstract application of a legal principle, since they were consistent in also assuming that the tenant regularly sold crops on the market. For example, Iavolenus made such an assumption when defining the *fructus* from a farm for the purpose of determining what should be included in a legacy (*D*. 33.2.42, 5 *ex poster. Lab.*). In treating this issue, Iavolenus considered not the physical status of the crops grown on the farm but the financial interests of the individual harvesting the crops.[60] This definition resolves the ambiguity of the meaning of the term *fructus*—whether it simply refers to "crops" grown on a farm or is to be understood as the "profit" from the farm. Iavolenus emphatically adopted the latter meaning of the term, and in doing so he made several interesting assumptions. He apparently envisioned that, under normal circumstances, the farm in question would be cultivated by a tenant. In any case, the tenant would have the same relationship to the market as would the landowner who cultivated the land directly. Iavolenus' definition took into account the changes in market prices for agricultural products over the course of the agricultural year, as well as the possibility that the tenant might seek to increase his profit in the same way as a landowner by waiting to sell his crops at the most opportune time. The tenant might even sell the crop in advance of the harvest. Wealthier landowners engaged in this practice to reduce their risk. Pliny, for example, regularly sold the vintages on his Tuscan estates to independent wine-dealers in advance of the harvest (*Ep.* 8.2), and Pliny's matter-of-fact description of his negotiations with these dealers suggests that it was a practice well known to upper-class landowners.

---

60. Iav. *D.* 33.2.42: "In fructu id esse intellegitur, quod ad usum hominis inductum est: neque enim maturitas naturalis hic spectanda est, sed id tempus, quo magis colono dominove eum fructum tollere expedit. itaque cum olea immatura plus habeat reditus, quam si matura legatur, non potest videri, si immatura lecta est, in fructu non esse."

The wealthy equestrian Aurelius Appianus also sold some of his grapes on his Egyptian estate to wine-dealers in advance of the vintage.[61] A tenant seeking to sell a crop in advance of the harvest, of course, required the cooperation of his landlord, since the crops remained the property of the landlord while they were attached to the soil, with the tenant only gaining ownership over them by *perceptio,* that is, by harvesting them.[62]

In his discussion of various types of crop theft, Africanus made the same assumption about the relationship of the tenant to the market (*D.* 47.2.62.8, 8 *quaest.*). In this case, a tenant sold crops "on the vine" [fructus pendentes] to a purchaser, who then removed them from the farm. The term *fructus pendentes* refers to vines or olives, so in this case again the tenant is envisioned as having the same relationship to the market as a larger landowner might. There are very few responses in the Digest that deal with the tenant's selling his crops; such sales involved few legal problems requiring them to be treated separately from the general law of sale. The major legal problem that the jurists addressed concerned the legal capacity of the tenant to sell crops. In advance of the harvest, as we have seen, the still growing crops formed part of the *fundus* and so belonged to the landowner. After the crops were harvested, the tenant owned them, but they remained pledged to the landowner as security for the rent. In the latter case, however, sales of crops fell under a more general category of alienation of property still pledged as security, so the problems arising from the tenant's selling crops on the market does not seem to have received any special treatment by the jurists.

Still, the description that Scaevola offered of one particular legacy suggests that active involvement of the tenant in the market was a con-

---

61. For the sales of Appianus' vintages to *karponai,* see my *Management and Investment,* 113–17, and, more generally on the marketing of wine on the estate of Appianus, see D. Rathbone, *Economic Rationalism and Rural Society in Third-Century* A.D. *Egypt* (Cambridge, 1991), 278–306.

62. The landowner's lien on the harvested crops is discussed in chap. 3. It was possible in Roman law to sell another's property, as well as to sell property pledged as security as long as the creditor agreed: cf. Ulp. *D.* 18.1.28 (41 *ad Sab.*), *D.* 20.6.4.1 (73 *ad ed.*), Gaius *D.* 20.6.7 pr. (*lib. sing. ad formam hypothecariam*), Marcianus *D.* 20.6.8.6–19 (*lib. sing. ad formam hypothecariam*), Paul. *D.* 20.6.10 pr. (3 *quaest.*). On these passages, see Kaser, *RP* I, 469 n. 74; cf. F. de Zulueta, *The Roman Law of Sale* (Oxford, 1945), 10–12, and Köhn, "Die Kolonen in den Rechtsbestimmungen," 195–99, 206.

ventional feature of the Roman estate economy (*D.* 33.1.21 pr., 22 *dig.*). An individual bequeathed to a freedman the right to one-fiftieth of the income from a group of estates (*praedia*) for as long as he should live. This income was to consist of whatever rent the tenants on the estate paid or, alternatively, of whatever income was realized from direct sales of the crops to purchasers.[63] In this case again, tenancy simply represents an alternative way to manage an estate producing for the market. The landowner himself or herself might sell the produce of the estate, or the tenants might market the crops themselves and provide the landowner with an income in that way.[64] The type of tenants envisioned in this case fulfilled the function of the tenants who provided an estate owner with a fixed and stable income in another case discussed by Scaevola, concerning the legacy of the income (*reditus*) from a *fundus* (*D.* 33.2.38, 3 *resp.*).[65] In this case, the widow of a deceased landowner received the legacy, but the heir was permitted to sell the *fundus* in question and instead offer the legatee the same annual income that the deceased testator had been accustomed to receiving by leasing the land out to tenants. The testator clearly intended to provide his wife with a fixed income, and that income could be achieved by leasing an estate out to tenants, who apparently paid a cash rent that remained very stable from one year to the next.

Leasing estates out to tenants who themselves were responsible for marketing the estate's produce clearly provided advantages for landowners seeking to stabilize their income and minimize their involvement in the management of their estates. But recognizing the existence of tenants capable of performing this task still does not explain how such tenants could be counted on to provide a stable rent year after year. Tenants might well be expected to generate revenues from selling crops on the market, but we are certainly justified in supposing that farm tenants were at least as averse to risk as their landlords. More important, farm tenants will have had much less flexibility than large landowners in overcoming problems posed by an inade-

---

63. Scaev. *D.* 33.1.21 pr.: "... 'quae [de *ins. Mo.*] praediis a colonis vel emptoribus fructus ex consuetudine domus meae praestantur.'"

64. This text is also discussed in chap. 2. The case had a provincial origin, since it refers to the interest rates prevalent in the province concerned.

65. See chap. 2 on this text and other responses indicating that the fixed income from an estate might be equated with the annual rent paid by the estate's tenants.

quate harvest or depressed market prices for their crops. Farm tenants facing the obligation to pay a cash rent had no choice but to produce for the market. Admittedly, in the long run larger landowners will have found themselves in the same difficulty if market prices for crops remained depressed, since they could not realistically be expected to transfer to other enterprises capital invested in agriculture. In the short run, however, landowners with larger cash reserves would have an easier time riding out a poor yield or depressed prices; in the latter circumstance, they could avoid selling at unfavorable prices and wait until the market for their produce improved. This option was probably not available to most tenants facing the obligation of making a yearly rent payment in cash.

The legal sources provide us with little direct information about this subject, but we can infer some basic characteristics of tenants' relationship to the market by considering again how the Roman legal authorities approached the principle of *remissio mercedis*. In defining the conditions under which tenants might be entitled to claim a remission of rent, the Roman legal authorities considered only conditions affecting the physical crop yield of the farm; they did not take into account at all market conditions or other factors making it difficult or even impossible for tenants to market their produce. As we will see later in this chapter, economic considerations often induced landowners to be more generous in granting remissions than this strict definition might require them to be. Let us consider again the principle of a remission granted on account of a poor crop, *remissio mercedis ob sterilitatem*, which the jurists in the second century A.D. developed.[66] A landlord might grant a remission of the rent during one year because of a poor crop. But as Papinian emphasized, this landlord still had a claim for the full rent for the entire lease period, and his tenant was required to pay the full rent if the following years produced a bountiful crop (apud Ulp. D. 19.2.15.4, 32 *ad ed*.).[67] This principle was reaffirmed in a rescript of Alexander Severus, who granted a petitioner's claim to a remission of

---

66. This discussion of the *remissio ob sterilitatem* is based on Sitzia, "Considerazione," 331–61.

67. This passage is quoted in n. 105 in this chapter. Mayer-Maly, *Locatio Conductio*, 144, in interpreting this passage as reflecting the imperial policy of enforcing on tenants an obligation to cultivate their land, argues that Papinian's doctrine presupposes that prices for crops could not have varied much; hence, the ruling reflects a more closely regulated economy of the early third century.

rent only if the petitioner's disastrous harvest caused by an unforeseen calamity was not balanced out by other "bountiful" years during the lease period (*CJ* 4.65.8, A.D. 231).⁶⁸

The terms used by the jurists to denote "poor" and "bountiful" years, *sterilitas* and *ubertas,* suggest the approach of the Roman authorities toward the problem of finding a way to mediate between adjudicating the landlord's claims and imposing too great a burden of risk on the tenant. The tenant was not legally entitled to a remission for the normal hazards involved in farming; these included a poor crop yield (Ulp. *D.* 19.2.15.5, 32 *ad ed.*). No source mentions poor market conditions as a reason for granting remissions of rent, and it seems clear that the Roman legal authorities would have viewed the changing market conditions for crops as one of the risks that the tenant would be expected to have taken into account before entering into the lease contract. We must not forget that the jurists reckoned in terms of the cash value of rents and losses by the tenant. That they did so seems clear from Ulpian's emphasis that a legal remission of rent on the basis of *vis maior* did not entitle the tenant for the damages that he would otherwise be able to claim if the landlord had for some other reason failed to provide him with a usable farm (*D.* 19.2.15.7, 32 *ad ed.*). The tenant received only a prorated reduction of his *merces,* and he also had to bear the loss of the seed.⁶⁹ The Roman legal authorities considered the physical crop yield to be the decisive factor determining whether the tenant would be held responsible to pay his full rent. The implication is that a tenant achieving a reasonable harvest could raise sufficient funds to

---

68. Cf. *CJ* 4.65.8: "Licet certis annuis quantitatibus fundum conduxeris, si tamen expressum non est in locatione aut [aut *VC cum Graecis,* ut *PR*] mos regionis postulat, ut, si qua labe tempestatis vel alio caeli vitio damna accidissent, ad onus tuum pertinerent, et quae evenerunt sterilitates ubertate aliorum annorum repensatae non probabuntur, rationem tui iuxta bonam fidem haberi recte postulabis, eamque formam qui ex appellatione cognoscet sequetur." Risk for *vis maior* can be allotted to the lessee, according to the local custom. For the reading *aut* and in general for the interpretation of this rescript, see Sitzia, "Considerazione," 357–58 and n. 92, with reference to other discussions. See also Mayer-Maly, *Locatio Conductio,* 145, who inserts *non* before *postulat.* For the same assertion of the lessor's right by Diocletian and Maximian, cf. *CJ* 4.65.18 (A.D. 290).

69. Ulp. *D.* 19.2.15.7: "Ubicumque tamen remissionis ratio habetur ex causis supra relatis, non id quod sua interest conductor consequitur, sed mercedis exonerationem pro rata: supra denique damnum seminis ad colonum pertinere declaratur."

pay his rent by selling produce; the legal sources provide no hint that landowners might even be compelled to grant remissions for any reasons other than the size of the harvest. In effect, the jurists took little account of prices, but not because they were unaware of the importance of prices to the tenant's interests. Rather, prices were not relevant to the legal issues under consideration, since the tenant had no choice but to sell produce on the market to raise the funds to pay the rent. By implication, in the long run at least, the tenant's prosperity depended first and foremost on the size of the harvest.[70]

In the real world, of course, matters were probably not so simple. It seems very likely that the vintages that Pliny arranged to sell from his Tuscan estates to *negotiatores* in advance of the harvest (*Ep.* 8.2) included those produced by his tenants.[71] Pliny was seeking to assure that his tenants would have access to sufficient funds to allow them to cultivate their farms productively and therefore to be able to pay him their rents at the end of the agricultural year. A case discussed by Ulpian suggests one method to which farm tenants had recourse to reduce their risk for market prices (*D.* 19.2.19.3, 32 *ad ed.*).[72] In this case, the lessor and tenant agreed that the tenant would pay the rent in the form of grain valued at a price specified when the contract was made, instead of the usual cash rent. Ulpian allowed the lessor nevertheless to demand cash as his rent, but the jurist ruled that the lessor had to compensate the tenant for his losses. The lessor and tenant in this case were taking very different approaches toward risk. The tenant was seeking to minimize his risk for the lower prices that might ensue at harvest

---

70. See Frier, *BIDR*, 3d ser., 31–32 (1989–90): 247–48, on the restricted range of circumstances considered by the jurists as grounds for a remission of rent.

71. On Plin. *Ep.* 8.2, see my "Approaches to Economic Problems in the 'Letters' of Pliny the Younger: The Question of Risk in Agriculture," *ANRW* II 33, no. 1 (1989): 555–90; I argue that the grapes that Pliny sold to the *negotiatores* in *Ep.* 8.2 included the production of his tenants. For a contrasting interpretation of this letter, see De Neeve, *Athenaeum* 78 (1990): 376–79.

72. Ulp. *D.* 19.2.19.3: "Si dominus exceperit in locatione, ut frumenti certum modum certo pretio acciperet, et dominus [dein *Mo.*] nolit frumentum accipere neque pecuniam ex mercede deducere, potest quidem totam summam ex locato petere, sed utique consequens est existimare officio iudicis hoc convenire, haberi rationem, quanto conductoris intererat in frumento potius quam in pecunia solvere pensionis exceptam portionem. simili modo et si ex conducto agatur, idem erit dicendum."

time, when grain would be most plentiful.[73] He clearly was settling for a lower price for his grain than he might have been able to obtain had he waited to sell it, but he was protecting himself against the risk that grain prices might collapse and leave him in no position to pay his rent. The landlord, by contrast, notwithstanding his subsequent change of heart, was gambling that he could sell the grain at a high price, possibly at some time in the future, well after the harvest, when grain would have again become more scarce. This case presupposes that the tenant under normal circumstance had to sell his produce on the market. In addition, the tenant's effort to reduce his risk for the market seems conventional, which suggests that changes in market prices posed more of a difficulty for tenants than the jurists' treatment of *remissio mercedis* might lead us to expect.

This state of affairs had important consequences for the way in which the tenant conducted his business. Given the uncertain nature of the market for agricultural goods and the harsh reality that he had no choice but to rely on this market to gather the funds needed for his rent, the tenant had every incentive to cultivate his farm as intensively as his resources would allow. Faced with an uncertain market, on the one hand, and fixed yearly rental charges, on the other, the tenant really had only one way to enhance his financial security, by maximizing as much as he could the marketable product from each unit of land leased.[74] In this way, the tenant could overcome short-term changes in market prices by selling more of his produce. As Kula has argued in a study of the peasant economy of feudal Poland, smaller farmers are less likely to be hurt by the depressed prices that accompany a bountiful harvest than by a small harvest. In the latter case, the higher prices for foodstuffs might not benefit the farmer achieving a small harvest, since he would have to sell too high a proportion of his harvest, leaving him

---

73. For fluctuations in wheat prices that affected the estate of Appianus, see Rathbone, *Economic Rationalism*, 464–66; wine prices also varied considerably within the year: see ibid., 466–68, and, more generally, H.-J. Drexhage, *Preise, Mieten/Pachten, Kosten und Löhne im römischen Ägypten bis zum Regierungsantritt Diokletians* (St. Katharinen, 1991), 58–73. In fourth-century Egypt, it was apparently common for loans in money to be repaid in crops at the harvest: see R.S. Bagnall, *Egypt in Late Antiquity* (Princeton, 1993), 73.

74. I am assuming a situation in which open land is not in unlimited supply, as it was, at least to some extent, on the imperial estates in the Medjerda Valley in North Africa.

with too little produce left over to supply his family and workforce, or to provide seed for the next year's crop. Although an unusually large harvest might result in depressed prices, a small farmer would have better chances of being able to pay his rent and other financial obligations if he could simply produce and sell more crops on the market.[75] This allocation of resources may not seem to be "rational," but farmers with no choice as to how they might secure their livelihoods had only one way to pursue security, by maximizing, in proportion to the labor and other productive resources that they had available, the harvest that they could potentially market.

As analysts of peasant agriculture have observed, the small farmer is likely to cultivate a given plot of land more intensively than a larger landowner. The small farmer generally seeks to achieve yearly targets of production; that is, he seeks to produce enough to meet his yearly financial obligations, including feeding his family and workforce, paying his rent (and, if applicable, taxes), and meeting any other socially imposed financial obligations that he has to bear. Such an individual has fixed costs that he has to meet, and the importance of producing enough to meet these fixed costs becomes even clearer for a tenant paying a fixed yearly rent in cash. Under these circumstances, the tenant can be expected to work harder, because he will not count as costs the labor supplied by himself and his family. For a larger landowner, in contrast, the costs of labor will always be a factor affecting planning, whether the labor consists of a combination of permanent salaried workers and wage laborers hired on a daily basis, as was the case on the estate of Aurelius Appianus, or is provided by slaves.[76] In any case, the small farmer, in many circumstances, will in effect be able to cultivate a given unit of land with lower labor costs than will a larger landowner.[77] This would not be the case if the large landowner were

---

75. See Kula, *An Economic Theory of the Feudal System*, 57–58, 62–68. See also W. Abel, *Agricultural Fluctuations in Europe from the Thirteenth to the Twentieth Centuries*, trans. O. Ordish (London, 1986) [= *Agrarkrisen und Agrarkonjunktur*, 3d ed. (Hamburg and Berlin, 1978)], 8–13.

76. For an analysis of the labor force on the estate of Aurelius Appianus, see Rathbone, *Economic Rationalism*, 88–174. The slave labor force would be kept to a minimum, since the owner would only have to feed, clothe, and house those slaves who could be kept busy throughout the year. The slave owner would have to balance the costs of maintaining slaves against the costs and inconvenience of hiring casual labor.

77. On the allocation of labor and capital by a small-scale farmer, see the analysis of peasant agriculture by the Russian economist A.V. Chayanov, "On

able to reduce labor and capital costs significantly through economies of scale, as did Aurelius Appianus on his huge estate, but it is likely that leasing to smaller cultivators, each responsible for the management and care of an individual farm, did help many Roman landowners to achieve incomes from their estates despite the difficulties involved in managing extensive and often scattered holdings.

Given the advantages that Roman landowners seem to have derived by allowing the tasks of managing and cultivating their farms to devolve on their tenants, we need to reconsider the economic significance of the short-term lease that was the norm for the Roman jurists. If the foregoing analysis of Roman tenancy is at all correct, landowner and tenant were dependent on one another. Both parties gained from the sharing of resources: the landowner provided land and significant fixed capital, and the tenant provided managerial expertise, labor, and movable capital resources, such as livestock. This mutualistic relationship, however, was not without conflict, as each party sought as much as possible to increase the contributions of the other. The short-term lease gave both parties added bargaining power. As comparative evidence shows, the short-term lease need not result in a large turnover of tenants from one lease period to the next, and it need not result in insecurity of tenure for tenants.[78] It is of course clearly possible that the improvements made by the tenant would allow the landowner to lease out the farm at a higher rent, as in the case addressed by Scaevola (*D.* 19.2.61 pr., 7 *dig.*). Theoretically, an unscrupulous landowner could take advantage of a tenant's expectations. He could entice the tenant to improve the farm and then use whatever leverage he could to force the

---

the Theory of Non-Capitalist Economic Systems" (1924), trans. C. Lane, in D. Thorner et al., eds., *The Theory of Peasant Economy* (Homewood, Ill., 1966), 1–28, and "Peasant Farm Organization" (1925), trans. R.E.F. Smith, in ibid., 29–269. For further refinement of Chayanov's theories, see J.R. Millar, "A Reformulation of A.V. Chayanov's Theory of Peasant Farm Economy," *Economic Development and Cultural Change* 18, no. 2 (1970): 219–29, and T.B. Wiens, "Uncertainty and Factor Allocation in a Peasant Economy," *Oxford Economic Papers*, n.s., 29 (1977): 48–60. For additional discussion of the economics of peasant agriculture, see D. Grigg, *The Dynamics of Agricultural Change* (London, 1982), 91–100.

78. See Currie, *The Economic Theory of Land Tenure*, 69–73, who discusses situations in which short-term leases do not involve insecurity of tenure for tenants. In his analysis of sharecropping in India, however, P.K. Bardhan, *Land, Labor, and Rural Poverty* (New York, 1984), 105–111, argues that the short-term lease provides the landlord with a means of enforcement but provides the tenant with a disincentive to invest.

tenant to pay a higher rent; if the current tenant would not pay more, he could find a new one.

But in view of the financial considerations that I have described, it would not generally be advantageous for landowners to proceed in this way. A landowner would of course be legally obligated to compensate the tenant for the value of unexhausted improvements. But even if the tenant was not able to vindicate his rights to this compensation by suing the landlord, the landlord would have every reason to respect the investments made by the tenant. Few tenants would invest more than they had to in order to fulfill the terms of their leases if they had no reason to be confident that they would be able to reap the fruits of their labors. The short-term lease provided a means of enforcement. At the end of the lease period, the landowner could expel an unsatisfactory tenant or, less dramatically, use the threat of taking this step or of renegotiating the lease contract as a way to enforce the tenant's obligation to cultivate his farm in accordance with norms negotiated or understood in the lease. The landowner's ability to use the limited duration of the lease contract would of course be restricted by the extent to which local custom accorded the tenant long-term rights to his land. It seems likely that, through the principle of *relocatio tacita*, the jurists subsumed under the normative Roman lease a variety of local tenure arrangements. Some of these may have accorded the tenant a de facto right of possession, even if in formal Roman law the tenant had no such right. This circumstance notwithstanding, the short-term lease might play a crucial role in the dynamic between landowner and tenant, offering the landowner one means of exercising bargaining power.

To some extent, however, the short-term lease also served the interests of the tenant, especially the tenant capable of contributing valuable resources of his own. The services of such tenants were especially valued, and such diverse experiences as that of Pliny the Younger in managing his own estates and that of the Roman government in managing the imperial estates in North Africa indicate that such tenants were in chronically short supply. Under this circumstance, the short-term lease afforded the tenant greater freedom of action; a tenant unable to gain satisfactory lease terms from one landowner would have the possibility of leasing from another. The difficulties that the Roman government experienced in administering the imperial estates in the Medjerda Valley in North Africa demonstrate this point. There, the tenants had perpetual rights to their land, but their terms of tenure did not bind them to the imperial estates. As a result, tenants on these estates used an implicit threat to migrate off the estates as a means of gaining better terms. These tenants complained to the imperial government about being forced to pay higher rent and perform additional labor services for the *conductores,* the middlemen who leased from the Fiscus the right to collect the rents from the tenants. The imperial government, despite benefiting, in the short run at least, from the *conductores'* ability to collect higher rents, nevertheless intervened on behalf of the tenants, reaffirming their traditional rights.[79] Absent the type of governmental measures binding tenants to their estates that characterized the later Roman Empire, private landowners would always be in competition with one another to secure and retain the service of tenants capable of contributing substantial resources.

Distribution of Risk between Landlord and Tenant

The Roman landowner was served best when the tenant had an interest in investing his managerial skills, labor, and other productive resources in the long-term productivity of his farm. To complete our model of the contribution made by farm tenancy in the Roman agrarian economy, we need to consider how Roman landowners and tenants shared the risks associated with agriculture. As I have argued, the landowner was restricted in his ability to bring to bear against the tenant the full range of legal options that Roman lease law seemingly provided the lessor for enforcing the tenant's contractual obligations. The lessor, theoretically at least, could sanction an unsatisfactory tenant in one of two principal ways. To enforce the tenant's obligation to make timely payments of rent and to keep the farm in good working order, the lessor could confiscate the property pledged by the tenant as security or keep the tenant in his debt.[80] The lessor could also use the threat of not renewing a short-term lease as a means to induce the tenant to cultivate his farm in

---

79. I discuss the petitions by *coloni* on North African estates in *The Economics of Agriculture,* 112–16, 146–53, 182–87; see also J. Kolendo, *Le colonat en Afrique sous le haut-empire,* 2d ed. (Paris, 1991), especially 47–74, as well as V. Weber, in K.-P. Johne et al., *Die Kolonen in Italien und den westlichen Provinzen des Römischen Reiches* (Berlin, 1983), 289–343.

80. For these means of enforcement in Roman tenancy, see Finley, "Private Farm Tenancy in Italy," 109–17; Finley emphasizes the bargaining power that tenant indebtedness afforded the landowner.

a satisfactory manner. The ability of the landowner to enforce the tenant's contractual obligations was made more complicated by the question of risk.

Droughts and crop-failures were and continue to be a characteristic feature of Mediterranean agriculture.[81] We should expect that one of the principal areas of dispute in the dynamic relationship between landowners and tenants was bearing the costs of such droughts. Indeed, the classical Roman farm lease imposed the bulk of the risk on the tenant. Because he paid his rent in cash, the tenant bore the full risk not only for the size of the harvest but for the market price of the crops as well. From this perspective, then, the landowner could take advantage of very favorable terms in such a way as to derive a steady cash income from his estates with little risk or managerial expense.

But the experiences of real-life landowners, as far as we are able to trace them, suggest that matters were far from being so clear-cut. Pliny, at any rate, saw neither the confiscation of the tenants' pledges nor the expulsion of unsatisfactory tenants as a realistic solution to the problem of chronic tenant indebtedness on the estate that he was considering purchasing at Tifernum Tiberinum (*Ep.* 3.19). Instead, he undertook to solve this problem by other means, all revolving around reducing the tenant's risk, first by granting substantial remissions of rent (*Ep.* 9.37.2, 10.8.5), and then, when these measures did not provide an adequate solution for the problem of tenant indebtedness, by replacing the time-honored system of cash rents with sharecropping (*Ep.* 9.37). Earlier, Columella had also dealt with the problem of the tenant's risk, if in a somewhat different fashion. Columella recommended that the landowner be most concerned that the tenants cultivate his farm productively; if this requirement was met, the farm would always provide a crop, and the tenant would be less likely to request a remission of rent (1.7.1).

In all likelihood, many Roman landowners with estates in Italy encountered the same problems that Pliny and Columella faced, even if

---

81. For the problem of drought in medieval Languedoc, see E. Le Roy Ladurie, *The Peasants of Languedoc*, trans. J. Day (Urbana, 1974; French ed., 1966), 133. For the effects of drought on the production of grain in ancient Thessaly, see P. Garnsey, T. Gallant, and D. Rathbone, "Thessaly and the Grain Supply of Rome during the Second Century B.C.," *JRS* 74 (1984): 30–44. For discussion of the problems of drought in the peasant economy of Greece, see T.W. Gallant, *Risk and Survival in Ancient Greece* (Stanford, 1991): 34–59, 101–12.

the solutions that they adopted might have been quite different from what Pliny did.⁸² It does not seem likely that landowners with little interest in supervising the management of their estates personally would have been as ready as Pliny to institute sharecropping, which imposes substantially higher managerial costs on the lessor than does leasing for a fixed rent in cash. Even so, they must have found some way to alleviate the tenant's risk.

Moreover, the problem of alleviating the tenant's risk was not confined to Italy during the early empire. For example, in Roman Africa the imperial government took extraordinary measures to reduce the risk of the *coloni* cultivating the imperial estates in the Medjerda Valley. The Fiscus collected the rent from the *coloni* as shares of the harvest, which of course provided the *coloni* with some protection against the danger of a small harvest and at the same time kept their ability to pay rent from being dependent on the market prices available for their crops. In this case, the imperial government offered the reduced risk inherent in sharecropping, but without reserving for itself the measures that landowners leasing to sharecroppers commonly implement to make sure that the sharecroppers cultivate their farms in a satisfactory manner. The *coloni* held their land under perpetual leasehold, so their lease rights could not be withdrawn and assigned to another party willing to pay a higher rent. At the same time, the *coloni* had unrestricted access to uncultivated land. This latter circumstance meant that the *coloni* were likely to cultivate their land less intensively than either tenants paying a fixed rent or tenants restricted in the amount of land that they could take under lease. Still, granting these concessions was an understandable policy for the imperial government, whose overriding goal was to secure the continued cultivation of the imperial estates. In the judgment of the imperial government, small-scale farmers with resources sufficient to allow them to cultivate their land independently could best be relied on to accomplish this goal for the long term.⁸³

Finally, papyrological evidence from Roman Egypt suggests that landowners there were regularly forced by circumstances to be flexible with their tenants. To cite one example, Soterichos son of Lykos, a peasant farmer in the first century A.D. at the village of Theadelphia in the

---

82. For the landowner's "social" obligations toward his tenants, see Capogrossi Colognesi, *Ai margini della proprietà fondiaria*, 194–97 and "Il regime," 191–93.

83. See my *The Economics of Agriculture*, chap. 5.

Fayum, continued to lease land from a certain Tamystha daughter of Antigonos for the better part of two decades, even though he was chronically behind on his rent.[84] It seems likely that this landowner was willing to be flexible about the rent to retain the services of a valuable tenant.

From this discussion it seems clear that for many landowners the long-term advantages gained from the tenant's continued ability to cultivate a farm productively far outweighed the short-term convenience of collecting a full rent in cash. As a result, economic considerations frequently forced landowners to go beyond their narrowly defined contractual requirements and assume a greater burden of risk. This hypothesis about the relationship between Roman landowners and their tenants draws support from the treatment by the Roman legal authorities of the problem of allotting risk in agriculture.

The jurists' treatment of this problem represents a key area for understanding how the institution of farm tenancy functioned in the early empire, since it had important consequences for defining the economic relationship between landowner and tenant. In treating this question, the Roman legal authorities applied the same conservatism that they displayed in other areas of tenancy law. The jurists seem to have recognized how economic considerations might force landowners to accede to their tenants' requests for remissions of rent, but at the same time they persisted in interpreting the legal basis for granting requests for remissions of rent in terms of the duties and rights of landlords and tenants as defined in the conventional short-term lease contract.

The classical Italian farm lease imposed the bulk of the risk on the tenant, who, since he paid a fixed rent in cash, bore the risk not only for the harvest but also for the market price of his crops. But, as we have seen, Roman law did grant the tenant relief from the obligation to pay rent if an unforeseeable disaster, or *vis maior*, made it impossible for the tenant to fulfill his contractual obligations (Servius apud Ulp. Dig. 19.2.15.2, 32 *ad ed.*; cf. Gaius 19.2.25.6, 10 *ad ed. prov.*).[85] There is no clear-

---

84. I discuss the relationship between Soterichos and Tamystha in *Management and Investment*, 142–47. For the affairs of Soterichos, see S. Omar, *Das Archiv des Soterichos* (Opladen, 1979).

85. On the development of the legal doctrine of *remissio mercedis*, see especially Sitzia, "Considerazione," 331–61; De Neeve, ZRG 100 (1983): 296–339; Frier, *BIDR*, 3d ser., 31–32 (1989–90): 237–70; Capogrossi Colognesi, *Ai margini*

cut explanation in the legal sources of the legal principle by which the tenant was exonerated from his obligation to pay rent, and this topic is quite controversial in contemporary scholarship. Ulpian explained the tenant's right as deriving from the lessor's failure to provide him with a farm that he could cultivate: "ut frui liceat" (Ulp. *D.* 19.2.15 pr.-1, 32 *ad ed.*).[86] Among modern scholars, De Neeve has accepted this explanation; in his view, a bad harvest could be viewed as an impairment of the tenant's enjoyment (*frui*) of the *fundus* and in certain circumstances could entitle the tenant to a remission of rent.[87] But Frier has recently questioned this explanation and argued instead that the Roman jurists, beginning with the Republican jurist Servius Sulpicius Rufus, would grant remissions of rent as a way of keeping the tenancy contract intact when circumstances made it impossible for the tenant to fulfill his obligations. These circumstances cannot in any way be considered a failure on the part of the lessor, but the lessor nevertheless bore the risk for them.[88] Capogrossi Colognesi, however, has challenged Frier's interpretation and has returned to the view that the tenant's right to a remission of rent derived from the landlord's not fulfilling his principal obligation under the lease, that is, his obligation to provide the tenant with a usable farm.[89]

---

della proprietà fondiaria, 143–99 and "Il regime," 163–94; Zimmermann, *The Law of Obligations,* 369–74; and, earlier, Mayer-Maly, *Locatio Conductio,* 140–47. On the classical Roman concept of *vis maior* in agricultural leases, see Mayer-Maly, 189–93 . See also I. Molnár, "Verantwortung und Gefahrtragung bei der *locatio conductio* zur Zeit des Prinzipats," *ANRW* II 14 (1982): 583–680, at 660–79. For general discussion of *vis maior* in Roman law, see Kaser, *RP* I, 507–8, 566–67, 571; cf. I. Molnár, "Die Ausgestaltung des Begriffes der vis maior im römischen Recht," *Iura* 32 (1981): 73–105.

86. Ulp. *D.* 19.2.15 pr.-1: "Ex conducto actio conductori datur. Competit autem ex his causis fere: ut puta si re quam conduxit frui ei non liceat . . . vel si quid in lege conductionis convenit, si hoc non praestatur, ex conducto agetur." Cf. ibid. sec. 2: ". . . oportere enim agrum praestari conductori, ut frui possit."

87. See De Neeve, *ZRG* 100 (1983): 296–339, especially 303. For discussion of this view, see Frier, *BIDR,* 3d ser., 31–32 (1989–90): 241–44.

88. See Frier, *BIDR,* 3d ser., 31–32 (1989–90): 237–70, especially 239–60.

89. Capogrossi Colognesi, *Ai margini della proprietà fondiaria,* 143–99, especially 143–48 and "Il regime," 163–94, especially 163–66. Cf. W. Ernst, "Das Nutzungsrisiko bei der Pacht," *ZRG* 105 (1988): 541–91, at 541–60, who argues that the tenant could claim remission of rent on the basis of *vis maior* because *vis maior* represented a circumstance for which the lessor alone could be held legally responsible.

The question for the jurists centered around defining the conditions that released the tenant from this obligation. Ulpian and apparently also Pomponius quoted Servius Sulpicius Rufus, who defined these conditions as "any force that cannot be resisted" [omnem vim, cui resisti non potest] (D. 19.2.15.2), such as an irresistible force of rivers or of jackdaws or starlings, disastrous and unforeseen weather conditions, an enemy attack, or an earthquake.[90]

The relief accorded the tenant derived apparently from a legal institutionalization of a widespread practice among landowners. The tenant could not be expected to pay a rent if external conditions made it impossible for him to cultivate his farm and to harvest the crop that he needed to pay his rent. Under this circumstance, landowners even as early as the late republic might grant remissions, and the jurists and the Roman imperial government were concerned with defining how such remissions affected the legal rights of landlords and tenants. In this sense, the allocation of risk for *vis maior* in farm tenancy was consistent with the treatment of this principle in other areas of Roman private law.[91] However, the tenant had no legal recourse to gain relief for losses resulting from hazards in farming that were foreseeable, *vitia ex re*, no matter how serious they were. These risks could of course be every bit as catastrophic to the interests of the tenant as those hazards classified by the jurists as *vis maior*. Accordingly, tenants were not legally entitled to relief on the basis of a poor crop alone. We have seen how the

---

90. Servius apud Ulp. *D.* 19.2.15.2: "Si vis tempestatis calamitosae contigerit, an locator conductori aliquid praestare debeat, videamus. Servius omnem vim, cui resisti non potest, dominum colono praestare debere ait, ut puta fluminum graculorum sturnorum et si quid simile acciderit, aut si incursus hostium fiat. . . ." Cf. "sed et si labes facta sit omnmeque fructum tulerit, damnum coloni non esse, ne supra damnum seminis amissi mercedes agri praestare cogatur. . . . sed et si ager terrae motu ita corruerit, ut nusquam sit, damno domini esse: oportere enim agrum praestari conductori, ut frui possit" (ibid.). In the papyrological attestation of this passage, Pomponius is cited as referring to Servius' discussion: "[E]t refert [Pomponius Servium existimasse omnem vim, cui resisti non potest, dominum colono praestare debere, . . . etc.]. The Pomponius text is preserved in *PSI* XIV 1449, quoted by Sitzia, "Considerazione," 331–32, with 331 n. 1 on the restoration of Pomponius, originally proposed by V. Arangio-Ruiz. For discussion of *D.* 19.2.15.2 and the possibility of postclassical changes, see Ernst, *ZRG* 105 (1988): 560–67, with 563 n. 79 on the restoration of Pomponius to the papyrus text.

91. See especially Frier, *BIDR*, 3d ser., 31–32 (1989–90): 246–48, and De Neeve, *ZRG* 100 (1983): 296–339.

emperor Antoninus Pius ruled against precisely this sort of request in a rescript and termed "revolutionary" a request for remission based on the advanced age of the vines (Ulp. D. 19.2.15.5). Finally, the relief provided to the tenant was limited; he was entitled only to a prorated exoneration from the requirement to pay rent and had to bear the loss of seed himself (Ulp. D. 19.2.15.7).[92] It goes without saying that the tenant received no compensation for any loss of income. The principle of granting a remission of rent on the basis of *vis maior*, then, did not relieve the tenant of any of the risk that he bore for the greatest danger affecting Mediterranean agriculture, namely, the severe variations in weather, especially rainfall, that made droughts a regular part of life.

Despite the limited nature of the protection offered to the tenant, the imperial government regularly had to adjudicate legal cases involving a tenant's claim for a remission of rent. That such claims were frequent is suggested by Columella's urgency in recommending ways for the landowner to avoid having to deal with a tenant making such a request (1.7.1). For example, Ulpian quoted a rescript (without naming the emperor) approving a tenant's claim for a remission of rent (D. 19.2.15.3).[93] In this case, the tenant claimed a remission because a fire had severely damaged the *fundus* held under lease. This case suggests at least in general how the tenant might make a successful claim. The tenant, having suffered a loss because of the fire, withheld his rent, and the lessor brought a successful suit against him. The tenant then submitted a petition to the emperor, who overturned the original ruling and released the tenant from the obligation of having to pay his rent.[94]

---

92. Ulp. D. 19.2.15.7: "Ubicumque tamen remissionis ratio habetur ex causis supra relatis, non id quod sua interest conductor consequitur, sed mercedis exonerationem pro rata: supra denique damnum seminis ad colonum pertinere declaratur."

93. Ulp. D. 19.2.15.3: "Cum quidam incendium fundi allegaret et remissionem desideraret, ita ei rescriptum est: 'Si praedium coluisti, propter casum incendii repentini non immerito subveniendum tibi est.'" On this passage, see Capogrossi Colognesi, *Ai margini della proprietà fondiaria*, 182–84 and "Il regime," 185—86: the phrase *si praedium coluisti* implies that the tenant was assumed not to have been negligent if he cultivated the farm in accordance with conventional practices. See also Sitzia, "Considerazione," 343–44.

94. For the importance of the rescript process in the development of the imperial policy of granting remissions of rent, see Mayer-Maly, *Locatio Conductio*, 142–43, with discussion of this particular passage. See my additional discussion in the subsequent text.

The tenant involved in this case is described as cultivating a *praedium;* he was in all likelihood a tenant leasing on a relatively large scale. There is no reason to believe that the emperor writing this rescript in any way sought to provide extra relief to the tenant; it is rather an interpretation (possibly a more generous interpretation from the point of view of the tenant) of the principle that the tenant could not be expected to pay rent if certain circumstances that had nothing to do with the way in which he cultivated the property made it impossible for him to meet his contractual obligations.[95]

We can detect a similar process in a rescript on this question issued by the emperor Alexander Severus (*CJ* 4.65.8, A.D. 231). In this case, the petitioner was also a tenant, and he claimed a remission of rent based on *sterilitas,* apparently a disastrously poor harvest (see my further discussion of *sterilitas* later in this section). The emperor defined the circumstances under which the tenant might be entitled to a remission of rent, then he instructed the judge who had the responsibility of hearing this case on appeal to decide the case accordingly.[96] The two petitions addressed to the emperor Antoninus Pius probably resulted from similar circumstances, with the tenants either appealing unfavorable court decisions or seeking an imperial rescript as support for the case that they would make before the provincial governor or other judge. Accordingly, a tenant faced with a situation that he felt would justify a remission of rent might petition the emperor, and the reply to this petition, in the form of an imperial rescript stating the legal principles according to which a remission might be granted, would provide the tenant powerful support for his case in an eventual hearing before an imperial official.[97] The decisions that a judge made in a court case involving a claim for a remission of rent, however, might have a wider application than simply for the individual tenant making the claim.[98] Indeed, the climatic conditions or other circumstances that prevented

---

95. The emperor insisted that the tenant cultivate the land: "si praedium coluisti"; on this point, see De Neeve, *ZRG* 100 (1983): 311–12.

96. The rescript is quoted in n. 68 in this chapter. For discussion of the details, see Frier, *BIDR,* 3d ser., 31–32 (1989–90): 256–57.

97. See De Neeve, *ZRG* 100 (1983): 330–36.

98. For the implications for other landowners in a given region when an individual landowner granted a remission, see Frier, *BIDR,* 3d ser., 31–32 (1989–90): 259–60. For the general applicability of some imperial rescripts, see Nörr, *ZRG* 98 (1981): 1–46, at 41–45. For the use of imperial constitutions as precedents by both jurists and emperors, see M. Peachin, *Iudex vice Caesaris* (Stuttgart, 1996), 21–24.

one tenant from cultivating his farm would also affect the interests of other tenants in the area. If one tenant were successful in claiming a remission, other tenants would have a stronger legal position as well. But just as important as the judge's decision was the attitude taken by an individual landowner. A landowner assenting to a tenant's claim for relief would put pressure on other landlords in the vicinity to do the same. The imperial government was aware of this circumstance, and the emperor Diocletian emphasized in a rescript that acts of generosity by landlords that went beyond their contractual obligations or the custom of the region in question should not be interpreted as affecting the rights of other landlords in the vicinity (CJ 4.65.19, A.D. 293).[99]

In the two cases ruled on by Antoninus Pius, the imperial government answered requests for remissions of rent based on the poor quality of the harvest. The involvement of the emperor in judging requests for remissions is significant, since normally legal issues involved in tenancy were settled by a private *iudex* or, in the provinces, by the governor.[100] This involvement of the emperor suggests how important this question was for Roman agriculture in the late principate and has led a number of commentators to view the principle of *remissio mercedis* not as a juristic doctrine but rather as an imperial policy. Under this view the principle of *remissio mercedis* provided relief for poor crops and is to be distinguished from the Servian scheme, which exonerated the tenant only in the event of *vis maior*. This policy would thus have been designed to address the deteriorating state of Roman agriculture in the late principate, particularly in Italy. According to Mayer-Maly, the purpose of this policy was to enforce the tenant's duty to cultivate the land. Thus remissions of rent would only be granted to tenants who, despite their efforts to cultivate their land, were prevented from fulfilling their contractual obligations by unforeseen disasters.[101] Such a policy would certainly have been consistent with a general imperial ideology that

---

99. Diocl., Max., *CJ* 4.65.19: "Circa locationes atque conductiones maxime fides contractus servanda est, si nihil specialiter exprimatur contra consuetudinem regionis. quod si alii remiserunt contra legem contractus atque regionis consuetudinem pensiones, hoc aliis praeiudicium non possit adferre."

100. Cases involving requests for *remissio mercedis* were apparently increasingly heard in *cognitio extraordinaria*. On the development of this institution, see I. Buti, "La 'cognitio extra ordinem': Da Augusto a Diocleziano," *ANRW* II 14 (1982): 29–59.

101. See Mayer-Maly, *Locatio Conductio*, 140–47, especially 142–43, with further literature. This position is persuasively refuted by De Neeve, *ZRG* 100 (1983): 296–339, especially 318–29.

considered it to be the duty of the empire's subjects to cultivate their farmland and thereby fulfill their fiscal obligations to the state.[102] The task of the jurists and imperial authorities of the Severan Age and afterward, then, was to define precisely the conditions under which a tenant could claim a remission.[103] Another approach is to view the imperial government as attempting to come to the relief of small farmers, who had become, according to this view, subject to increasing pressures that threatened to destroy their economic independence. At the very least, the imperial policy of *remissio mercedis* would have established a more convenient procedural framework for tenants seeking remissions, since the *cognitio extraordinaria* could potentially give tenants more direct relief than the cumbersome traditional formulary system.[104]

In addition to the rescripts issued by Antoninus Pius, the principal evidence for this view about *remissio mercedis* comes from a response of Papinian, quoted by Ulpian, concerning remissions granted for poor crops, *ob sterilitatem* (D. 19.2.15.4). Papinian ruled that a remission granted under this circumstance did not cancel the landlord's claim on the full rent if the following years produced a "bountiful" crop.[105] Pap-

---

102. This ideology found expression in many places, for example, in the famous edict of Caracalla expelling Egyptians from Alexandria. Egyptians were to return to their villages of origin to fulfill their duty toward the state: ἐ[κεῖνοι] κωλ[ύ]εσθαι ὀφε[ί]λουσι οἵτινες φεύγουσι τὰς χώρας τὰς ἰδίας ἵνα μὴ / ἔρ[γον] ἀγροικον ποιῶσι . . . (*Sel.Pap.* II 215, *P.Giss.* 40 col. ii.23–24, *W.Chr.* 22, A.D. 215).

103. See Mayer-Maly, *Locatio Conductio*, 143–47.

104. For this view, see especially Ankum, *RIDA* 19 (1972): 219–38. De Neeve, *ZRG* 100 (1983): 330–36, is skeptical that the *cognitio extraordinaria* would have provided the average tenant with any substantial advantages over the formulary system, but he does allow that the increased possibility of submitting petitions to the emperor would have broadened the class of tenants able to use the imperial legal system to defend their right to a remission of rent.

105. Papin. apud Ulp. D. 19.2.15.4: "Papinianus libro quarto responsorum ait, si uno anno remissionem quis colono dederit ob sterilitatem, deinde sequentibus annis contigit uber[i]tas, nihil obesse domino remissionem, sed integram pensionem etiam eius anni quo remisit exigendam. hoc idem et in vectigalis damno respondit." For a full discussion of this passage, see Frier, *BIDR*, 3d ser., 31–32 (1989–90): 253–56, with further literature: the *quis* in this response is apparently an imperial official, not a landowner, so the remission was imposed on the landowner by a legal decision. Cf. De Neeve, *ZRG* 100 (1983): 321–24, who views the *quis* as lessor, as does Ankum, *RIDA* 19 (1972): 229–30. On the reading *uber[i]tas*, see Frier, 254 n. 76.

inian established the primacy of this claim by defining that any remission, even if granted as an outright gift, *verbo donationis*, would be treated as a contractual transaction. Only a remission granted in the final year of the lease would not be compensated by later bountiful harvests, and in this circumstance the remission would hold only when the landlord granting it was aware that the previous years had produced plentiful harvests.[106] Strictly speaking, this principle could not have done much to reduce the risk for the tenant, except in the case of an absolute disaster, but the concept *ubertas* must have been subjective, as was no doubt its converse, *sterilitas*.[107] Given this situation, there must have been a great deal of room for negotiation about what conditions would justify the tenant's claim for a remission of rent and, more important, a greater deal of room for negotiating under what circumstances the landlord could demand back the full rent for the year remitted in the years that follow. This same principle of balancing bad years against good ones appears in the rescript of Alexander Severus that I have already discussed (*CJ* 4.65.8). In this rescript, the emperor upheld the right of tenants to seek remission for *sterilitates* when neither the lease contract nor local custom imposed all risk for poor harvests on the tenant and when the poor year was not compensated by "bountiful" crops in other years.[108] Regulating the remission of rent remained an imperial concern at least until the reign of Diocletian, when, as we have seen, that emperor dealt with the problem of landlords going beyond their own contractual requirement or local custom by granting remissions (*CJ* 4.65.19). Diocletian upheld the primacy of the lease contract and local custom by ruling that such remissions did not prejudice the contractual rights of other landlords.

It seems very doubtful, however, that the imperial government would have developed such a policy to protect the interests of the small

---

106. Papin. apud Ulp. *D.* 19.2.15.4: "sed et si verbo donationis dominus ob sterilitatem anni remiserit, idem erit dicendum, quasi non sit donatio, sed transactio. quid tamen, si novissimus erat annus sterilis, in quo ei remiserit? verius dicetur et si superiores uberes fuerunt et scit locator, non deberi eum ad computationem vocari." Mayer-Maly, *Locatio Conductio*, 144–45, with further literature, considers this part of the Ulpian text to be interpolated.

107. See Sitzia, "Considerazione," 347–51, on the subjectivity of *sterilitas* and *ubertas*.

108. *CJ* 4.65.8 is quoted in n. 68 in this chapter. The phrase *ubertate aliorum annorum* can refer to preceding as well as to following years, although there is some ambiguity: see Sitzia, "Considerazione," 358–59 and nn. 93, 94.

farmer, since it would represent an unprecedented intrusion by the state into the private contractual relationships between landowners and tenants. Instead, I consider much more convincing the now prevalent view that the Roman legal authorities sought, in the second and third centuries A.D., to define more precisely how grants of remissions affected the continuing obligations of landowners and tenants toward one another.[109] The term *sterilitas* used by Papinian as grounds for remission is admittedly not found in the Servian scheme discussed by Pomponius and Ulpian, but this term does not represent an altogether new criterion for exonerating the tenant from his obligation to pay rent. Rather, the Servian scheme leaves some ambiguities about what constituted suitable grounds for claiming a relief. A climatic disaster that wiped out the crop canceled the tenant's obligations, but less dramatic climatic conditions might also reduce the crop and leave the tenant no more able to pay his rent. In some cases these conditions would be distinguishable from the types of disaster enumerated by Servius only by degree. We can see how subjective claims for remission on the basis of *vis maior* might be from the discussion of this subject by the second-century jurist Gaius (*D.* 19.2.25.6, 10 *ad ed. prov.*).[110] In Gaius' formulation, *vis maior* could be defined as a situation in which the tenant's crops were damaged beyond a tolerable point: "si plus, quam tolerabile est, laesi fuerint fructus." Such a formulation implies that the judgment of what constituted "intolerable damage" to the tenant's crops was always a matter of interpretation. As Sitzia argues, however, economic reasons might compel landlords not to dispute their tenants' requests for relief but to grant remissions. Papinian then established that remissions granted under such circumstances did not compromise the landlord's future claim to his full rent. Papinian was therefore clarifying existing legal principles concerning *remissio mercedis*, rather than creating new ones.[111]

109. This view is shared by Sitzia, De Neeve, and Frier.

110. Gaius 19.2.25.6: "Vis maior, quam Graeci θεοῦ βίαν appellant non debet conductori damnosa esse, si plus, quam tolerabile est, laesi fuerint fructus: alioquin modicum damnum aequo animo ferre debet colonus, cui immodicum lucrum non aufertur. apparet autem de eo nos colono dicere, qui ad pecuniam numeratam conduxit: alioquin partiarius colonus quasi societatis iure et damnum et lucrum cum domino fundi patitur."

111. See Sitzia, "Considerazione," 344–51. It was also possible for a provision about remissions in the event of *sterilitas* to be part of the lease contract. Such an arrangement, involving the judgment of a third party, is discussed in

We can detect, then, that the jurists' approach in their treatment of the distribution of risk in farm tenancy is similar to their approach in the other areas of law that we have examined. The jurists remained very conservative in terms of the types of contractual relationships that they would recognize, and accordingly they undertook to fit a wide range of very complicated economic relationships into the legal framework of the normative Roman lease. This conservatism on the part of the Roman legal authorities can best be seen in the rescript of Diocletian concerning remissions of rent (CJ 4.65.19, already mentioned earlier in this section); there, the emperor emphasized his interest in maintaining the legal validity of the lease contract: "circa locationes atque conductiones maxime fides contractus servanda est, si nihil specialiter exprimatur contra consuetudinem regionis." According to Diocletian's principle, the generosity of landowners who went beyond their strict contractual requirements did not compromise the contractual rights of other landowners. In all likelihood, however, many, if not most, lease relationships were based to a considerable extent on local customs and traditions; the rescripts of Diocletian characterize how the Roman imperial authorities endeavored to understand and adjudicate these relationships in terms of conventional legal categories. In doing so, they were taking the same approach that they took to describe agreements that were apparently intended to be long-term tenancy relationships when they used the terms of the normative five-year lease, in effect asserting the claim of normative lease law to regulate a wide variety of economic relationships.[112]

To return to the economic situation of the farm tenant, the treatment of remission of rent by the Roman legal authorities again suggests the considerable bargaining power that many tenants must have enjoyed. To preserve an economic system from which they benefited, Roman landowners had to balance asserting their own rights against the need to maintain the services of tenants capable of making a significant contribution in terms of resources and managerial expertise. In many lease relationships, the cash rent due in any given year must have been a

---

the *Opiniones* attributed to Ulpian (*D.* 50.8.3.2, 3 *opinionum*): "Sed si in locatione fundorum pro sterilitate temporis boni viri arbitratu in solvenda pensione cuiusque anni pacto comprehensum est, explorata lege conductionis fides bona sequenda est." On this passage, see Mayer-Maly, *Locatio Conductio*, 146–47; Sitzia, 355–56; and Frier, *BIDR*, 3d ser., 31–32 (1989–90): 257 n. 83.

112. For the normative function of rescripts, see n. 98 in this chapter.

matter for negotiation, depending on the nature of the contribution that the tenant made to the landowner's financial interest. Tenants nominally leasing on a short-term basis but in fact remaining on their farms year after year were not likely to see their property confiscated by the landowner if they proved unable or unwilling to pay their full rent. A landowner taking such steps would be acting against his own interest, because he would be compromising the productivity of the very individuals on whom he depended in the long term for his income. A wiser course for such landowners was to remit the tenant part of the rent due; the doctrine of Papinian sought to uphold the principle that in doing so the landowner was not renouncing any of his rights, in particular his eventual legal claim to the full rent as stipulated in the lease contract.

The Roman legal authorities, then, recognized the importance of the tenant's long-term investment to Roman agriculture, but their concern was to define the legal rights of the parties involved in a lease relationship, rather than to formulate economic policy. In this effort, they remained steadfastly conservative, seeking at all times to describe the rights and duties of landowners and tenants in terms of the existing normative lease contract that remained the basis for Roman lease law.[113] The insistence of the Roman legal authorities on the landowner's rights, however, does not imply that landowners were particularly successful at vindicating them. The legal right that he might have for back rents from his tenants, for example, would have been a purely academic question for Pliny, who believed that he had to take extraordinary measures to reduce his tenants' burden of risk. Columella sought to solve the problems of remissions and vindicating his full legal rights by avoiding even being put in the position of having to grant a remission of rent. In the understanding of both landowners, the continued productive cultivation of the farm was the most important service offered by their tenants, and similar considerations would have characterized the actions of numerous other comparable landowners.

Conclusion

I have argued that their dependence on the productivity of tenants placed Roman landowners in a somewhat contradictory position. On

---

113. This type of legal conservatism differs from that attributed to the jurists by A. Watson, *The Spirit of Roman Law* (Athens, Ga., 1995), who views the jurists as isolated from societal concerns.

the one hand, the normative Roman farm lease seemed designed to serve the interests of the landowner, since the tenant provided steady cash rents and bore the bulk of the risk for the harvest and the market prices of the crops. In addition, tenants made a valuable contribution by investing their resources in maintaining the productivity of their individual farms. Landowners, moreover, had a full range of legal remedies at their disposal to enforce the tenant's satisfactory performance of his contractual obligations. They could confiscate the tenant's property or use the threat of canceling the tenant's lease as a means of enforcement. On the other hand, economic realities made it very difficult for landowners to make use of these remedies; landowners regularly had to assume a much greater burden of the risk than might be suggested by the terms of the normative Roman lease, and they had to be flexible in dealing with their tenants to maintain their access to the tenant's investment, which was the preeminently important advantage offered by farm tenancy.

The Roman legal authorities were forced to take account of the economic relationship between landowner and tenant to be able to judge cases in a consistent manner. The jurists persisted in envisioning the landlord-tenant relationships in conventional terms, even when these terms had long since ceased to describe realistically the conditions under which most tenants occupied their land. But the jurists did not adopt this strategy because they were ignorant of the economic factors surrounding farm tenancy or simply chose not to take them into account. Instead, the jurists—and the imperial chancellery in answering petitions—practiced this conservatism because they maintained the primacy of Roman private law as the means to regulate the relationships between landowners and tenants in the Roman world.[114]

This model of the relationships on an estate derives from the assumptions that the jurists made about the interests of landowners and tenants when they formulated legal rules by which to judge cases involving farm tenancy. Clearly the jurists' assumptions offer at best an "idealized" picture of the Roman economy, and it cannot be argued that this idealized picture exactly describes estates in Italy or in any of the provinces during the early empire. Nevertheless, the jurists'

---

114. This point is made by Frier, *ZRG* 96 (1979): 227–28, who argues that the Roman jurists sought to maintain "the influence of Roman private law over lease relations" (227).

assumptions are consistent enough for us to infer the existence of basic economic relationships. How these relationships worked themselves out in the real world depended on the individual geographical and historical conditions under which landowners owned and tenants cultivated their land. The idealized picture of the landlord-tenant relationship that emerges from the Digest would most nearly describe the management of estates in areas under which private legal relationships were conducted in accordance with Roman law; these areas would include Italy and probably some of the western provinces.[115] But this limitation of the evidence need not represent a crucial stumbling block for analyzing agricultural conditions in the early empire. Rather, the legal evidence allows us to infer the existence of relationships that characterized the Roman economy generally, even if the form in which these relationships manifested themselves varied considerably from one region to another. By analyzing these relationships, we can come to a better understanding of the essential characteristics of the Roman economy.

---

115. See D. Liebs, "Römische Provinzialjurisprudenz," *ANRW* II 15 (1976): 288–362, on the provincial background of the writings of certain jurists; most of the classical jurists seemed to have written their legal opinions with the conditions of the city of Rome or at least of Italy in mind.

Conclusion: The Jurists and the Roman Agrarian Economy

In this study, I have endeavored to analyze the goals and planning of upper-class Romans in a crucial sector of the Roman economy, agriculture. Given the importance of agriculture as the basis of wealth in the Roman economy, the goals that upper-class Romans sought to achieve from their agricultural holdings and the methods that they used to pursue them were a critical factor in defining the fundamental economic relationships of the early Roman Empire. The problem for the historian has always been to find some means of identifying and analyzing how upper-class Romans understood economic problems.

Literary evidence from arguably representative sources, such as Pliny the Younger and to a lesser extent the Roman agronomists, allows us to trace the way in which individual members of the upper-classes understood their economy. We are justified in drawing broader conclusions from these sources insofar as we can infer that the writers shared basic assumptions about the Roman economy with their reading public. Similarly, papyrological evidence from Roman Egypt informs us about the organization of agricultural properties in that province and the methods used to manage them, allowing us to extrapolate more general conclusions about landownership in the Roman Empire. When examined against comparative evidence for economic relationships from better documented periods of history, the literary and papyrological evidence provides a solid foundation for a debate about the nature of the Roman economy.[1] We can further strengthen this foundation by considering the potentially rich source of evidence for the Roman economy provided by Roman legal literature. While there was undoubtedly

---

1. See H.W. Pleket, "Wirtschaft," in *Handbuch der Europäischen Wirtschafts- und Sozialgeschichte*, vol. I (Stuttgart, 1990), 25–160, and "Agriculture in the Roman Empire in Comparative Perspective," in H. Sancisi-Weerdenburg et al., eds., *De Agricultura* (Amsterdam, 1993), 317–42, for discussion of the Roman economy in a comparative historical perspective.

a close relationship between the development of Roman private law and that of the Roman economy, analyzing this complex relationship is itself a difficult process that raises many intricate questions. To what extent does law enshrine economic relationships? Given the limited information that exists concerning economic relationships, to what extent can we learn about these by studying Roman private law?

In this study, I have attempted to make use of the legal sources and to explore the ways in which they could enhance our understanding of the Roman economy. Specifically, I have examined the legal regulations that the Roman jurists developed for areas of the law impinging directly on the financial interests of upper-class Romans, including tutorship, legacies and trusts, and tenancy. Recent scholarship has argued persuasively that the Roman jurists did respond to the needs of the upper-classes as they formulated principles in crucial areas of the law.[2] In my view, one can make a strong case that, in the areas of the law concerning private property that I have singled out, the jurists were concerned to respond to the needs of the class to which they themselves belonged. My discussion has shown that the jurists made a consistent set of assumptions about the financial interests of upper-class Romans and about the general conditions of the Roman economy.

My purpose in examining the economic assumptions of the Roman jurists is not to provide a comprehensive account of the economic interests and activities of upper-class Romans but rather to develop a model that defines the most important relationships affecting their decision making. Certainly, we must grant that individual upper-class Romans will have invested and managed their wealth in more diverse ways than ever could be envisioned in the legal sources. Even so, the economic principles underlying the rules formulated by the Roman jurists are significant for understanding the Roman economy. Admittedly, we should expect the jurists to have made very conservative assumptions about the Roman economy as they regulated tutorship. Likewise, in the law of legacy, the jurists' concern to develop conventional descriptions of estates necessarily simplified matters because they were intended to be general rules to settle potential disputes. As I have sought to demonstrate, however, the jurists' rules and definitions did respond primarily to the needs of upper-class Romans, since relevant literary and papyro-

---

2. See especially the works of B.W. Frier and S.D. Martin discussed in my introduction.

logical material reveals that wealthy Romans acted on precisely the types of economic considerations that the jurists envisioned. At the very least, the legal material, when viewed against relevant literary and papyrological sources, indicates that the jurists shared broadly held views about the nature of the Roman economy and the constraints that it imposed on upper-class Romans in managing their wealth. Consequently, I believe the case can be made that the abundant legal material provides an important source for understanding the economic mentality of upper-class Romans.

This analysis of the legal material reveals several basic relationships that characterized the Roman economy. The first is the nature of an agrarian estate as an investment. The assumptions that underlie the legal regulations for tutorship and the bequest of estates suggest that an estate essentially represented security and that the overriding goal for many landowners was to manage their estates in such a way as to maintain a stable income. This concern for stability put upper-class Romans in a somewhat contradictory situation. While the resources available to them seemingly gave upper-class Romans tremendous economic power, their concern was to avoid any risk that might threaten their ability to maintain their position within Rome's social hierarchy. This aversion to risk must have affected the willingness of Romans to try to take advantage of any available commercial possibilities, but more important for our purposes, it reduced the economic power that landowners could exercise in managing their own estates. For many landowners, tenancy offered a ready means to keep as low as possible their risks and expenses in managing their estates. But this concern for stability weakened the bargaining power of landowners leasing their land to tenants. Despite the wide social and economic gulf that separated landowners from most tenants, landowners could not simply impose conditions on tenants but instead had to contend with a continuing shortage of suitable tenants and had to make adjustments in the terms of land tenure to keep their estates cultivated and productive for the long term.

The Letters of Pliny the Younger document the difficulties that Roman landowners faced in managing estates leased out to tenants, as they also demonstrate the considerable bargaining power that tenants enjoyed. The jurists' treatment of farm tenancy, however, indicates that the difficulties that Pliny experienced were not an isolated phenomenon but characteristic of the Roman economy. Though we can detect this dynamism between landowner and tenant in various areas of the law on lease, it is especially evident in the jurists' treatment of allocating the risk in agriculture. In the development of the legal principle of *remissio mercedis*, we can trace the Roman legal authorities' concern to respond to a basic aspect of Mediterranean agriculture while preserving what was economically most advantageous for the landowner, namely, the continuing presence of productive tenants. What is remarkable in the Roman doctrine of *remissio mercedis* is that the jurists were responding not simply to the needs of a relatively small number of elite tenants who were able to relate to their landowners on a socially and legally equal basis but rather to the needs of landowners and tenants of strikingly divergent economic and social status.

The difficulty in analyzing the legal regulations for farm tenancy as evidence for economic history is that the jurists were concerned not with making economic policy, but with defining the legal relationships between landowners and tenants arising from changing economic conditions. And insofar as the consistent and expected conservatism of the jurists led them to define a variety of tenure arrangements in terms of conventional Roman private law, the task of tracing economic changes is made all the more difficult. Indeed, an analysis of the legal sources for farm tenancy does not allow us to describe in great detail the particular conditions under which any group of tenants occupied their land. We learn instead how the jurists responded to legal issues arising from the dynamic economic relationship between landowners and tenants, one that was shaped to a large extent by the former's desire for economic security. The jurists' treatment of the law of tenancy confirms how landowners were circumscribed by an overriding economic conservatism in their dealings with their tenants. Again, it must be emphasized that my conclusions do not in and of themselves describe the relationships between any individual landowner and his or her tenants. But through examining the legal evidence we do gain a deeper appreciation of the constraints affecting all landowners, and understanding these general constraints provides us with the necessary background against which to analyze the Roman economy in all its complexity.

Bibliography

Abel, W. *Agricultural Fluctuations in Europe from the Thirteenth to the Twentieth Centuries*. Trans. O. Ordish. London, 1986. [= *Agrarkrisen und Agrarkonjunktur*. 3d ed. Hamburg and Berlin, 1978.]
Alcock, S. *Graecia Capta: The Landscapes of Roman Greece*. Cambridge, 1993.
Alston, R. *Soldier and Society in Roman Egypt: A Social History*. London and New York, 1995.
Amelotti, M. *Il testamento romano attraverso la prassi documentale*. Vol. I, *Le forme classiche di testamento*. Florence, 1966.
Ankum, H. "*Remissio mercedis*." *RIDA* 19 (1972): 219–38.
Ash, H.B., ed. and trans. *Columella de Re Rustica*. Vol. I. Loeb Edition. London and Cambridge, Mass., 1977. First ed., 1941.
Aubert, J.-J. *Business Managers in Ancient Rome: A Social and Economic Study of Institores, 200 B.C. – A.D. 250*. Columbia Studies in the Classical Tradition 21. Leiden, 1994.
Aymard, A. "Les capitalistes romaines et la viticulture italienne." *Annales ESC* 2 (1947): 257–65.
Bagnall, R.S. "Landholding in Late Roman Egypt: The Distribution of Wealth." *JRS* 82 (1992): 128–49.
———. *Egypt in Late Antiquity*. Princeton, 1993.
———. "Managing Estates in Roman Egypt: A Review Article." *BASP* 30, nos. 3–4 (1993): 127–35.
Bagnall, R S., and B.W. Frier. *The Demography of Roman Egypt*. Cambridge Studies in Population, Economy, and Society in Past Time 23. Cambridge, 1994.
Bardhan, P.K. *Land, Labor, and Rural Poverty: Essays in Development Economics*. New York, 1984.
Berger, A. *Encyclopedic Dictionary of Roman Law*. Transactions of the American Philosophical Society, n.s., vol. 43, pt. 2. Philadelphia, 1953. Reprint, 1980.
Bingen, J. "Documents de l'Egypte romaine." *BASP* 22 (1985): 14–21.
Biscardi, A. "L'*Oratio Severi* e il divieto di *obligare*." In *Studi in onore di Giuseppe Grosso*, vol. III, 245–66. Turin, 1970.
Bossu, C. "L'objectif de l'institution alimentaire: Essai d'évaluation." *Latomus* 48 (1989): 372–82.
Bowman, A.K. *The Town Councils of Roman Egypt*. Am.Stud.Pap. 11. Toronto, 1971.
Bowman, A.K., and D. Rathbone. "Cities and Administration in Roman Egypt." *JRS* 82 (1992): 107–27.

Brockmeyer, N. "Arbeitsorganisation und ökonomisches Denken in der Gutswirtschaft des römischen Reiches." Ph.D. diss., Ruhr-Universität Bochum, 1968.
Bürge, A. "Vertrag und personale Abhängigkeiten im Rom der späten Republik und der frühen Kaiserzeit." *ZRG* 97 (1980): 105–56.
———. "Fiktion und Wirklichkeit: Soziale und rechtliche Strukturen des römischen Bankwesens." *ZRG* 104 (1987): 465–558.
Buti, I. "La 'cognitio extra ordinem': Da Augusto a Diocleziano." *ANRW* II 14 (1982): 29–59.
Capogrossi Colognesi, L. "Grandi proprietari, contadini e coloni nell'Italia Romana (I–III d.C.)." In A. Giardina, ed., *Società romana e impero tardo-antico*, vol. I, *Istituzioni, ceti, economie*, 325–65, 703–23. Rome and Bari, 1986.
———. "Il regime degli affitti agrari." *Scienze dell'antichità, Storia, Archeologia, Antropologia* 6–7 (1992–93): 163–253.
———. *Ai margini della proprietà fondiaria*. Rome, 1995.
Carandini, A. "Columella's Vineyard and the Rationality of the Roman Economy." Trans. E. Fentress. *Opus* 2 (1983): 172–204.
Carlsen, J. *Vilici and Roman Estate Managers until A.D. 284*. ARID Suppl. 24. Rome, 1995.
Cervenca, G. "Studi sulla *Cura Minorum*." Pt. 1. "*Cura Minorum* e *Restitutio in Integrum*." *BIDR*, 3d ser., 14 (1972): 235–317.
———. "Studi sulla *Cura Minorum*." Pt. 3. "L'estensione ai minori del regime dell'*Oratio Severi*." *BIDR*, 3d ser., 21 (1979): 41–94.
Chalon, G. *L'édit de Tiberius Julius Alexander: Étude historique et exégétique*. Bibliotheca Helvetica Romana 5. Olten and Lausanne, 1964.
Champlin, E. *Final Judgments: Duty and Emotion in Roman Wills, 200 B.C.–A.D. 250*. Berkeley and Los Angeles, 1991.
Chastagnol, A. *Le sénat romain à l'époque impériale: Recherches sur la composition de l'Assemblée et le statut de ses membres*. Paris, 1992.
Chayanov, A.V. "On the Theory of Non-Capitalist Economic Systems." 1924. Trans. C. Lane in D. Thorner, B. Kerblay, and R.E.F. Smith, eds., *The Theory of Peasant Economy*, 1–28. Homewood, Ill. 1966.
———. "Peasant Farm Organization." 1925. Trans. R.E F. Smith in D. Thorner, B. Kerblay, and R.E.F. Smith, eds., *The Theory of Peasant Economy*, 29–269. Homewood, Ill. 1966.
Cheung, S.N.S. *The Theory of Share Tenancy*. Chicago, 1969.
Chiusi, T.J. "Landwirtschaftliche Tätigkeit und actio institoria." *ZRG* 108 (1991): 155–86.
Christes, J. *Sklaven und Freigelassene als Grammatiker und Philologen im antiken Rom*. Forschungen zur antiken Sklaverei 10. Wiesbaden, 1979.
Cipolla, C.M. *Before the Industrial Revolution: European Society and Economy, 1000–1700*. 3d ed. New York and London, 1994.
Clark, C., and M. Haswell. *The Economics of Subsistence Agriculture*. 4th ed. London, 1970.
Cotton, H. "The Guardianship of Jesus Son of Babatha: Roman and Local Law in the Province of Arabia." *JRS* 83 (1993): 94–108.

Criniti, N. *La Tabula Alimentaria di Veleia*. Fonti e studi, 1a ser., 14. Parma, 1991.
Crook, J. "Classical Roman Law and the Sale of Land." In M.I. Finley, ed., *Studies in Roman Property*, 71–83. Cambridge, 1976.
———. "Women in Roman Succession." In B. Rawson, ed., *The Family in Ancient Rome: New Perspectives*, 58–82. London and Ithaca, 1986.
———. "Feminine Inadequacy and the *Senatusconsultum Velleianum*." In B. Rawson, ed., *The Family in Ancient Rome: New Perspectives*, 83–92. London and Ithaca, 1986.
Currie, J.M. *The Theory of Land Tenure*. Cambridge, 1981.
D'Arms, J.H. *Commerce and Social Standing in Ancient Rome*. Cambridge, Mass., 1981.
Delia, D. *Alexandrian Citizenship during the Roman Principate*. American Classical Studies 23. Atlanta, 1991.
De Ligt, L. *Fairs and Markets in the Roman Empire: Economic and Social Aspects of Periodic Trade in a Pre-Industrial Society*. Dutch Monographs on Ancient History and Archaeology 11. Amsterdam, 1993.
———. "The Nundinae of L. Bellicius Sollers." In H. Sancisi-Weerdenburg, R.J. van der Spek, H.C. Teitler, and H.T. Wallinga, eds., *De Agricultura: In Memoriam Pieter Willem de Neeve (1945–1990)*. Dutch Monographs on Ancient History and Archaeology 10, 238–62. Amsterdam, 1993.
De Martino, F. "Coloni in Italia." *Labeo* 41 (1995): 35–65.
De Neeve, P.W. "Remissio Mercedis." *ZRG* 100 (1983): 296–339.
———. "Fundus as Economic Unit." *RHD* 52 (1984): 3–19.
———. *Colonus: Private Farm-Tenancy in Roman Italy during the Republic and the Early Principate*. Amsterdam, 1984.
———. "The Price of Agricultural Land in Roman Italy and the Problem of Economic Rationalism." *Opus* 4 (1985): 77–109.
———. "A Roman Landowner and His Estates: Pliny the Younger." *Athenaeum* 78 (1990): 363–402.
de Pachtère, F.G. *La table hypothécaire de Veleia: Étude sur la propriété foncière dans l'Appenin de Plaisance*. Paris, 1920.
de Zulueta, F. *The Roman Law of Sale*. Oxford, 1945.
Di Porto, A. "Impresa agricola ed attività collegate nell'economia della 'villa': Alcune tendenze organizzative." In *Sodalitas: Scritti in onore di Antonio Guarino*, vol. VII, 3235–77. Naples, 1984.
Dixon, S. "Family Finances: Terentia and Tullia." In B. Rawson, ed., *The Family in Ancient Rome: New Perspectives*, 93–120. London and Ithaca, 1986.
Drexhage, H.-J. *Preise, Mieten/Pachten, Kosten und Löhne im römischen Ägypten bis zum Regierungsantritt Diokletians*. Vorarbeiten zu einer Wirtschaftsgeschichte des römischen Ägypten 1. St. Katharinen, 1991.
Duncan-Jones, R. *The Economy of the Roman Empire: Quantitative Studies*. 2d ed. Cambridge, 1982.
———. "Review." *JRS* 76 (1986): 296–97.
———. *Structure and Scale in the Roman Economy*. Cambridge, 1990.
———. *Money and Government in the Roman Empire*. Cambridge, 1994.

Eck, W. *Die staatliche Organisation Italiens in der hohen Kaiserzeit*. Vestigia 28. Munich, 1979.
el Abbadi, M.H. "P.Flor. 50: Reconsidered." *Proceedings of the XIV International Congress of Papyrologists, Oxford, 24–31 July 1974*, 91–96. Egypt Exploration Society, Graeco-Roman Memoirs 61. London, 1975.
Ernst, W. "Das Nutzungsrisiko bei der Pacht." *ZRG* 105 (1988): 541–91.
Finley, M.I. "Private Farm Tenancy in Italy before Diocletian." In M.I. Finley, ed., *Studies in Roman Property*, 103–21. Cambridge, 1976.
———. *The Ancient Economy*. 2d ed. Berkeley and Los Angeles, 1985.
———. *Ancient History: Evidence and Models*. London, 1985.
Flach, D. *Römische Agrargeschichte*. Handbuch der Altertumswissenschaft 3 no. 9. Munich, 1990.
Foxhall, L. "The Dependent Tenant: Land Leasing and Labour in Italy and Greece." *JRS* 80 (1990): 97–114.
Frank, T. *Rome and Italy of the Empire*. In *An Economic Survey of Ancient Rome*, vol. V. Baltimore, 1940.
Frayn, J. *Markets and Fairs in Roman Italy: Their Social and Economic Importance from the Second Century B.C. to the Third Century A.D.* Oxford, 1993.
Frederiksen, M.W. "Theory, Evidence and the Ancient Economy." *JRS* 65 (1975): 164–71.
Frier, B.W. "Law, Technology, and Social Change: The Equipping of Italian Farm Tenancies." *ZRG* 96 (1979): 204–28.
———. *Landlords and Tenants in Imperial Rome*. Princeton, 1980.
———. "Roman Life Expectancy: Ulpian's Evidence." *HSCP* 86 (1982): 213–51.
———. "Review of *Servus Quasi Colonus: Forme non Tradizionali di Organizzazione del Lavoro nella Società Romana*, by G. Giliberti, and *Colonus: Privégrondpacht in Romeins Italie tijdens de Republiek en het vroege Principaat*, by P.W. de Neeve." *ZRG* 100 (1983): 667–76.
———. "Review of *Die Kolonen in Italien und den westlichen Provinzen des Römischen Reiches: Eine Untersuchung der literarischen, juristischen und epigraphischen Quellen vom 2. Jahrhundert v.u.Z. bis zu den Severern*," by K.-P. Johne, J. Köhn, and V. Weber." *ZRG* 102 (1985): 564–69.
———. *The Rise of the Roman Jurists: Studies in Cicero's* pro Caecina. Princeton, 1985.
———. "Law, Economics, and Disasters down on the Farm: 'Remissio Mercedis' Revisited." *BIDR*, 3d ser., 31–32 (1989–90): 237–70.
———. "Subsistence Annuities and Per Capita Income in the Early Roman Empire." *CP* 88 (1993): 222–30.
Gallant, T.W. *Risk and Survival in Ancient Greece: Reconstructing the Rural Domestic Economy*. Stanford, 1991.
Gardner, J.F. *Women in Roman Law and Society*. London, 1986.
Garnsey, P. *Social Status and Legal Privilege in the Roman Empire*. Oxford, 1970.
———. "Non-Slave Labour in the Roman World." In P. Garnsey, ed., *Non-Slave Labour in the Greco-Roman World*, Cambridge Phil. Soc. Suppl. 6, 34–47. Cambridge, 1980.

Garnsey, P., T. Gallant, and D. Rathbone. "Thessaly and the Grain Supply of Rome during the Second Century B.C." *JRS* 74 (1984): 30–44.
Garnsey, P., K. Hopkins, and C.R. Whittaker, eds. *Trade in the Ancient Economy.* London, 1983.
Garnsey, P., and R. Saller. *The Roman Empire: Economy, Society and Culture.* Berkeley and Los Angeles, 1987.
Giliberti, G. *Servus Quasi Colonus: Forme non tradizionali di organizzazione del lavoro nella società romana.* 2d ed. Naples, 1988.
Goffart, W. *Caput and Colonate: Towards a History of Late Roman Taxation.* Phoenix Suppl. 12. Toronto, 1974.
Goldsmith, R.W. "An Estimate of the Size and Structure of the National Product of the Early Roman Empire." *Review of Income and Wealth* 30, no. 3 (1984): 263–88.
———. *Premodern Financial Systems: A Historical Comparative Study.* Cambridge, 1987.
Greene, K. *The Archaeology of the Roman Economy.* Berkeley and Los Angeles, 1986.
Griffin, M. *Seneca: A Philosopher in Politics.* Oxford, 1976.
Grigg, D. *The Dynamics of Agricultural Change.* London, 1982.
Hahn, J. *Der Philosoph und die Gesellschaft: Selbstverständnis, öffentliches Auftreten und populäre Erwartungen in der hohen Kaiserzeit.* Heidelberger Althistorische Beiträge und Epigraphische Studien 7. Stuttgart, 1989.
Hanson, A.E. "Two Copies of a Petition to the Prefect." *ZPE* 47 (1982): 233–43.
———. "The Archive of Isidoros of Psophthis and P. Ostorius Scapula, *Praefectus Aegypti.*" *BASP* 21 (1984): 77–87.
Harl, K.W. *Coinage in the Roman Economy, 300 B.C. to A.D. 700.* Baltimore and London, 1996.
Harris, W.V. "Between Archaic and Modern: Problems in Roman Economic History." In W.V. Harris, ed., *The Inscribed Economy: Production and Distribution in the Roman Empire in the Light of Instrumentum Domesticum, JRA* Supplementary Series 6, 11–29. Ann Arbor, 1993.
Heitland, W.E. *Agricola: A Study of Agriculture and Rustic Life in the Greco-Roman World from the Point of View of Labour.* Cambridge, 1921. Reprint, Westport, Conn., 1970.
Helmholz, R.H. "The Roman Law of Guardianship in England, 1300–1600." *Tulane Law Review* 52, no. 2 (1978): 223–57.
Hennig, D. "Untersuchungen zur Bodenpacht im ptolemäisch-römischen Ägypten." Ph.D. diss., Ludwig-Maximilians-Universität zu München, 1967.
Herrmann, J. *Studien zur Bodenpacht im Recht der graeco-aegyptischen Papyri.* Münch. Beitr. 41. Munich, 1958.
Hodkinson, S. "Animal Husbandry in the Greek Polis." In C.R. Whittaker, ed., *Pastoral Economies in Classical Antiquity,* Cambr. Phil. Soc. Suppl. 14, 35–74. Cambridge, 1988.
Holtheide, B. "Matrona Stolata–Femina Stolata." *ZPE* 38 (1980): 127–34.
Honoré, T. *Ulpian.* Oxford, 1982.
———. *Emperors and Lawyers.* 2d ed. Oxford, 1994.

Hopkins, K. "Taxes and Trade in the Roman Empire (200 B.C.–A.D. 400)." *JRS* 70 (1980): 101–25.

Hopkins, K., and G. Burton. "Political Succession in the Late Republic (249–50 B.C.)." In K. Hopkins, *Death and Renewal. Sociological Studies in Roman History* 2, 31–119. Cambridge, 1983.

Howgego, C. "The Supply and Use of Money in the Roman World, 200 B.C. to A.D. 300." *JRS* 82 (1992): 1–31.

Johne, K.-P., V. Weber, and J. Köhn. *Die Kolonen in Italien und den westlichen Provinzen des Römischen Reiches: Eine Untersuchung der literarischen, juristischen und epigraphischen Quellen vom 2. Jahrhundert v.u.Z. bis zu den Severern.* Berlin, 1983.

Johnston, D. "Munificence and *Municipia*: Bequests to Towns in Classical Roman Law." *JRS* 75 (1985): 105–25.

———. "Prohibitions and Perpetuities: Family Settlements in Roman Law." *ZRG* 102 (1985): 220–90.

———. *The Roman Law of Trusts*. Oxford, 1988.

———. "Successive Rights and Successful Remedies: Life Interests in Roman Law." In P. Birks, ed., *New Perspectives in the Roman Law of Property: Essays for Barry Nicholas,* 153–67. Oxford, 1989.

Jongman, W. *The Economy and Society of Pompeii.* Dutch Monographs on Ancient History and Archaeology 4. Amsterdam, 1988.

Jördens, A. *Vertragliche Regelungen von Arbeiten im späten griechischsprachigen Ägypten.* [= *P.Heid.* V.] Heidelberg, 1990.

Kaser, M. *Das römische Privatrecht.* Vol. I, *Das altrömische, das vorklassische und klassische Recht.* Handbuch der Altertumswissenschaft 3 no. 3.3.1. 2d. ed. Munich, 1971.

———. *Zur Methodologie der römischen Rechtsquellenforschung.* SAWW (Phil.-hist. Kl.) 277, no. 5. Vienna, 1972.

Kaster, R.A. *Guardians of Language: The Grammarian and Society in Late Antiquity.* Berkeley and Los Angeles, 1988.

Keenan, J.G. "The Will of Gaius Longinus Castor." *BASP* 31 (1994): 101–7.

Kehoe, D.P. "Allocation of Risk and Investment on the Estates of Pliny the Younger." *Chiron* 18 (1988): 15–42.

———. *The Economics of Agriculture on Roman Imperial Estates in North Africa.* Hypomnemata 89. Göttingen, 1988.

———. "Approaches to Economic Problems in the 'Letters' of Pliny the Younger: The Question of Risk in Agriculture." *ANRW* II 33, no. 1 (1989): 555–90.

———. *Management and Investment on Estates in Roman Egypt during the Early Empire.* PTA 40. Bonn, 1992.

———. "Economic Rationalism in Roman Agriculture." *JRA* 6 (1993): 476–84.

———. "Investment in Estates by Upper-Class Landowners in Early Imperial Italy: The Case of Pliny the Younger." In H. Sancisi-Weerdenburg, R.J. van der Spek, H.C. Teitler, and H.T. Wallinga, eds., *De Agricultura: In Memoriam Pieter Willem de Neeve (1945–1990),* Dutch Monographs on Ancient History and Archaeology 10, 214–37. Amsterdam, 1993.

———. "Approaches to Profit and Management in Roman Agriculture: The Evidence of the Digest." In J. Carlsen, P. Ørsted, and J.E. Skydsgaard, eds., *Landuse in the Roman Empire*, ARID Suppl. 22, 45–58. Rome, 1994.

———. "Legal Institutions and the Bargaining Power of the Tenant in Roman Egypt." *APF* 41, no. 2 (1995): 232–62.

Kelly, J.M. *Roman Litigation*. Oxford, 1966.

Köhn, J. "Die Kolonen in den Rechstbestimmungen." In K.-P. Johne, V. Weber, and J. Köhn, *Die Kolonen in Italien und den westlichen Provinzen des Römischen Reiches: Eine Untersuchung der literarischen, juristischen und epigraphischen Quellen vom 2. Jahrhundert v.u.Z. bis zu den Severern*, 167–257. Berlin, 1983.

Kolendo, J. *Le traité d'agronomie des Saserna*. Archiwum Filologiczne 29. Wroclaw, 1973.

———. *Le colonat en Afrique sous le haut-empire*. 2d ed. Paris, 1991.

———. "Ostentation sociale et grande propriété." In *Du Latifundium au Latifondo: Un héritage de Rome, une création médiéavale ou moderne?* Actes de la table rond internationale du CNRS organisée à l'Université Michel de Montaigne Bordeaux III les 17–19 décembre 1992, 425–36. Talence, 1995.

Krause, J.-U. *Witwen und Waisen im römischen Reich*. Vol. III, *Rechtliche und soziale Stellung von Waisen*. Heidelberger Althistorische Beiträge und Epigraphische Studien 18. Stuttgart, 1995.

Kraut, B. "Seven Heidelberg Papyri concerning the Office of Exegetes." *ZPE* 55 (1984): 167–90.

Kula, W. *An Economic Theory of the Feudal System: Towards a Model of the Polish Economy, 1500–1800*. Trans. L. Garner. London, 1976.

Kunkel, W. *Herkunft und soziale Stellung der römischen Juristen*. 2d ed. Graz, Vienna, and Cologne, 1967.

Lenel, O. "Die cura minorum in der klassischen Zeit." *ZRG* 35 (1914): 129–213.

Le Roy Ladurie, E. *The Peasants of Languedoc*. Trans. J. Day. Urbana, 1974. French ed., 1966.

Lewis, N., and M. Reinhold. *Roman Civilization: Selected Readings*. 3d ed. Vol. II. New York, 1990.

Liebs, D. "Römische Provinzialjurisprudenz." *ANRW* II 15 (1976): 288–362.

———. *Die Jurisprudenz im spätantiken Italien (240–640 n.Chr.)*. Berlin, 1987.

Lo Cascio, E. "Gli *Alimenta*, l'agricoltura italica e l'approvvigionamento di Roma." *RAL* 33 (1978): 311–52.

———. "Gli *Alimenta* e la 'politica economica' di Pertinace." *RFIC* 108 (1980): 264–88.

———. "Forme dell'economia imperiale." In A. Schiavone, ed., *Storia di Roma*, vol. II, *L'impero mediterraneo*, pt. 2, *I principi e il mondo*, 313–65. Turin, 1991.

———. "Fra equilibrio e crisi." In A. Schiavone, ed., *Storia di Roma*, vol. II, *L'impero mediterraneo*, pt. 2, *I principi e il mondo*, 701–31. Turin, 1991.

———. "L'affitto agrario in Italia nella prima età imperiale: A proposito di alcuni lavori recenti." *Scienze dell'antichità, Storia, Archeologia, Antropologia* 6–7 (1992–93): 257–68.

———. "Considerazioni sulla struttura e sulla dinamica dell'affitto agrario in età imperiale." In H. Sancisi-Weerdenburg, R.J. van der Spek, H.C. Teitler, and H.T. Wallinga, eds., *De Agricultura: In Memoriam Pieter Willem de Neeve (1945–1990)*, Dutch Monographs on Ancient History and Archaeology 10, 296–316. Amsterdam, 1993.

———. "The Size of the Roman Population: Beloch and the Meaning of the Augustan Census Figures." *JRS* 84 (1994): 23–40.

———, ed. *Dall'affitto agrario al colonato tardo-antico: Continuità o frattura?* Naples, 1997.

Łoś, A. "Les intérêts des affranchis dans l'agriculture italienne." *MEFRA* 104 (1992): 709–53.

Love, J. "The Character of the Roman Agricultural Estate in the Light of Max Weber's Economic Sociology." *Chiron* 16 (1986): 99–146.

———. *Antiquity and Capitalism: Max Weber and the Sociological Foundations of Roman Civilization*. London and New York, 1991.

MacCormack, G. "The Liability of the Tutor in Classical Roman Law." *Irish Jurist* 5 (1970): 369–90.

MacMullen, R. *Corruption and Decline of Rome*. New Haven, 1988.

Macve, R.H. "Some Glosses on Ste. Croix's 'Greek and Roman Accounting.'" In P. Cartledge and F.D. Harvey, eds., *Crux: Essays Presented to G.E.M. de Ste. Croix on his 75th Birthday*, 233–64. Exeter, 1985.

Martin, R. *Recherches sur les agronomes latins et leurs conceptions économiques et sociales*. Collection d'études anciennes. Paris, 1971.

Martin, S.D. *The Roman Jurists and the Organization of Private Building in the Late Republic and Early Empire*. Collection Latomus 204. Brussels, 1989.

Mattingly, D.J. "Oil for Export? A Comparison of Libyan, Spanish and Tunisian Olive Oil Production in the Roman Empire." *JRA* 1 (1988): 33–56.

———. "The Olive Boom: Oil Surpluses, Wealth and Power in Roman Tripolitania." *Libyan Studies* 19 (1988): 21–41.

Mayer-Maly, T. *Locatio Conductio: Eine Untersuchung zum klassischen römischen Recht*. Vienna, 1956.

Melillo, G. *Economia e giurisprudenza a Roma: Contributo al lessico economico dei giuristi romani*. Forme materiali e ideologie del mondo antico 10. Naples, 1978.

Mette-Dittmann, A. *Die Ehegesetze des Augustus: Eine Untersuchung im Rahmen der Gesellschaftspolitik des Princeps*. Historia Einzelschriften 67. Stuttgart, 1991.

Migliardi Zingale, L. *I testamenti romani nei papiri e nelle tavolette d'Egitto: Silloge di documenti dal I al IV secolo d.C.* 2d ed. Turin, 1991.

Millar, F. *The Emperor in the Roman World (31 B.C.–A.D. 337)*. Ithaca, 1977.

Millar, J.R. "A Reformulation of A.V. Chayanov's Theory of Peasant Economy." *Economic Development and Cultural Change* 18, no. 2 (1970): 219–29.

Mitteis, L., and U. Wilcken. *Grundzüge und Chrestomathie der Papyruskunde*. 2 vols. Leipzig, 1912. Reprint, Hildesheim, 1963.

Molnár, I. "Die Ausgestaltung des Begriffes der vis maior im römischen Recht." *Iura* 32 (1981): 73–105.

———. "Verantwortung und Gefahrtragung bei der *locatio conductio* zur Zeit des Prinzipats." *ANRW* II 14 (1982): 583–680.

Mrozek, S. "Die Privaten Alimentarstiftungen in der römischen Kaiserzeit." In H. Kloft, ed., *Sozialmassnahmen und Fürsorge: Zur Eigenart antiker Sozialpolitik,* 155–66. Grazer Beiträge, Supplementband 3. Graz and Horn, 1988.

Muth, R.F. "Real Land Rentals in Early Roman Egypt." *Explorations in Economic History* 31 (1994): 210–24.

Nicolet, C. "La pensée économique des Romains, République et Haut-Empire." In *Rendre à César: Économie et société dans la Rome antique.* Bibliothèque des Histoires, 117–219. Mesnil-sur-L'Estrée, 1988. = "Il pensiero economico dei Romani." In L. Firpo, ed., *Storia delle idee politiche, economiche e sociali.* Vol. I, 877–960. Turin, 1982.

Nörr, D. "Zur Reskriptenpraxis in der hohen Prinzipatszeit." *ZRG* 98 (1981): 1–46.

Oliver, J.H. *Greek Constitutions of Early Roman Emperors from Inscriptions and Papyri.* Mem. Amer. Phil. Soc. 178. Philadelphia, 1989.

Omar, S. *Das Archiv des Soterichos.* Pap.Colon. 8. Opladen, 1979.

Osborne, R. "Social and Economic Implications of the Leasing of Land and Property in Classical and Hellenistic Greece." *Chiron* 18 (1988): 279–323.

Oxford Latin Dictionary. Ed. P.G.W. Glare. Oxford, 1982.

Parkin, T. *Demography and Roman Society.* Baltimore, 1992.

Parsons, P.J. "Petitions and a Letter: The Grammarian's Complaint." In A.E. Hanson, ed., *Collectanea Papyrologica: Texts Published in Honor of H. C. Youtie,* vol. II, PTA 20, 409–46. Bonn, 1976.

Patterson, J.R. "Crisis: What Crisis? Rural Change and Urban Development in Imperial Appennine Italy." *PBSR* 55 (1987): 115–46.

Pavis-d'Escurac, H. "Aristocratie sénatoriale et profits commerciaux." *Ktema* 2 (1977): 339–55.

Peachin, M. "Consultation with a Magistrate in Justinian's *Code.*" *CQ* 42, no. 2 (1992): 448–58.

———. *Iudex vice Caesaris: Deputy Emperors and the Administration of Justice during the Principate.* Heidelberger Althistorische Beiträge und Epigraphische Studien 21. Stuttgart, 1996.

Pleket, H.W. "Wirtschaft." In *Handbuch der Europäischen Wirtschafts- und Sozialgeschichte,* vol. I, F. Vittinghoff, ed., *Europäische Wirtschafts- und Sozialgeschichte in der Kaiserzeit,* 25–160. Stuttgart, 1990.

———. "Agriculture in the Roman Empire in Comparative Perspective." In H. Sancisi-Weerdenburg, R.J. van der Spek, H.C. Teitler, and H.T. Wallinga, eds., *De Agricultura: In Memoriam Pieter Willem de Neeve (1945–1990),* Dutch Monographs on Ancient History and Archaeology 10, 317–42. Amsterdam, 1993.

Purcell, N. "Wine and Wealth in Ancient Italy." *JRS* 75 (1985): 1–19.

Rathbone, D. "Italian Wines in Roman Egypt." *Opus* 2 (1983): 81–98.

———. "The Ancient Economy and Graeco-Roman Egypt." In L. Criscuolo and G. Geraci, eds., *Egitto e storia antica dall'ellenismo all'età araba: Bilancio di un confronto,* Atti del colloquio internazionale (Bologna, 31 agosto–2 settembre 1987), 159–76. Bologna, 1989.

———. *Economic Rationalism and Rural Society in Third-Century* A.D. *Egypt: The Heroninos Archive and the Appianus Estate*. Cambridge, 1991.
———. "More (or Less?) Economic Rationalism in Roman Agriculture." *JRA* 7 (1994): 432–36.
Robert, J., and L. Robert. "Bulletin épigraphique." *REG* 61 (1948): 137–212.
Rosafio, P. "Rural Labour Organization in Pliny the Younger." *ARID* 21 (1993): 67–79.
———. "Slaves and *Coloni* in the Villa System." In: J. Carlsen, P. Ørsted, and J.E. Skydsgaard, eds., *Landuse in the Roman Empire*, ARID Suppl. 22, 145–58. Rome, 1994.
Rostovtzeff, M. *The Social and Economic History of the Roman Empire*. 2d ed. Rev. P.M. Fraser. 2 vols. Oxford, 1957. Reprint, 1979.
Rowlandson, J. *Landowners and Tenants in Roman Egypt: The Social Relations of Agriculture in the Oxyrhynchite Nome*. Oxford Classical Monographs. Oxford, 1996.
Rupprecht, H.-A. "Zum Ehegattenerbrecht nach den Papyri." *BASP* 22 (1985): 291–95.
———. "Die Beendigung von Vertragsverhältnissen, Überlegungen zur Rechtswirklichkeit anhand der Pacht." *JJP* 20 (1990): 119–28.
Sachers, E. "Tutela." *RE* 7 (1948): 1497–1599.
Saller, R.P. *Personal Patronage under the Early Empire*. Cambridge, 1982.
———. "Roman Dowry and the Devolution of Property in the Principate." *CQ* 34, no. 1 (1984): 195–205.
———. *Patriarchy, Property and Death in the Roman Family*. Cambridge Studies in Population, Economy and Society in Past Time 25. Cambridge, 1994.
Sanfilippo, C. *Pauli Decretorum Libri Tres*. Milan, 1938.
Scheidel, W. "Pächter und Grundpacht bei Columella (*Colonus*-Studien II)." *Athenaeum* 81 (1993): 391–439.
———. "Sklaven und Freigelassene als Pächter und ihre ökonomische Funktion in der römischen Landwirtschaft (Colonus-Studien III)." In H. Sancisi-Weerdenburg, R.J. van der Spek, H.C. Teitler, and H.T. Wallinga, eds., *De Agricultura: In Memoriam Pieter Willem de Neeve (1945–1990)*, Dutch Monographs on Ancient History and Archaeology 10, 182–96. Amsterdam, 1993.
———. *Grundpacht und Lohnarbeit in der Landwirtschaft des römischen Italien*. Europäische Hochschulschriften, Reihe III, Geschichte und ihre Hilfswissenschaften 624. Frankfurt, 1994.
———. "Finances, Figures and Fiction." *CQ* 46, no. 1 (1996): 222–38.
Schleich, T. "Überlegungen zum Problem senatorischer Handelsaktivitäten." Pts. 1 and 2. *MBAH* 2, no. 2 (1983): 65–90; 3, no. 1 (1984): 37–76.
Schubert, P. *Les archives de Marcus Lucretius Diogenes et textes apparentés*. PTA 39. Bonn, 1990.
Schuller, W. "Zum *Pignus Tacitum*." *Labeo* 15 (1969): 267–84.
Schwartz, J. *Les archives de Sarapion et ses fils: Une exploitation agricole aux environs d'Hermoupolis Magna (de 90 à 133 p.C.)*. Bibliothèque d'étude 29. Cairo, 1961.
Seidl, E. *Rechtsgeschichte Ägyptens als römischer Provinz: Die Behauptung des ägyptischen Rechts neben dem römischen*. St. Augustin, 1973.

Shaw, B.D. "Rural Markets in North Africa and the Political Economy of the Roman Empire." *AntAfr* 17 (1981): 37–83.
Sherwin-White, A.N. *The Letters of Pliny: A Social and Historical Commentary.* Oxford, 1966. Rev. ed., 1985.
Sijpesteijn, P. J. "Further Remarks on Some Imperial Titles in the Papyri." *ZPE* 45 (1982): 177–96.
Simon, H.A. *Models of Man, Social and Rational.* New York, 1957.
———. *Administrative Behavior.* 3d ed. New York, 1976.
———. *The Sciences of the Artificial.* 2d ed. Cambridge, Mass., 1981.
———. *Reason in Human Affairs.* Stanford, 1983.
Sirago, V. *L'Italia agraria sotto Traiano.* Université de Louvain, Recueil de travaux d'histoire et de philologie, 4th ser., 16. Louvain, 1958.
Sitzia, F. "Considerazioni in tema di *periculum locatoris* e *di remissio mercedis.*" In *Studi in memoria di Giuliana D'Amelio,* vol. I, *Studi storico-giuridici,* 331–61. Milan, 1978.
Solazzi, S. *La minore età nel diritto romano.* Rome, 1912.
———. *Curator impuberis.* Rome, 1917.
———. "Tutele e Curatele." In *Scritti di diritto romano,* vol. II, 1–66. Naples 1957. Originally published in two parts in *RISG* 53 (1913): 263–97; 54 (1914): 17–70, 273–94.
Steinwenter, A. *Fundus cum instrumento: Eine agrar- und rechtsgeschichtliche Studie. SAWW* (Phil.-hist. Kl.) 221, no. 1. Vienna and Leipzig, 1942.
Strobel, K. "Zu Fragen der frühen Geschichte der römischen Provinz Arabia und zu einigen Problemen der Legionslokation im Osten des Imperium Romanum zu Beginn des 2. Jh. n.Chr.." *ZPE* 71 (1988): 251–80.
Talbert, R. *The Senate of Imperial Rome.* Princeton, 1984.
Taubenschlag, R. *The Law of Greco-Roman Egypt in Light of the Papyri, 332 B.C.–640 A.D.* 2d ed. Warsaw, 1955.
———. "Die Alimentationspflicht im Licht der Papyri." In *Opera Minora,* vol. II, 539–55. Warsaw, 1959.
Tchernia, A. *Le vin de l'Italie romaine: Essai d'histoire économique d'après les amphores.* Bibliothèque des écoles françaises d'Athène et de Rome 261. Rome, 1986.
Tellegen, J. W. *The Roman Law of Succession in the Letters of Pliny the Younger.* Vol. I. Stud.Amst. 21. Zutphen, 1982.
Tozzi, G. *Economisti greci e romani.* Milan, 1961.
Treggiari, S. *Roman Freedmen in the Late Republic.* Oxford, 1969.
———. *Roman Marriage: Iusti Coniuges from the Time of Cicero to the Time of Ulpian.* Oxford, 1991.
Tripolitis, A. "Some Princeton Papyri Reconsidered." *BASP* 5 (1968): 7–16.
Turpin, W. "Imperial Subscriptions and the Administration of Justice." *JRS* 81 (1991): 101–18.
Vera, D. "Strutture agrarie e strutture patrimoniali nella tarda antichità: L'aristocrazia romana fra agricoltura e commercio." *Opus* 2 (1983): 489–533.
———. "Simmaco e le sue proprietà: Struttura e funzionamento di un patrimo-

nio aristocratico del quarto secolo d.C." In F. Paschoud, G. Fry, and Y. Rütsche, eds., *Colloque Génevois sur Symmaque*, 231–76. Paris, 1986.

———. "Schiavitù rurale e colonato nell'Italia imperiale." *Scienze dell'antichità, Storia, Archeologia, Antropologia* 6–7 (1992–93): 291–339.

———. "Dalla 'villa perfecta' alla villa di Palladio: Sulle trasformazioni del sistema agrario in Italia fra principato e dominato." Pts. 1 and 2. *Athenaeum* 83, no.1 (1995): 189–211; 83, no. 2 (1995): 331–56.

Veyne, P. "La table des Ligures Baebiani et l'institution alimentaire de Trajan." Pts. 1 and 2. *MEFRA* 69 (1957): 81–135; 70 (1958): 177–241.

———. "Vie de Trimalcion." *Annales ESC* 16 (1961): 213–47.

———. "Mythe et réalité de l'autarcie à Rome." *REA* 81 (1979): 261–80.

Waszynski, S. *Die Bodenpacht: Agrargeschichtliche Papyrusstudien*. Vol. I, *Die Privatpacht*. Leipzig and Berlin, 1905.

Watson, A. *The Spirit of Roman Law*. Athens, Ga., 1995.

White, J.L. *Light from Ancient Letters*. Philadelphia, 1986.

White, K.D. *Roman Farming*. Ithaca, 1970.

Whittaker, C.R. "Trade and the Aristocracy in the Roman Empire." *Opus* 4 (1985): 49–75.

Wieacker, F. "Textkritik und Sachforschung: Positionen in der gegenwärtigen Romanistik." *ZRG* 91 (1974): 1–40.

———. *Römische Rechtsgeschichte: Quellenkunde, Rechtsbildung, Jurisprudenz und Rechtsliteratur*. First Section: *Einleitung, Quellenkunde, Frühzeit und Republik*. Handbuch der Altertumswissenschaft 10 no. 3.1.1. Munich, 1988.

Wiens, T.B. "Uncertainty and Factor Allocation in a Peasant Farm Economy." *Oxford Economic Papers*, n.s., 29 (1977): 48–60.

Woolf, G. "Food, Poverty and Patronage: The Significance of the Epigraphy of the Roman Alimentary Schemes in Early Imperial Italy." *PBSR* 58 (1990): 197–228.

Worp, K. *Corpus Papyrorum Raineri*. Vol. XVIIA, *Griechische Texte XIIA, Die Archive der Aurelii Adelphios und Asklepiades*. Vienna, 1991.

Zahrnt, M. "Antinoopolis in Ägypten: Die hadrianische Gründung und ihre Privilegien in der neueren Forschung." *ANRW* II 10, no. 1 (1988): 669–706.

Zimmermann, R. *The Law of Obligations: Roman Foundations of the Civilian Tradition*. Cape Town, 1990.

List of Ancient Sources

Inscriptions

BGU
  VII 1563: 163–64n. 61

CIL
  III 6998, 13652: 86–87
  V 4489: 85
  V 5262: 78
  VI 10229: 94n. 41
  VIII 1641: 79–80
  IX 1455: 83
  IX 5845: 84–85
  X 114: 87
  X 444: 84
  X 1880: 85n. 19
  X 5853: 84
  X 6328: 80
  XI 379: 85
  XI 419: 85
  XI 1147: 79, 83

FIRA III
  48: 94n. 41
  53: 86–87
  55d: 80
  118: 85

ILS
  2927: 78
  3456: 84
  3775: 84–85
  6271: 84
  6278: 80
  6328a: 85n. 19
  6468–71: 87–88
  6469: 87, 88
  6509: 83

  6663: 85
  6664: 85
  6675: 79, 83
  6818: 79–80
  7196: 86–87
  8370: 85
  8376: 85

REG 61 (1948): 168 no. 106: 85–86

OGIS 669: 163–64n. 61

Legal Sources

Coll.
  2.7.7: 149–50n. 26
  12.7.9: 149–50n. 26

CJ
  4.14.5: 168
  4.51.1: 63n. 82
  4.65.5: 149
  4.65.8: 148n. 22, 214–15, 228, 231
  4.65.9: 186
  4.65.11: 163
  4.65.15: 185n. 6
  4.65.16: 207
  4.65.18: 215n. 68
  4.65.19: 229, 231, 233
  4.65.21: 148n. 22
  5.12.18: 105
  5.13.1.15: 67n. 90
  5.32.1: 24
  5.36.3: 24, 145
  5.37.3: 29
  5.37.4: 39
  5.37.9: 33
  5.37.20: 51n. 58

CJ (continued)
  5.37.22: 59n. 72
  5.37.22.3a: 64–65n. 85
  5.37.22.5a: 52–53n. 60
  5.37.24.1: 40
  5.40.2: 25n. 4
  5.50.1 pr.: 30
  5.50.2 pr.: 30
  5.50.2.1: 30
  5.51.2: 50
  5.51.3: 39, 63–64
  5.56.1: 51
  5.56.3: 40, 52
  5.62.11.1: 24, 145
  5.62.2: 24
  5.70.2: 55n. 65, 58–59n. 72
  5.71–74: 55n. 65
  5.71.1: 67n. 89
  5.71.4 pr.: 58–59
  5.71.16 pr.: 59n. 72, 62n. 76
  5.72.2: 66
  5.73.2: 61n. 76
  5.74.1: 61n. 76
  5.75.5: 61n. 76
  6.38.2 pr.–1: 98n. 47
  10.53 (52) 6 pr.–1: 109n. 70
  11.48.8.1: 142n. 10
  11.50.1: 207n. 55

CTh
  2.4.1.1: 25n. 4
  3.30.1: 51n. 58

D.
  1.21.2.1: 55n. 64
  2.14.56: 153
  4.4.47 pr.: 62n. 76
  7.1.9 pr.: 118
  7.1.9.1–6: 118
  7.1.9.2: 118n. 88
  7.1.10: 118n. 88
  7.1.13.4: 98n. 47, 118n. 88
  7.1.13.5: 118
  7.1.13.6: 118n. 88
  7.1.15.6: 117–18
  7.1.34.1: 190–91, 193–94, 202

  7.1.50: 104n. 57
  7.1.57.1: 90
  7.1.65 pr.: 118n. 88
  7.4.29 pr.: 154
  7.4.29.2: 118
  9.2.28.9: 149–50n. 26
  9.2.28.11: 149–50n. 26
  12.1.4.1: 147n. 19, 207n. 53
  15.3.16: 168
  16.1.18.1: 69
  18.1.28: 212n. 62
  18.1.34.7: 63
  18.1.68 pr.: 165–66
  18.1.75: 153n. 35
  18.1.79: 154–55, 164
  18.2.6.1: 188n. 15
  19.1.13.6: 83n. 11
  19.1.13.11: 166
  19.1.13.16: 166n. 66
  19.1.13.30: 187
  19.1.17.2: 114n. 82
  19.1.21.4: 152–53, 155
  19.1.49 pr.: 165
  19.2: 138
  19.2.4: 190
  19.2.9 pr.: 147n. 18
  19.2.9.1: 147n. 19
  19.2.9.1: 201–2
  19.2.11.2: 148n. 20
  19.2.11.4: 150n. 26
  19.2.13.11: 147, 147n. 19, 152n. 33, 206, 207n. 53
  19.2.14: 206, 207n. 53
  19.2.15 pr.–1: 225
  19.2.15.1: 150, 151
  19.2.15.1–2: 147n. 18
  19.2.15.1–3: 203–4
  19.2.15.2: 203–4, 224, 226
  19.2.15.3: 148n. 20, 227–28
  19.2.15.4: 214, 230–31
  19.2.15.5: 203, 204–5, 215, 226–27
  19.2.15.7: 215, 227
  19.2.15.9: 154
  19.2.18: 154
  19.2.19.2: 147n. 17, 149, 194–95
  19.2.19.3: 148n. 21, 216–17

List of Ancient Sources 255

19.2.21.4: 164n. 62
19.2.24.1: 150n. 28
19.2.24.2: 147n. 19
19.2.24.2–3: 199–200n. 39
19.2.24.3: 199–200
19.2.24.4: 147n. 18, 147n. 19
19.2.24.5: 147n. 19, 190–91, 191–92
19.2.25.1: 185–86
19.2.25.3: 147n. 17, 148n. 20, 151
19.2.25.6: 11, 142, 147n. 17, 148, 224, 232
19.2.30.4: 149–50n. 26
19.2.31: 152
19.2.32: 147n. 19, 163, 185
19.2.33: 147n. 18
19.2.51 pr.: 148n. 20, 153
19.2.52: 153
19.2.53: 150n. 28, 152n. 34
19.2.54 pr.: 152
19.2.54.1: 153–54, 185, 187
19.2.54.2: 149
19.2.55.1: 199
19.2.61 pr.: 199n. 38, 200–1, 219
20.1.32: 165, 176–77
20.2.7 pr.: 150n. 28
20.6.4.1: 212n. 62
20.6.7 pr.: 212n. 62
20.6.8.6–19: 212n. 62
20.6.10 pr.: 212n. 62
20.6.14: 149, 152
23.3.69.4: 32n. 20
23.4.22: 104–5, 154
23.5.4: 67
23.5.4: 67–68
23.5.5: 68
23.5.7 pr.–1: 67–68
23.5.11: 68
23.5.13 pr.: 68
23.5.13.4: 68
23.5.18 pr.: 69
24.3.25.4: 147n. 19, 186–87
24.3.7.8: 147n. 18, 148n. 21
26.5.27 pr.: 24
26.7: 2
26.7.1 pr. f.: 54–55n. 62
26.7.3.2: 38, 38–39n. 33

26.7.3.4: 24
26.7.4: 24
26.7.5 pr.: 38, 39–40
26.7.5.9: 62
26.7.7.2: 40, 40–41n. 39
26.7.7.3: 39
26.7.7.4: 51
26.7.7.5: 51n. 59
26.7.7.6: 51
26.7.7.7: 39
26.7.7.10: 39, 50
26.7.7.11: 50
26.7.7.12: 51
26.7.9.3: 34n. 26
26.7.10: 29, 34
26.7.12.1: 29
26.7.12.2: 29
26.7.12.3: 28, 32
26.7.12.4: 52
26.7.13 pr.: 31
26.7.13.1: 52, 53
26.7.13.2: 31–32, 33
26.7.15: 50, 52
26.7.16: 53
26.7.18 pr.: 34n. 26
26.7.32.2: 41
26.7.32.3: 41
26.7.33 pr.: 34
26.7.35: 53
26.7.39 pr.: 53n. 61
26.7.39.3: 24
26.7.39.8: 24
26.7.40 pr.: 25n. 4
26.7.43.1: 32
26.7.44 pr.: 53
26.7.46 pr.: 41–42, 154
26.7.46.2: 50
26.7.46.2–3: 51
26.7.46.7: 51–52
26.7.47.1: 29
26.7.47.2: 24
26.7.47.4: 35–37, 91
26.7.49: 39n. 34, 50
26.7.51: 24
26.7.52: 32–33
26.7.54: 51

D. (continued)
  26.7.57 pr.: 50n. 57
  26.7.57.1: 41n. 40
  26.7.58.1: 50
  26.7.58.3: 50n. 55
  26.10.3.14: 29
  26.10.3.16: 39n. 34
  26.10.7.1: 29
  26.10.7.2: 29
  27.2.2 pr.: 30
  27.2.2 pr.–2: 31
  27.2.2.1: 30
  27.2.2.5: 30n. 17
  27.2.3 pr.–1: 30
  27.2.3.2: 30
  27.2.3.3: 30n. 17
  27.2.3.6: 30
  27.2.4: 33
  27.3.1 pr.: 34n. 27
  27.3.1.2: 33
  27.3.1.4: 33, 33n. 22
  27.4: 34n. 27
  27.4.1 pr.: 30
  27.4.3.2–3: 50n. 55, 51
  27.4.3.4: 50
  27.4.3.6: 39–40n. 36
  27.4.3.7: 34, 42
  27.5.4: 34n. 26
  27.8.1.15: 38n. 32
  27.8.2: 61n. 76
  27.9: 2, 55, 55n. 65
  27.9.1 pr.–2: 55
  27.9.1.1–2: 56–58
  27.9.1.1–2: 58n. 72
  27.9.1.2: 60
  27.9.1.3: 60
  27.9.1.4: 59–60
  27.9.3.1: 61
  27.9.3.4: 59
  27.9.3.5: 59
  27.9.3.6: 60
  27.9.4: 60
  27.9.5 pr.–1: 60
  27.9.5.3: 66
  27.9.5.9: 62
  27.9.5.10: 62

  27.9.5.13: 62–63
  27.9.5.14: 62
  27.9.8.1: 66n. 88
  27.9.9: 64
  27.9.13 pr.: 63
  27.10.5: 25n. 4
  28.5.35.3: 99–100
  30.92 pr.: 111
  30.120.2: 185
  31.77.5: 67n. 90
  32.27.2: 153, 174, 193
  32.28 pr.: 70
  32.37 pr.: 101
  32.38 pr.: 70
  32.38.1: 70n. 97
  32.38.2: 70n. 97
  32.38.3: 70n. 97
  32.43: 32
  32.44: 122
  32.60.3: 115–16
  32.64: 122
  32.65 pr.: 175
  32.78.1: 122
  32.78.3: 123, 171–72
  32.91 pr.: 120, 169n. 73, 173
  32.91.1: 97n. 44, 123
  32.92 pr.: 97n. 45, 120
  32.97: 120–22, 169n. 73, 170
  32.101 pr.: 81–82
  32.101.1: 172n. 79
  33.1.18 pr.: 90
  33.1.19 pr.: 90
  33.1.21 pr.: 90, 101–2, 212–13
  33.1.21.1: 89–90, 101
  33.1.21.3: 82
  33.1.21.5: 91–92
  33.2.17: 81
  33.2.25: 71
  33.2.30.1: 153, 190–91
  33.2.32.4: 93–94
  33.2.32.5: 103–4
  33.2.32.7: 102
  33.2.38: 71, 104
  33.2.38: 102–3, 213
  33.2.42: 211
  33.4.1.15: 186–87

33.7: 2, 113
33.7.8 pr.: 114, 115
33.7.8.1: 97
33.7.8.1: 115
33.7.9: 115
33.7.10: 115
33.7.11: 115
33.7.12 pr.: 114
33.7.12 pr.–14: 114
33.7.12.1: 116
33.7.12.3: 114–15, 167
33.7.12.4: 169–70
33.7.12.12–13: 115
33.7.12.27: 98, 173
33.7.12.28–41: 98
33.7.15.2: 97
33.7.18.2: 115
33.7.18.4: 169, 170
33.7.18.6: 115
33.7.20 pr.: 97n. 45, 172
33.7.20.1: 169, 171, 173
33.7.20.3: 120n. 93, 123, 152, 169n. 73, 173
33.7.20.6: 97n. 44
33.7.24: 150
33.7.26.1: 115
33.7.27 pr.–1: 172
33.7.27 pr.: 172n. 80
33.7.27.1: 172
33.10.7.2: 100
34.1.4 pr.: 94–95, 107
34.1.8: 91
34.1.9 pr.: 38, 91
34.1.12: 94
34.1.15 pr.: 37, 91
34.1.16.2: 37–38, 91n. 34
34.1.20.1: 90
34.1.20.2: 90
34.1.22 pr.: 31
34.1.22.1: 95
34.3.16: 147, 147n. 19, 190, 191
34.3.18: 191n. 21
34.3.28 pr.: 35n. 28
35.2.3.2: 36n. 30
35.2.25.1: 89n. 32
39.3.4.2–3: 202n. 44

39.3.5: 202n. 44
40.7.40 pr.: 170–71
41.1.44: 151
41.3.33.1: 154
41.4.7.3: 54–55
42.1.15.12: 39n. 35
43.26.4.4: 207n. 53
43.26.6 pr.: 207n. 53
43.32.1.4: 147n. 19
43.33.1 pr.: 149
43.33.2: 149
45.1.89: 147n. 19
46.1.52.2: 152
46.1.58 pr.: 152n.33
46.6.11: 38n. 32
47.2.26.1: 148
47.2.68.5: 207n. 53
47.2.62.8: 150n. 28, 212
47.2.86: 152n. 33
49.14.3.6: 163n. 61
49.14.50: 187–89
50.8.3.2: 232–33n. 111
50.8.5 pr.: 152n. 34
50.8.11.2: 69n. 95

Frag.Vat.
 13: 164–65, 165n. 64
 44: 185
 45: 59n. 72
 158: 55n. 65, 55n. 66
 212–14: 55n. 65, 55n. 66
 258: 105n. 62

Gaius, *Inst.*
 2.63: 67n. 90

Iust., *Inst.*
 2.8 pr.: 67n. 90

[Paul.] *Sent.*
 2.18.2: 148n. 20
 2.18.4: 199n. 38
 2.30: 55n. 65

PSI XIV 1449: 226n. 90

Literary Sources

Aulus Gellius
  14.2: 49n. 54

Cato
  Agr.
    1: 42n. 43, 75n. 109
    1.1: 156n. 42
    2.1 f.: 156n. 42
    4: 156n. 42

Cicero
  2 *Verr.* 3.119: 107n. 67

Columella
  1 pref. 7: 72n. 101
  1 pref. 12: 156n. 42
  1.1.18: 156n. 42
  1.1.18–20: 156n. 42
  1.1.20: 157n. 46
  1.7.1: 222
  1.7.1: 227
  1.7.1–2: 148n. 21, 156, 197
  1.7.1–4: 156–57
  1.7.1–7: 156–57
  1.7.2: 148n. 21
  1.7.3–4: 174–75
  1.7.3–7: 197
  1.7.5: 15
  1.7.6: 157n. 45
  1.7.6–7: 157, 168
  1.7.7: 157n. 45
  1.8.20: 156n. 42
  3.3: 198
  3.3.3: 72n. 102
  3.13.6–7: 88n. 27

Petronius
  48.2–3: 73
  53.4: 73–74
  53.5–8: 74
  76.2–10: 73
  76.8–9: 73
  76.9: 46n. 48, 47n. 51
  77.3: 73

Pliny (the Elder)
  Nat.
    14.48: 72n. 99
    14.49–51: 71–72
    14.51: 72
    18.31: 156n. 42
    18.35: 156n. 42, 157n. 46
    18.37: 71
    18.43: 156n. 42

Pliny (the Younger)
  Ep.
    1.8.9–10: 78n. 2
    1.8.10: 78n. 2
    1.8.10–11: 79n. 3
    1.24: 146n. 14
    2.17: 152
    2.20.13: 27n. 8
    2.4: 25–28, 47–48
    2.4.2: 26, 48
    2.4.3: 26
    3.11.2: 48
    3.19: 26, 42, 46–47, 48, 72–73, 111–12, 164, 177n. 82, 189–90, 196, 222
    3.19.1: 4n. 4
    3.19.2: 176
    3.19–6–7: 172n. 79, 173, 177, 195
    3.19.7: 111, 112n. 77
    3.19.8: 46
    4.6.1: 198–99
    4.13.5–8: 81n. 8
    5.4: 210
    5.6: 152
    5.13: 210
    6.3: 107, 116–17, 146n. 15
    6.3.1: 107n. 66
    6.8: 48–49
    6.8.5: 49
    6.8.5–6: 28n. 10
    6.19.4: 74
    6.30: 146n. 14
    7.18: 80, 112
    7.18.1: 80
    7.18.2: 78n. 2
    7.18.2–3: 82

List of Ancient Sources

7.18.2–4: 112
7.18.3: 83
7.18.4: 83
7.30.3: 26
7.30.3: 146n. 13
8.2: 145, 198–99, 211, 216
9.36: 146n. 13
9.36.6: 146n. 13
9.37: 12, 26, 145–46, 190, 195–96, 222
9.37.2: 190, 196n. 30, 222
9.37.2–4: 148
9.37.3: 196n. 31
10.8.5: 26, 177–78n. 93, 189n. 17, 190, 199, 222
10.54: 74–75n. 108, 80–81
10.58.5: 109n. 70
10.70.2: 81n. 8

Seneca
  Ep.
    41.7: 45

SHA
  Alex. Sev.
    32.3: 11
    40.2: 117n. 85
  Marc.
    10.11: 55n. 62
    11.8: 74n. 108
  Pert.
    9.3: 84n. 15

Suetonius
  Gram.
    23: 72n. 99

Tacitus
  Ann.
    6.16.3: 45–46
    13.30.2: 85

Papyri

BGU I 136: 31n. 18
BGU I 326: 125–26
C.Pap.Jud. II 450: 86n. 22
CPR VI 76: 126n. 107
CPR XVIIA 17a, b: 109
FIRA III 6: 127n. 110
FIRA III 47: 126n. 107
FIRA III 50: 125–26
FIRA III 51: 127–29
M.Chr. 86: 31n. 18
M.Chr. 189: 111, 111n. 76
M.Chr.: 274: 158–59
M.Chr. 317: 127–29
M.Chr. 316: 125–26
New Primer 50: 125–26
Oliver, *Greek Constitutions,* no. 247: 86
Pap.Lugd.Batav. VI (= *P.Fam.Tebt.*) 33: 82n. 9
Pap.Lugd.Batav. VI (= *P.Fam.Tebt.*) 53: 31n. 18
P.Amh. II 71: 93
P.Amh. II 85: 158–59
P.Amh. II 86: 158–59
P.Coll. Youtie II 66: 107–8
P.Diog. 10: 124–25
P.Diog. 29: 161
P.Fam.Tebt. (= *Pap.Lugd.Batav.* VI) 33: 82n. 9
P.Fam.Tebt. (= *Pap.Lugd.Batav.* VI) 53: 31n. 18
P.Flor. I 50: 129–31
P.Flor. I 58: 106
P.Gen. I 31: 105–6
P.Giss. 40: 230n. 102
P.Heid. IV 336: 160–61
P.Heid. IV 337: 159n. 50
P.Lips. 10: 111
P.Lond. II 383: 82n. 9
P.Mich. VII 422: 106
P.Mich. VIII 562: 162
P.Mich. IX 525: 35n. 27
P.Mich. IX 562: 162–63
P.Oxy. I 71: 107n. 67
P.Oxy. IV 705: 86
P.Oxy. IV 707: 152n. 34
P.Oxy. IV 727: 161–62
P.Oxy. VI 898: 31n. 18, 65n. 86
P.Oxy. VI 907: 92, 127–29

P.Oxy. VIII 1102: 109–111
P.Oxy. IX 1208: 45n. 47
P.Oxy. X 1279: 209n. 57
P.Oxy. XII 1541: 132n. 122
P.Oxy. XIV 1630: 152n. 34
P.Oxy. XXVII 2474: 92, 131–34
P.Oxy. XXXIV 2713: 35n. 27
P.Oxy. XXXIV 2723: 44–45
P.Oxy. XXXVIII 2848: 44n. 46, 86, 86n. 22
P.Oxy. XLIII 3113: 31n. 18
P.Oxy. XLVII 3366: 107–8
P.Oxy. XLVIII 3921–2: 31n. 18, 50n. 55
P.Oxy. LI 3638: 43–44
P.Princ. II 38: 129
PSI X 1102: 34–35n. 27, 132n. 122
PSI X 1121: 132n. 122
PSI XII 1258: 31n. 18, 95
PSI XIII 1325: 71n. 98
P.Select. (= Pap.Lugd.Bat. XIII) 14: 126–27
P.Soter. 1–2: 88n. 28
P.Strasb. 511: 208

P.Test.Roma.^{2} 5: 126n. 107
P.Test.Roma.^{2} 7: 92, 126–27
P.Test.Roma.^{2} 9: 71n. 98
P.Test.Roma.^{2} 12: 71n. 98, 125–26
P.Test.Roma.^{2} 18: 126n. 107
P.Test.Roma.^{2} 20: 124–25
P.Test.Roma.^{2} 23: 129
P.Test.Roma.^{2} 24: 92, 127–29
P.Test.Roma.^{2} 26: 92, 131–34
P.Turner 24: 45n. 47
P.Yadin 15: 34n. 24
SB I 5217: 127n. 110
SB I 5761: 65
SB V 7630: 71n. 98
SB VI 9049: 159n. 50
SB X 10533: 209n. 57
SB XVI 12557: 159n. 50
SB XVIII 13300: 106
Sel.Pap. I 85: 125–26
Sel.Pap. II 215: 230n. 102
W.Chr. 22: 230n. 102
W.Chr. 407: 86n. 22

General Index

Accounting: by Columella, 74n. 106, 198n. 36; on estate of Appianus, 19, 121–22
Acilius Sthenelus, investing in viticulture, 72n. 99
Actio aquae pluviae arcendae, 202n. 44
Actio de peculio, 168n. 71
Actio ex conducto, 150, 151, 185n. 6, 199n. 38, 202
Actio ex empto, 165
Actio ex testamento, 192
Actio ex vendito, 164, 187
Actio in rem verso, 168n. 71
Actio institoria, 113n. 81, 119n. 90
Actio rationibus distrahendis, 61
Actio Serviana, 149n. 24
Actio suspecti, 61
Actio tutelae, 56, 60–61
Actor, as manager of an estate, 165, 169n. 73, 173, 176
Aesthetic aspects of estates, 28n. 10, 97–98, 117
Alexander, Tiberius Iulius, edict of, 163–64n. 61
Alienation of property: restricted by *Oratio Severi,* 54–67; restricted by testamentary prohibitions, 70–71, 90; restricted in dotal law, 67–69; restricted in municipal law, 69n. 95, 86
Alimenta: bequests of, 37, 38, 70n. 97, 90–91, 95; bequeathed in Roman Egypt, 95, 129, 132–33; provided by tutors to pupils, 29–31
Alimentary foundations, 78–84, 103, 112

Antichretic leases, 111, 111n. 76, 163n. 60
Antinoopolis: alimentary foundation, 82n. 9; civic privileges, 124
Appianus, Aurelius, management of his estates, 4, 19, 121–22, 130, 212
Auctoritas, of tutor, 23

Babatha, archive of, 33–34n. 24
Bellicius Sollers, L., petitioning senate to establish periodic market, 210

Calendarium, bequest of, 120, 122
Calpurnii Firmi, their landholdings in Egypt, 44–45
Caninius Rebilus, civic benefactor, 85–86
Capitalism, in Roman economy, 18
Cash, significance to management of an estate, 119–23
Cato the Elder, on investing in land, 42n. 43, 73n. 104, 75n. 109
Cautio, in tenancy, 123, 152
Cautio rem pupili salvam fore, 38n. 32
Chayanov, A.V., on peasant economy, 218–19n. 77
Cibaria, bequest of, 94
Civic benefactions, funding of, 77–88, 96
Cognitio extraordinaria: in trusts, 55n. 62; in tutorship, 54–55n. 62, 64n. 83; protecting tenants, 229n. 100, 230

Columella: calculations about viticulture, 74n. 106, 198; recommendations about leasing, 148n. 21, 156–57, 168, 174–75, 197–98, 222, 227, 234; views about agriculture, 72n. 101

Commerce, significance to Roman economy, 16–17, 73, 76, 239

Conductores, on African imperial estates, 73n. 105, 176n. 89, 178n. 93, 221

Conservatism: of jurists in law of legacy, 113n. 79; in law of tenancy, 10–11, 143, 184, 206–7, 224, 233, 234, 235, 240; in law of tutorship, 71, 75–76, 138n. 2

Consilium, hearing cases, 120n. 94, 188

Contrarium tutelae iudicium, 34n. 27, 50n. 56

Cornelius Fabatus, role in managing his estates, 146

Credit market: as envisioned by jurists, 52; in Roman empire, 45–49. *See also* Loans

Crop prices: excluded as grounds for remission of rent, 215, 216; risk for in tenancy, 213–14, 217–18, 224

Crops: disposition when tenant is evicted, 185; pledged as security in tenancy, 155, 212. *See also* Harvest, tenants' risk for

Cura minorum: assimilated with tutorship, 23, 39n. 35, 57; general features of, 22–23. *See also* Curators

Curatores impuberum, duties of, 57

Curator furiosi, affected by *Oratio Severi*, 66n. 88

Curator rei publicae, 69n. 95, 81n. 8

Curators: administering minors' property, 57–58; subject to *Oratio Severi*, 56–60

Curatorship. See *Cura minorum*, assimilated with tutorship; Curators

Custom, affecting lease relationships, 229, 231, 233

Debt: incurred by *servi quasi coloni*, 168, 170–71, 173; incurred by upper-class Romans, 25–26, 27, 47–49, 110, 111; of Pliny's tenants, 81, 195–96; of pupils, 56, 60–65, 69. *See also* Loans; *Reliqua colonorum*

Decreta of Paul, 120n. 94

Dependents, financial support of, 89–96, 123–34

Discretionary expenses, of pupils, 31–33. *See also* Pupils

Donatio, affecting lease obligations, 231

Dotal property: leased by tenant, 154, 186–87; restrictions on its alienation, 67–69. *See also* Dowry

Dowry: including land, 104–5, 110; provided by Pliny, 26, 48; provided by tutors for pupils, 32–33; in wills from Egypt, 128. *See also* Dotal property

Drought, 42–43, 222, 227. See also *Sterilitas*, as basis for remission of rent

Economic planning: by Pliny, 95–96, 111–12; by upper-class Romans, 1–7, 15–21, 75, 178–79; in funding civic benefactions, 77–88; in providing for dependents, 89–96, 123–34

Economies of scale, 44, 219

Economy of Roman empire, 15–20, 237–40. *See also* Jurists, assumptions about Roman economy

Emphyteutic rights, affected by *Oratio Severi*, 59

Emptio tollit locatum, 184

Enforcement, by lessor in tenancy, 155–56, 200, 219, 221–24

Epitropoi. See Tutorship, in Egypt

Estates: bequeathed in Egypt, 123–34; economic value, 6–7, 106–7, 111,

117, 119–20, 135, 239; economic value as conceived by jurists in law of legacy, 6–7, 96–112, 113–16, 119; economic value in Egypt, 107–11, 124, 129–31, 134; in law of usufruct, 117–19; organization of, 3–4, 176–77, 179n. 95. *See also* Investment in land

Eviction, of tenants, 155, 156, 184, 185, 189, 200–201. *See also* Possession

Fertilizer, in legacy of *fundus cum instrumento,* 114n. 82, 115–16
Fideiussores. See Guarantors, in tenancy
Fiscus, selling estate with tenant, 187–89
Flavius Archippus, financial support of, 108–9
Fraud: by curators, 63–64; by tutors, 50–51, 64, 65
Freedmen/women: in agriculture, 71–72; as beneficiaries of wills, 70, 89, 90, 94–95, 101–2, 106–7, 120–21, 125, 192; serving pupils, 28; as tenants (*see* Tenants)
Fructus: bequest of, 71; defined in context of estate, 115–16, 147, 211; in *Oratio Severi,* 61–62n. 78
Fundus, as object of lease, 3, 141, 147, 151, 177
Fundus cum instrumento: aesthetic aspects of, 97–98; as an "ideal" type, 116; as defined by jurists, 113–17, 134–35, 150–51, 167
Fundus instructus, distinguished from *fundus cum instrumento,* 97–98, 173

Grammarians, petitioning for means of support, 107–8
Guarantors, in tenancy, 152

Harvest, tenants' risk for, 213–14, 216–18, 226–27
Heroninos archive, 16n. 25, 19, 20, 24

Horion, Aurelius, benefactor at Oxyrhynchus, 44n. 46, 86
Hypallagma, 111
Hypotheca, in law of tenancy, 149n. 24

Ideal types: in *fundus cum instrumento,* 116; in law of tenancy, 11n. 15, 15, 138, 235
Imperial estates, in North Africa, 12, 142n. 10, 176n. 89, 178n. 93, 182, 217n. 74, 220–21, 223
In diem addictio, 187–88
Income: affected by harvest, 42–43, 213–14, 216–18; as defined for an estate, 41–42, 54, 71, 96–112, 113, 119, 133–34, 213
Infamia, imposed on tutors, 61
Instrumentum: as defined in a legacy of *fundus cum instrumento,* 113–16, 120n. 91, 123; in farm leases, 150–51, 195; in legacy of usufruct, 118; as regulated by *Oratio Severi,* 60n. 74
Interest rates, 36, 74n. 106, 102
Interpolation: affecting interpretation of legal texts, 9–10; in law of tutorship, 23, 57, 58n. 72
Invecta aut illata, 149. *See also* Pledges
Investment: factors limiting in Roman economy, 1, 2, 3, 6–7, 36–38, 75, 89, 107; of the pupil's wealth, 35–54
Investment in farm improvements: by beneficiaries of wills, 100–101, 102, 103–4; by owner on estate under *servi quasi coloni,* 167–68; in tenancy, 4–5, 144, 145, 179–80, 181–82, 193–209, 219–20, 233–34, 235; by tenants, 4–5, 8, 144, 145, 179–80, 193–209, 219–20, 233–34, 235; landowner's aversion to, 109, 202; tenant's right to compensation for, 182, 199–202, 220
Investment in land: by Appianus, 19; as conceived by Cato, 42n. 43, 75n. 109; in Egypt, 43–45; as envisioned

Investment in land *(continued)*
  in dotal law, 68; to fund civic benefactions, 78–88; literary evidence for, 71–75; by Pliny, 42, 46, 72–73, 111–12, 164; of the pupil's wealth, 38–45, 54, 63; for security, 54, 63, 64, 74, 76, 77, 239; by Trimalchio, 73–74
Irrigation, in Egypt, 108n. 69, 130, 134
Iudicium tutelae, 30, 34n. 26, 36–37, 52, 53
Iulius Apollinarius, C., management of his land, 162–63

Jurists: assumptions about credit market, 52; assumptions about Roman economy, 6–8, 10, 12, 14, 20–21, 71, 77, 138, 238–39; economic assumptions in law of legacy, 5–6, 99, 100–1, 134–45, 158; economic assumptions in law of tenancy, 141, 143–44, 178, 179, 210–11, 217, 235–36; economic assumptions in law of tutorship, 22, 38–43, 52–54, 63–64, 71, 76, 179; geographical application of their rulings, 142–43, 236; interpreting *Oratio Severi*, 60, 62–64, 65–67; methods in dotal law, 68; methods in law of legacy, 14, 96–101, 113–16, 134–35, 238; methods in law of tenancy, 8, 10–12, 137, 138, 143, 147–49, 177, 182–83, 206–8, 210–11, 233, 234, 235, 240; methods in law of tutorship, 14, 40–41, 49–50, 51–52, 75–76, 238; regulating remission of rent, 214, 224, 232–34, 240; social concerns of, 2–3, 8, 10–15, 238, 240; upper-class focus of, 13, 23–25, 65–66, 138–40. *See also* Conservatism, of jurists in law of legacy
Justinian Code: as evidence for economic history, 14–15; responding to petitions from provinces, 142–43

Kula, W., views on Polish economy, 217–18

Labor: on estate of Appianus, 218–19; in peasant agriculture, 218
Land. *See* Estates; Investment in land
Land prices, as a function of rents, 42, 111–12
Lata neglegentia, in tutorship, 40
Law courts, tenants' access to, 140–42, 210–11
Leases: compulsory, 163, 163–64n. 61, 208–9; long-term, 165, 183–94, 205; perpetual, 84n. 17, 197–98, 198n. 34, 220, 223; prodomatic, 65, 163n. 60. *See also* Normative farm lease; Tenants; Tenancy
Legacy, significance to upper classes, 13, 89, 96–97. *See also* Economic planning; Jurists; Trusts; Wills
Legal sources, as evidence for economic history, 2–3, 7, 9–15, 20–21, 22, 77, 137–43, 179, 181, 237–38, 240. *See also* Jurists
Lex Falcidia, 85n. 18, 89n. 32
Lex Iulia de fundo dotali, 67–69
Lex Plaetoria, 54
Liability: of trustee, 91; of tutors (*see* Tutors)
Livestock: on estate of Appianus, 122; in legacy of *fundus cum instrumento*, 114, 115–16; in legacy of *fundus instructus*, 173; provided by tenants, 149–50, 177, 181, 195; provided to *servus quasi colonus*, 168
Loans: as asset on an estate, 119, 122–23; as fund to support civic benefactions, 82, 83, 86–87; of municipal funds, 74–75n. 108; by Pliny, 46–48; of the pupil's money, 36, 40, 45, 49–54; risks involved in, 52–54; by upper-class Romans, 45–49
Locatio-conductio, formal aspects of, 138, 147–49. *See also* Leases; Normative farm lease; Tenancy

Lodging, bequest of, 93n. 39, 133
Lollianus, grammarian, 107–8
Longinus Castor, C., will of, 71n. 98, 125–26

Management of estates: difficulties faced by Roman landowners, 117, 137, 144–46, 179; facilitated by tenancy, 144–46, 155, 156–63, 166–73
Manumission, testamentary, 121, 125, 132, 170
Market conditions, affecting finances of upper-class Romans, 19, 43, 113, 131, 135, 145. *See also* Sale of crops
Markets, tenants' dependence on, 18, 155, 209–12, 213–14, 215–16, 217
Mill, in legacy of *fundus cum instrumento*, 115
Misthokarpeia, 111
Models, their use in economic history, 14, 20–21, 143, 221, 235, 238
Monetization, of Roman rural economy, 16
Municipal governments, problems in managing funds, 81, 82
Municipal law, restricting alienation of property, 69n. 95
Mutuum, 46

Normative farm lease: allocation of risk in, 222, 224; as envisioned by jurists, 11, 104, 138, 140, 147–55, 178, 183, 184, 194–95, 206–7, 210–11, 233, 234, 235

Opera rustica, performed by tenant, 147–48
Oratio Severi: affecting upper classes, 65–66; applying to curators, 56–58; economic principles behind, 54–67, 75; extended by Constantine to urban and movable property, 52–53n. 60; as interpreted by jurists, 58–67; sources for, 55; strengthening protection offered to pupils, 60–62

Ousiai, in Egypt, 163–64n. 61, 176n. 89

Papyri, as evidence for economic history, 8, 30, 77, 124, 143, 158, 208, 237, 238–39
Peasant agriculture: characteristics of, 217–19; in Egypt, 223–24; in Roman economy, 18; sales of land in, 75n. 109
Peculium, of *servus quasi colonus*, 41, 121, 166–67, 168–69, 170
Perceptio, 212
Periodic markets (*nundinae*), 209–10
Perpetual leases. *See* Leases
Petitions: from African *coloni*, 221; concerning inheritance, 105–6; concerning tenancy, 208; concerning tutorship, 34–35n. 27, 65n. 86; to establish periodic market, 210; from a grammarian, 107–9; in Justinian Code, 15, 142–43; against property manager, 107n. 67; for remission of rent, 203–6, 214–15, 226–27, 228, 229, 235; to sell pupil's property, 63
Phrontistes, managing land in Egypt, 161
Pignus, in law of tenancy, 149n. 24. See also *Invecta aut illata*; Pledges
Pledges: in law of tenancy, 149–50, 155, 156, 189–90, 195, 206, 207, 212, 221, 222; in loans, 176; in *relocatio tacita*, 207
Pledging of property: in alimentary programs, 83–84; prohibited by *Oratio Severi*, 58, 58–59n. 72, 59–60, 63n. 82; restricted in dotal law, 67; to support benefactions, 94
Pliny the Younger: attitude toward wealth, 25–28; borrowing money, 46–47; characteristics of his estates, 4, 176, 177; funding alimentary foundation, 78, 80–83, 95–96; income from viticulture, 198–99; investing in estates, 42, 46, 72–73, 111–12, 164; leasing to tenants, 12, 26, 73, 81, 111–12, 145–46, 156, 173,

Pliny the Younger *(continued)* 177, 177–78n. 93, 182, 189–90, 195–96, 222, 234, 239–40; lending money, 46–48; providing for his nurse, 107, 116–17, 125, 146n. 15; selling vintages, 211, 216; as source for economic history, 20, 25, 237, 239–40; sources of income, 26–27, 46; visiting his estates, 145–46

Poland, feudal economy of, 196n. 31, 217–18

Pompeii, economy of, 17–18

Possession: tenants' de facto right of, 220; tenants lacking right of, 184–89, 192

Praedia instructa, 173. See also *Fundus instructus,* distinguished from *fundus cum instrumento*

Praetor tutelarius, 54

Prices. *See* Crop prices, Market conditions, affecting finances of upper-class Romans; Markets, tenants' dependence on; Risk

Profit: in Roman agrarian economy, 2, 3, 6–7, 28n. 10, 77, 89, 93, 95–96, 100–111, 113, 116–17, 123–24, 133–34, 135, 239; of tenants, 209–21

Pupils: lending their money, 49–54; management of their property, 34–45, 54–67; recourse against tutors, 36–37, 38n. 32, 39, 40, 60–62; their use of wealth, 28–33

Quarries: on estates held under usufruct, 118–19; on *fundi dotales,* 69; on property of pupils, 60

Quod vi aut clam interdict, 202n. 44

Rationality, in Roman economy, 18, 19, 20, 218

Reditus: bequest of, 84, 90, 102–3, 213; as product of *fundus cum instrumento,* 113

Regulus, M. Aquilius, 27n. 8

Reliqua colonorum: bequest of, 120–21, 122–23, 152, 153, 154, 171–72, 173; excluded from legacies of estates, 122–23

Relocatio tacita, 200, 206–9, 220

Remissio mercedis. See Remission of rent

Remission of rent: Columella's efforts to avoid, 197n. 33, 222, 227, 234; as legal doctrine, 141, 142, 148, 202–6, 214–16, 217, 221–34, 240; by Pliny, 26, 190, 196, 222, 234

Remmius Palaemon, Q., investing in land, 71–72

Rents, in lease law, 11, 142, 147–49, 156, 160, 183, 222

Rescripts: answering petitions for remissions of rent, 203–6, 214–15, 227–31; legal significance of, 139n. 4, 204n. 50, 227n. 94, 228. *See also* Petitions

Restitutio in integrum, 58, 61, 61–62n. 78

Risk: allocation in Egyptian agriculture, 223–24; allocation in farm tenancy, 8, 221–34, 235; allocation on African imperial estates, 223; associated with viticulture, 107, 198; burden assumed by landowners, 214, 222, 228–29, 232, 235; for crop prices, 213–14, 216–18; efforts of landowners to avoid, 4, 19, 75–76, 96, 136, 180, 239; for harvest, 42–43, 95–96, 204, 213–14, 214–16, 222, 226–27; in loans, 52–54; in various contracts, 155n. 40; tenants' aversion to, 213–14

Roman economy. *See* Economy of Roman Empire

Sale of crops: from *fundus cum instrumento,* 116; by tenants (*see* Markets). *See also* Market conditions, affecting finances of upper-class Romans

Sale of land: affecting tenancy, 109, 164–66, 184, 186, 187–89; among upper-class Romans, 74–75; in

peasant agriculture, 75n. 109. *See also* Investment in land; Possession
Sarapion, archive of, 175n. 87
Scaevola, P. Mucius, on legacy, 99
Security: from bequests of land in Egypt, 127–28, 131; sought by landowners, 6–7, 15, 54, 63, 64, 74, 76, 77, 96, 135–36, 137, 179, 181, 240
Security of tenure: as treated by jurists, 141, 184, 190–94, 206–9; of tenants, 182, 185, 189–90, 200, 219, 220; of tenants in Egypt, 208–9, 223–24
Senatus Consultum Pegasianum, 85n. 18
Senatus Consultum Velleianum, 69
Seneca, investing in land, 72, 74
Servi quasi coloni, 41n. 42, 114–15, 121, 166–73, 179, 195n. 29
Servitudes: affected by *Oratio Severi*, 59; on dotal property, 67
Servius Sulpicius Rufus: on legacy, 100; on remission of rent, 226, 229, 232
Sharecropping, 11–12, 142, 145–46, 148, 196, 223
Shepherds, in legacy of *fundus cum instrumento*, 115
Short-term lease: as a means of enforcement of tenant's obligations, 200, 219–20, 221–22; serving interests of tenant, 220
Slaves: attending pupils, 30, 31; bequeathed in Roman Egypt, 129; cultivating land managed by *servi quasi coloni*, 167, 170; as labor force on estates, 3, 5, 136, 171–72, 176–77, 218; in legacies of estates, 98, 99–100, 114, 115, 171–72, 173; provided by tenants (*see* Tenants)
Soldiers: as landowners, 162–63, 186n. 9; as testators, 35–36, 126n. 10
Soterichos, archive of, 88n. 28, 223–24
Standard of care, of tutors. *See* Tutors
Sterilitas, as basis for remission of rent, 204, 214–15, 228, 229, 230–31, 232, 232–33n. 111
Subleasing, 177
Suetonius Tranquillus, role in managing estates, 146
Suppellex, legacy of, 100

Tabularium, bequest of, 120
Tacit pledging, 150n. 28
Tacit renewal of the lease. *See Relocatio tacita*
Tarius Rufus, L., investing in land, 71
Taxes: effect on economy, 16n. 25, 210n. 58; in Egyptian farm leases, 159; in Egyptian land sales, 109
Tenancy: advantages for landowners, 3, 4–6, 7–8, 15, 77, 96, 136, 137, 144–46, 155, 157, 162–66, 171, 179–80, 181, 196, 219, 221, 224, 233–34, 234–35, 240; advantages for tenant, 202–3, 209–21; Columella's recommendations concerning (*see Columella*); contractual terms, 104, 147–49; on dotal property, 154, 186–87; duration of the lease, 147; enforcement by lessor (*see* Enforcement, by lessor in tenancy); in Egypt, 104–5, 133–34, 160–63, 182, 183, 208–9, 223–24; on estate held under usufruct, 154, 201–2; forms of rent (*see* Rents, in lease law); on imperial estates (*see* Imperial estates, in North Africa); jurists' treatment of (*see* Jurists); on land belonging to town, 87–88; on land received through bequest, 104–6, 133–34, 160–62, 163; long-term, 165, 183–94, 197–98, 205, 233; in long-term leases (*see* Leases); on Pliny's estates (*see* Pliny the Younger); on pupils' land, 43, 158–62; at Pompeii, 18; on vineyards, 88, 199–201, 203–5. *See also* Leases, Sharecropping, Tenants
Tenants: access to law courts, 140, 141–42, 211; aversion to risk,

Tenants (*continued*)
213–14; bargaining power, 15, 189, 194, 233–34, 239; as beneficiaries of lessor's will, 153, 190–92, 193, 202; benefiting from leasing (*see* Tenancy); compelled to remain on farms (*see* Leases); duties under lease, 147–49, 151; duty to cultivate land, 229–30; employing slaves, 104, 149, 152, 174–75, 195; on estates with slaves, 171–72; of free and servile status on same estate, 173; freedmen serving as, 192–93; incentives to cultivate intensively, 217–19; investing in farm improvements (*see* Investment in farm improvements); lacking right of possession (*see* Possession); as landowners, 154–55, 175; large-scale, 153, 154, 175–76, 176n. 89, 177, 177–78n. 93, 183; resources of, 149–52, 153–54, 174–79, 183, 195, 195n. 28; selling crops on the market (*see* Markets, tenants' dependence on); status as envisioned by jurists, 139–41, 147–55. *See also* Security of tenure; Sharecropping; Tenancy

Testamentary prohibitions, against alienating property, 70–71

Testamentum Dasumii, 94n. 41

Tributum, funded by a trust, 91

Trimalchio, investing in land, 73–74

Trusts: confused with legacies, 103n. 56; envisioned as funds, 103n. 56; to fund *alimenta*, 90–91; to fund civic foundation, 81n. 8; to invest pupil's funds, 35–37; prohibiting alienation of property, 70; to support dependents, 89–96, 132; with tenant as beneficiary, 191–92. *See also* Economic planning; Legacy, significance to upper classes; Wills

Tutela. *See* Tutors; Tutorship

Tutor suspectus, 29, 61

Tutors: administering property of upper-class pupils, 23–25, 65–66; duties in managing pupils' property, 34–45, 49–54, 58–67, 75–76; financial liability, 35, 36–37, 38n. 32, 40, 50–52, 60–62, 63, 64; obligations in later empire, 40; providing for pupils' social needs, 28–33; standard of care, 29, 34, 40, 42, 51–52. *See also* Tutorship

Tutorship: affecting upper classes, 23–25, 65–66; in Egypt, 31n. 18, 50n. 55, 65, 65n. 86, 95, 128–29, 132, 133–34, 158–61; general features of, 12–14, 22–23. *See also* Pupils

Ubertas, affecting claim for remission of rent, 214–15, 230–31

Unexhausted improvements, 193n. 26, 199, 220

Unwritten contracts, in tenancy, 208–9

Urbanus colonus, in Columella, 174–75

Usucapio, in disputes involving tutorship, 60

Usufruct: beneficiary leasing to owner, 154; bequeathed to tenant, 153, 190–91, 193, 202; jurists' treatment of, 98n. 47, 117–19; leasing of, 105, 154, 201; legacy of, 90, 93–94, 102, 103–4, 117–19

Usurae pupillares, 39, 53

Ut frui liceat, in tenancy, 147, 225

Varro, on investing in land, 73n. 104

Vectigal, on Pliny's estates, 82–83, 112

Veleia: alimentary foundation, 79, 83–84; organization of estates there, 4

Vestiaria, bequest of, 94

Veterans: as landowners in Egypt, 106, 161; serving as curators in Egypt, 129; as testators, 71n. 98, 125, 126–27

Vilici: Columella's views about, 156–57; in legacy of *fundus cum*

instrumento, 167; managing estates, 3, 107n. 67, 145n. 12, 146n. 15, 119n. 90, 165, 169n. 73, 172, 173, 176, 177

Villa: in *fundus cum instrumento,* 97; maintenance of, 102–4; part of *fundus* held under lease, 139, 141, 151–52, 154–55, 172, 177; on Pliny's estates, 152

Vineyards: bequest of, 87–88, 95, 126–27, 128, 129; Columella's recommendations concerning their leasing, 157; divided among members of Hermopolite family, 129–31; maintenance of, 88, 199–201, 203–5; mortgage of, 65n. 86; purchase of, 43–45, 71–72. *See also* Investment in farm improvements

Vintages, sold in advance, 145, 211, 212, 216

Vis maior, as basis for remission of rent, 142, 203–4, 215, 224, 226, 229, 232

Vitia ex re, in law of tenancy, 203–4, 226

Viticulture: Columella's calculations about, 72n. 101, 74n. 106, 198; in Egypt, 43–44; on land cultivated by tenant, 187–89, 194–95, 198–200, 203–5; speculative investments in, 71–72; risks in, 107, 198

Volusius Saturninus, L., recommendations about tenancy, 174

Wealth: defined as access to yearly income, 134; Pliny's attitude toward, 25–28; of senators, 27n. 8; as used by pupils, 28–33

Wills: in Egypt, 92–93, 123–29, 131–34; as means to provide for dependents, 13, 89, 96, 192; in *Oratio Severi,* 56; provisions of, 35–37, 70, 71, 77, 81–82, 89–96, 101–4, 105—7, 120–21, 190–93, 212–13. *See also* Economic planning; Legacy; Trusts

Written contracts, in tenancy, 208–9

DATE DUE	
FEB 27 1998	

UPI   PRINTED IN U.S.A.